WAVE FORCES ON OFFSHORE STRUCTURES

A thorough understanding of the interaction of waves and currents with offshore structures has now become a vital factor in the safe and economical design of various offshore technologies. There has been a significant increase in the research efforts to meet this need. Although considerable progress has been made in the offshore industry and in the understanding of the interaction of waves, currents, and wind with ocean structures, most of the available books concentrate only on practical applications without a grounding in the physics. This text strives to integrate an understanding of the physics of ocean–structure interactions with numerous applications. This more complete understanding will allow the engineer and designer to solve problems heretofore not encountered and to design new and innovative structures. The intent of this book is to serve the needs of future generations of engineers designing more sophisticated structures at ever-increasing depths.

Dr. Turgut "Sarp" Sarpkaya is an internationally recognized authority in fluid mechanics research and was named by Cambridge University as one of the world's one thousand greatest scientists. "Sarp," as he is known to friends and colleagues, is the recipient of the Turning Goals into Reality Award by NASA, and he was selected Freeman Scholar by the American Society of Mechanical Engineers (ASME). Sarpkaya received his Ph.D. from The University of Iowa, followed by postdoctoral work at the Massachusetts Institute of Technology. He was the Thomas L. Fawick Distinguished Professor at the University of Nebraska and taught at the University of Manchester. He was named Professor and Chairman of Mechanical Engineering at the Naval Postgraduate School in 1967 and Distinguished Professor in 1975. His research over the past 50 years has covered the spectrum of hydrodynamics. His oscillating flow tunnel and the vortex-breakdown apparatus are two among several unique research facilities he has designed. Sarpkaya has published more than 200 papers and has explored for the Defense Advanced Research Projects Agency (DARPA) numerous classified projects dealing with the hydrodynamics and hydroacoustics of submarines.

He served as chairman of the Executive Committee of the Fluids Engineering Division of ASME and the Heat Transfer and Fluid Mechanics Institute. He is a Fellow of the Royal Society of Naval Architects and Marine Engineers, Fellow of the ASME, and Associate Fellow of the American Institute of Aeronautics and Astronautics.

WAVE FORCES ON OFFSHORE STRUCTURES

Turgut "Sarp" Sarpkaya

CAMBRIDGE
UNIVERSITY PRESS

CAMBRIDGE
UNIVERSITY PRESS

32 Avenue of the Americas, New York NY 10013-2473, USA

Cambridge University Press is part of the University of Cambridge.

It furthers the University's mission by disseminating knowledge in the pursuit of education, learning and research at the highest international levels of excellence.

www.cambridge.org
Information on this title: www.cambridge.org/9781107461161

First published 2010
First paperback edition 2014

A catalogue record for this publication is available from the British Library

Library of Congress Cataloguing in Publication data

Sarpkaya, Turgut.
Wave forces on offshore structures / Turgut Sarpkaya.
    p.   cm.
Includes bibliographical references and indexes.
ISBN 978-0-521-89625-2 (hardback)
1. Offshore structures – Hydrodynamics.   2. Wave resistance
(Hydrodynamics)   3. Ocean waves – Mathematical models.   I. Title.
TC1665.S27   2010
627'.98–dc22   2009034862

ISBN 978-0-521-89625-2 Hardback
ISBN 978-1-107-46116-1 Paperback

The book is dedicated to the memory of Herman Schlichting,
who shared with me his time and inspiration for an unforgettable year at
AVA/Göttingen in 1972.

Contents

Preface

In 1857, William Thomson (Lord Kelvin) wrote to Sir George Gabriel Stokes, "Now I think hydrodynamics is to be the root of all physical science, and is at present second to none in the beauty of its mathematics." However, 117 years later, Sir Geoffrey Ingram Taylor had a thoughtful reminder (1974): "Though the fundamental laws of the mechanics of the simplest fluids, which possess Newtonian viscosity, are known and understood, to apply them to give a complete description of any industrially significant process is often far beyond our power."

The ultimate requirement for the understanding and quantification of the ocean–structure interaction is the understanding of the physics of turbulence, separation, and computational fluid dynamics. This is particularly true for the extraction of oil and gas. Computers (virtual simulations) and experiments (i.e., what can be calculated and what can be measured) have and will continue to have unsurpassable limitations. Even if one were able to calculate a given fluid–structure interaction, with sufficient time and resources, one will not be able to cover the entire parameter space. Thus, approximations, experiments, experience, successes and failures, empirical equations, and virtual modeling will continue to define the limits of our power to design structures that serve the intended purposes. Experience, physical insights, computers, past regrets, and future failures will continue to define the form and the substance of ocean–structure interactions. Developing the technology and carrying out the physical processes to discover and extract oil and gas, particularly from greater depths at increased risks and cost, have become one of the major challenges facing the industrialized countries, notwithstanding the efforts to develop alternate sources of energy.

The impetus for the offshore industry came from the tragic failure of the Texas Tower No. 4 (a U.S. Air Force radar platform) in January 1961, which was quite recent, considering the fact that humans have occupied this planet for millions of years. Such tragedies have precipitated the need to understand wave–structure interaction, tower dynamics in random seas, and gusts at ever-increasing depths in friendly as well as hostile environments. The pre-computer era and unsophisticated electronic age had to go through their own evolutions to alleviate the frustrations of the offshore engineer. Now the attention of the offshore engineer is engaged with the problems of how to study, measure, and observe the ocean environment and how to exploit its riches. These tasks require not only a periodic assessment of the state-of-the-art and definition of research goals for the foreseeable future, but also the understanding and use of *virtual modeling*, because the fundamental and physical research and numerical simulations are cheaper than regret and unknown conservatism.

The principal difficulties stem from the problems associated with the determination of the state of the sea, the three-dimensional and random nature of the ocean–structure interactions, quantification of the forces exerted on structures, and the lack of fluid-mechanically satisfying closure model(s) for turbulence. The collective creative imagination of many paved the way to physics-based laws, approximate models, and enlightening explanations of infinitely complex (but beautiful) phenomena.

Considerable progress has been made during the past few decades. The seminal experimental studies of Keulegan and Carpenter in 1956 inspired much experimental work and led, in the mid-1970s, to a better understanding of the vortex trajectories about bluff bodies, evolution of the transverse vortex street, three-dimensional instabilities and damping at very low K values, effects of the aspect ratio and three-dimensionality of flow along a cylinder, flow about tube bundles, vortex-induced oscillations, dynamic response, separation-point excursions, and to the demonstration (by this writer) of the dependence of the force coefficients not only on K (the Keulegan–Carpenter number) but also on $\beta = Re/K = D^2/\nu T$ (first conceived and used by Sarpkaya in 1976). This brought a new age of experiments in the laboratories and in the field, reliable data analysis, model tests, new journals, and ever-increasing practical and engineering research with the help of computers (starting with memories as large as 264 bytes in 1956 with an IBM 650). This writer is fortunate to have been an active part of this evolution.

Since the 1970s, the dependence of the force coefficients on K, k/D, and β has led to a better understanding of the vortex trajectories about bluff bodies, evolution of the transverse vortex street, separation–point excursion, three-dimensional instabilities, flow separation and damping at very low K values, effects of aspect ratio and three-dimensionality of flow along a cylinder, coherent and quasi-coherent structures, flow about multi-tube bodies, vortex-induced vibrations, numerical simulations, large-eddy simulation (LES) (an ingenious compromise based on "multilayer" filtering), Reynolds-averaged Navier–Stokes (RANS) equations, and detached eddy simulation (DES), which emerged to alleviate the difficulties that are inherent in both LES and RANS solutions; see, e.g., Mittal and Moin

(1997) and Breuer (1998). Other models, such as DESIDER (detached eddy simulation for industrial aerodynamics), UFAST (unsteady effects in shock-wave-induced separation), POD (proper orthogonal decomposition), unsteady-RANS, and others are relatively more recent entries into the art of modeling.

The friendly and enlightening input of many of my colleagues (Drs. C. Dalton, G. E. Karniadakis, C. C. Mei, D. J. Newman, O. H. Oakley, B. M. Sumer, J. K. Vandiver, M. M. Zdravkovich, and many others) is sincerely appreciated. I have had the privilege to be assisted by several colleagues with specific talents and information. Professor Charles Dalton (a longtime friend) read each chapter and provided valuable input and encouragement. Many graduate students have worked with me: Dr. Gerardo Hiriart Le Bert, Dr. Farhad Rajabi, Dr. Ray L. Shoaff, Dr. Frank Novak, and others working toward their M.S. or Ph.D. degrees. Mr. Jack McKay's ingenuity permeated the design and construction of numerous research facilities. I express my sincere appreciation for the extensive research support and willing cooperation extended to me through the years by the Office of Naval Research and the National Science Foundation. Finally, I am truly grateful to Mrs. Irma Fink, Ms. Greta Marlatt, and Mr. Jeff Rothal, who have found and delivered with incredible speed hundreds of references. For all of these I express to them and the Naval Postgraduate School my special gratitude. I owe my special thanks to Peter C. Gordon, Senior Editor of Cambridge University Press, for his friendly advice and encouragement during the past two years.

I am deeply indebted to Professor Günther F. Clauss for his gracious permission to use the drawings and photographs of his fixed platforms and guyed and compliant towers.

This book is dedicated to the designers and builders of offshore structures and to the researchers in this field. Their concern for the advancement of the state of the art motivated my work. I sincerely hope that my efforts, modest relative to their monumental achievements, will meet their approval.

T. "Sarp" Sarpkaya, California

www.omae2006.com

A video is presented by Sarpkaya in the above Web site in black and white. It depicts the motion of vortices for about two minutes and may be played over and over again.

In addition, there is a PowerPoint presentation by Sarpkaya on fluid damping (Chapter 7). It is provided *as a courtesy of Dr.-Ing. Walter L. Kuehnlein (advice@sea2ice.com), and Sarah@seatoskymeetings.com and ian@seatoskymeetings.com. To them, I express my gratitude.*

1

Introduction

It is appropriate to begin this introduction with a thoughtful reminder by Sir Geoffrey Ingram Taylor (1974): "Though the fundamental laws of the mechanics of the simplest fluids, which possess Newtonian viscosity, are known and understood, to apply them to give a complete description of any industrially significant process is often far beyond our power." This is particularly true for oil-hungry mankind's desire for hydrocarbons, with pollution as a bonus. The principal difficulties stem from the problems associated with the determination of the state of the sea, the three-dimensional and random nature of the ocean–structure interaction, quantification of the forces exerted on structures, and the lack of a fluid-mechanically satisfying closure model for turbulence. The past hundred years have shown that "almost-randomness" is the law of the physics of turbulence. One can quantify the consequences of turbulence and probe into its behavior for a given event only through approximate models and physical and numerical experiments (provided that the wideband of relevant scales is fully resolved). Even if the direct numerical simulations (DNSs) at Reynolds numbers as high as 10^7 were possible in the centuries to come, the large parameter space in any application precludes a purely numerical solution.

The past four decades have seen an explosion of interest in the broad subject of ocean hydrodynamics. This interest led to an improved and more realistic understanding of the physical characteristics of some time-dependent flows about bluff bodies and their mathematical formulation and experimental exploration. On the one hand, attention has been focused on controlled laboratory experiments, which allow for the understanding of the separate effects of the governing and influencing parameters, and, on the other hand, on mathematical and numerical methods, which allow for the nearly exact solution of some wave-loading situations (large bodies, whose volumes are as large as 10 times that of the great pyramid of Khufu). For many practically significant fluid–structure interactions involving flow separation, vortex motion, turbulence, dynamic response, and structural and fluid-dynamical damping, however, direct observations and measurements are continuing to provide the needed information, whereas theory has not yet played an important role.

The hydrodynamic loading situations that are well understood are those that do not involve flow separation. Thus they are amenable to nearly exact analytical treatment. These concern primarily the determination of the fluid forces on large objects in the diffraction regime where the characteristic dimension of the body relative to the wavelength is larger than about 0.2. The use of various numerical techniques is sufficient to predict accurately the forces and moments acting on the body, provided that the viscous effects and the effects of separation for bodies with sharp edges are ignored as secondary.

The understanding of the fluid–structure interactions that involve extensive separation and dependence on numerous parameters, such as the Reynolds number and the Keulegan–Carpenter number $K = 2\pi(A/D)$, a parameter that does not depend on time (it is a simple length ratio). There are several reasons for this. First, although the physical laws governing motion (the Navier–Stokes equations) are well known, valid approximations necessary for numerical and physical model studies are still unknown. Even the unidirectional steady flow about a bluff body remains theoretically unresolved. Fage and Johansen's (1928) pioneering work and Gerrard's (1965) vortex-formation model, followed by a large number of important experiments, have provided extremely useful insights into the mechanism of vortex shedding. It became clear that a two-dimensional ambient flow about a two-dimensional body does not give rise to a two-dimensional wake, and only a fraction (about 60% for a circular cylinder) of the original circulation survives the vortex formation.

Offshore technology has experienced a remarkable growth since the 1940s, when offshore drilling platforms were first used in the Gulf of Mexico. At the present time, a wide variety of offshore structures are being used, even under severe environmental conditions. These are predominantly related to oil and gas recovery, but they are also used in other applications such as harbor engineering and ocean energy extraction. Difficulties in design and construction are considerable, particularly as structures are being located at ever-increasing depths and are subjected to extremely hostile environmental conditions. The discovery of major oil reserves in the North Sea has accelerated such advances, with fixed platforms in the North Sea, now located in water depths up to about 185 m and designed to withstand waves as high as 30 m. In more recent years the depths to be reached for more hydrocarbon resources have increased to 1600 m or more. In fact, the depths reached during the past 55 years increased as $h \approx (1/540)N^{3.5}$, where h is the depth and N is the number of years, starting with $N = 0$ in 1949.

The potential of major catastrophic failures, in terms of both human safety and economic loss, underlines the critical importance of efficient and reliable design. In January 1961, the collapse of Texas Tower No. 4 off the New Jersey coast involved the loss of 28 lives. In March 1980, the structural failure and capsizing of the mobile rig *Alexander Keilland* in the Ekofisk field in the North Sea involved the loss of over 100 lives. The *Piper Alpha* oil and gas platform caught fire in 1988, leading to the loss of 167 lives. The *Petrobras* (a floating production system) sank in the Campos Basin in 2001 and cost 10 lives.

1.1 Classes of offshore structures

It is appropriate at the outset to provide some perspective to what follows by classifying briefly the wide variety of offshore structures that are in current use or that have been seriously proposed. The major offshore structures used in the various stages of oil recovery include both mobile and fixed drilling platforms, as well as a variety of supply, work, and support vessels.

The various offshore structures currently in use have been described in detail in the trade and technical literature. Mention is made of Bruun (1976), who summarized the offshore rigs used in the North Sea, and Watt (1978), who reviewed the design and analysis requirements of fixed offshore structures used in the oil industry. Ships and moored shiplike marine vessels are also used extensively, but they are treated within the field of naval architecture and are not of primary consideration in this book.

In earlier years the development and production activities at an offshore site were primarily carried out with fixed platforms. The jacket or template structures, and extensions to them, were the most common platforms in use. A jacket platform comprises a space frame structure, with piles driven through its legs. An extension to this concept includes the space frame structures that employ skirt piles or pile clusters. Some platforms contain enlarged legs to provide for self-buoyancy during installation. Jacket platforms are located throughout the world, including the North Sea, where they may be exposed to waves with heights approaching 100 ft.

Figure 1.1 shows numerous structures with a variety of names [fixed platforms: gravity-based structures (GBSs) and the jacket]. These are followed by guyed and compliant towers. Under the general title of floating structures, there are tension leg platforms (TLPs), SPAR-buoys, and floating production systems (FPSs). The largest platform until 1980 was one installed in the Cognac field off the Louisiana coast in a water depth of just over

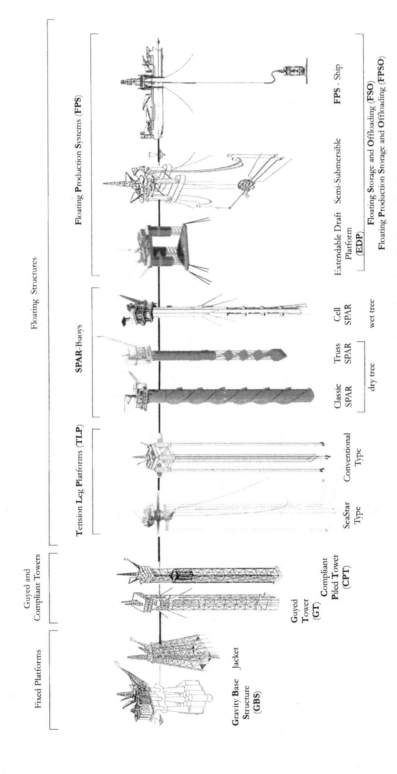

Figure 1.1. Representative offshore platforms (after Günther F. Clauss 2007).

4

300 m. This platform contained 60000 metric tons of steel and was fabricated in three sections, which were joined under water on site.

Gravity platforms depend on excessive weight, rather than on piles, for their stability. They are thus suited to sites with overconsolidated soils and have been used primarily in the North Sea. The most usual gravity platforms comprise a large base, which has the capacity for significant oil storage. In addition to their being located in depths of several hundred feet, gravity platforms are characterized by large horizontal dimensions. For example, a typical platform may be 180 m high, its base may have a diameter of 90 to 120 m, and it may have the capacity to store one million barrels of oil.

In more recent times, exploratory drilling is usually carried out with mobile drilling rigs. These include submersible platforms, rendered stationary with chain and wire mooring systems (Barltrop, 1998). They are limited to relatively shallow water, jackup platforms, drill ships or drill barges, and semisubmersible (SS) drilling platforms. The sketches of representative SS platforms are shown in Fig. 1.1 (GBS, jacket, guyed tower, and compliant structures). Such platforms are capable of operating at large depths. The major buoyancy members are placed well below the mean water level under operating conditions to minimize the wave action and to withstand severe weather conditions. They have a relatively low metacentric height (GM) that tends to reduce their pitch and roll motions. With careful design, counteracting inertial forces on surface-piercing columns and on submerged hulls may be made to cancel each other at a certain wave frequency. However, one must be beware of the fact that their high GM limits their variable load. In recent years, their primary station keeping by chain and wire has been augmented by azimuthing thrusters to assist the mooring system and the SS when in transit. The advances in dynamic positioning systems enabled them to operate and to transit at deeper waters without mooring.

What may be called a cousin of SS's is the tension leg platform (TLP). They are similar to SSs in a number of ways. TLPs have a greater water plane area, typically three to six surface-piercing columns, taut vertical mooring tethers, and a complete set of pontoons. All this is made possible because they are stationary.

Spar platforms (Halkyard, 1996) are essentially vertical, almost submerged, circular cylinders kept on station by lateral catenary anchor lines. Their center of gravity is below the center of buoyancy because of the fixed ballast concentrated at their bottom. This gives rise to a relatively large GM, i.e., enhances stability. However, as Barltrop (1998) noted, even though the large draft of a spar significantly reduces heave, its vertical length gives rise to a number of ocean–structure interaction problems: overturning moments that are due to wind, offloading forces, and current-induced forces. The spars are often fitted with spiral strakes to suppress vortex-induced vibrations. If the effect of the strakes is not experimentally optimized, they may give rise to massive separation at sufficiently high Reynolds numbers and enhance fluid loading and instability.

It should be noted in passing that the center of gravity of a spar may be lowered further (relative to that of a circular cylinder), and giving it a conical shape may further reduce the wave- and current-induced forces: round or hemispherical at the bottom and narrowing toward the free surface (resembling a pear). Above the free surface it may be reduced to a cylinder of constant radius. Such a "pear"-shaped spar has never been used.

The sequence of calculation procedures needed to establish the structural loading generally involves (a) establishing the wave climate in the vicinity of a structure, either on the basis of recorded wave data or by hindcasting from available meteorological data; (b) estimating design wave conditions for the structure; (c) selecting and applying a wave theory to determine the corresponding fluid particle kinematics; (d) using a wave-force formulation to determine the hydrodynamic forces on the structure (often very difficult near the mean water level where wave motion, currents, and strong gusts cannot be quantified); (e) calculating the structural response; and (f) calculating the structural loading, which includes base shear and moment, stresses, and bending moments. These steps may serve only as a rough indicator. The most important fact is that the design of a structure is based on computational fluid dynamics and virtual modeling, as pointed out earlier.

2

Review of the Fundamental Equations and Concepts

2.1 Equations of motion

The equations for an incompressible Newtonian fluid (known as the Navier–Stokes equations) may be written as

$$\frac{\partial u}{\partial t} + u\frac{\partial u}{\partial x} + v\frac{\partial u}{\partial y} + w\frac{\partial u}{\partial z} = X - \frac{1}{\rho}\frac{\partial p}{\partial x} + \nu\left(\frac{\partial^2 u}{\partial x^2} + \frac{\partial^2 u}{\partial y^2} + \frac{\partial^2 u}{\partial z^2}\right) \quad (2.1.1a)$$

$$\frac{\partial v}{\partial t} + u\frac{\partial v}{\partial x} + v\frac{\partial v}{\partial y} + w\frac{\partial v}{\partial z} = Y - \frac{1}{\rho}\frac{\partial p}{\partial y} + \nu\left(\frac{\partial^2 v}{\partial x^2} + \frac{\partial^2 v}{\partial y^2} + \frac{\partial^2 v}{\partial z^2}\right) \quad (2.1.1b)$$

$$\frac{\partial w}{\partial t} + u\frac{\partial w}{\partial x} + v\frac{\partial w}{\partial y} + w\frac{\partial w}{\partial z} = Z - \frac{1}{\rho}\frac{\partial p}{\partial z} + \nu\left(\frac{\partial^2 w}{\partial x^2} + \frac{\partial^2 w}{\partial y^2} + \frac{\partial^2 w}{\partial z^2}\right) \quad (2.1.1c)$$

The equation of continuity is expressed as

$$\frac{\partial u}{\partial x} + \frac{\partial v}{\partial y} + \frac{\partial w}{\partial z} = \text{div } q = 0 \quad (2.1.2)$$

in which u, v, w represent the velocity components in the x, y, z directions, respectively; X, Y, Z, are the components of the body force per unit mass in the corresponding directions; p is the pressure; and v is the kinematic viscosity of the fluid. The terms like Du/Dt denote the substantive acceleration. They are also known as the Eulerian derivative, material derivative, or comoving derivative of velocity.

The substantive acceleration consists of a local acceleration (in unsteady flow) that is due to the change of velocity at a given point with time and a convective acceleration that is due to translation, as, for example, in steady flow in a diverging (converging) pipe. The operator D/Dt may be applied to density, temperature, etc., to determine its respective Eulerian derivatives. Obviously, the difference between the substantive acceleration and the local

7

acceleration in a given direction accounts for the nonlinear convective accelerations, which do not vanish in *nonparallel* flows. In some cases (very slow
or creeping motions) the convective accelerations may be neglected. The
resulting linear equations are more amenable to analytical and numerical
solutions whose upper limit of validity can be determined by experiments
and, to some extent, by direct numerical simulations.

   The Navier–Stokes equations evolved over a period of 18 years starting
with Navier (1827), undergoing different derivations by Poisson (1831), and
de Saint Venant (1843), and culminating with Stokes in 1845. However, the
boundary conditions (no slip, no penetration) were not established unam-
biguously. It took another 6 years to firmly set the *hypothesis* that there is
no slip on a boundary or, more precisely, the fluid immediately adjacent to
the boundary acquires the velocity of the boundary. This was a monumental
achievement, *based mostly on heuristic reasoning*, as noted by Stokes (1851):
"I shall assume, therefore, as the conditions to be satisfied at the boundaries
of the fluid, that the velocity of a fluid particle shall be the same, both in magnitude and direction, as that of the solid particle with which it is in contact."
The derivation of the Navier–Stokes equations for an incompressible Newtonian fluid with constant viscosity may be found in many basic reference
texts (see, e.g., Schlichting 1968, or later issues) and is not repeated here.

   When gravity is the only body force exerted, a body-force potential may
be defined such that $\Omega = -gh$ and

$$X = \frac{\partial \Omega}{\partial x}, \quad Y = \frac{\partial \Omega}{\partial y}, \quad Z = \frac{\partial \Omega}{\partial z}$$

where h is height above a horizontal datum. Then Eqs. (2.1.1) reduce to

$$\frac{Du}{Dt} = -\frac{1}{\rho}\frac{\partial(p + \rho g h)}{\partial x} + v\,\nabla^2 u \tag{2.1.3a}$$

$$\frac{Dv}{Dt} = -\frac{1}{\rho}\frac{\partial(p + \rho g h)}{\partial y} + v\,\nabla^2 v \tag{2.1.3b}$$

$$\frac{Dw}{Dt} = -\frac{1}{\rho}\frac{\partial(p + \rho g h)}{\partial z} + v\,\nabla^2 w \tag{2.1.3c}$$

or, in more convenient vector notation, we have

$$\frac{\partial \boldsymbol{q}}{\partial t} + (\boldsymbol{q}\,\mathrm{grad})\boldsymbol{q} = -\frac{1}{\rho}\nabla(p + \rho g h) + v\,\nabla^2 \boldsymbol{q} \tag{2.1.4}$$

where $\boldsymbol{q}$ is the velocity vector and may be written as $\boldsymbol{q} = iu + jv + kw$.

   If L is a characteristic length scale and U is a reference velocity, then
(2.1.1) can be expressed in dimensionless form by resorting to L and U. Then

it is seen that the ratio UL/v, which is a Reynolds number, represents the ratio of the inertial to viscous forces. In a wide class of flows, the Reynolds number is very large and the viscous terms in the preceding equations are much smaller than the remaining inertial terms over most of the flow field. A notable exception is the *boundary layer* (a brilliant concept discovered by Prandtl 1904) in which the velocity gradients are steep and the viscous stresses are significant.

Through the use of an order-of-magnitude analysis, Prandtl has shown that, for large Reynolds numbers ($Re = U/v$, where x is the distance along the boundary), the equations of motion and continuity for a two-dimensional (2D) flow may be reduced to

$$\frac{\partial u}{\partial t} + u\frac{\partial u}{\partial x} + v\frac{\partial u}{\partial y} = -\frac{1}{\rho}\frac{\partial p}{\partial y} + v\frac{\partial^2 u}{\partial y^2} \qquad (2.1.5)$$

$$\frac{\partial u}{\partial x} + \frac{\partial v}{\partial y} = 0 \qquad (2.1.6)$$

A thorough discussion of all types of steady and unsteady boundary layers is given in Schlichting (1979). As noted earlier, the boundary conditions to be satisfied on the surface of a rigid body are that there will be *no slip* and *no penetration*. The boundary conditions at a free surface are discussed following the introduction of the velocity potential.

2.2 Rotational and irrotational flows

The rates of rotation of a fluid particle about the x, y, and z axes are given by (see, e.g., Schlichting 1968b)

$$\omega_x = \frac{1}{2}\left(\frac{\partial w}{\partial y} - \frac{\partial v}{\partial z}\right), \quad \omega_y = \frac{1}{2}\left(\frac{\partial u}{\partial z} - \frac{\partial w}{\partial x}\right), \quad \omega_z = \frac{1}{2}\left(\frac{\partial v}{\partial x} - \frac{\partial u}{\partial y}\right) \qquad (2.2.1)$$

They are components of the rotation vector $\omega = 1/2$ curl q. The flows for which curl $q \neq 0$ are said to be rotational because each fluid particle undergoes a rotation as specified by Eq. (2.2.1), in addition to translations and pure straining motions. The absence of rotation, i.e., $\omega_x = \omega_y = \omega_z = 0$, does not, however, require that the fluid be inviscid. In other words, in the regions of flow where curl $q = 0$, a real fluid exhibits an irrotational or inviscid-fluid-like behavior because the shear stresses vanish.

Rotation is related to two other fundamental concepts, namely, circulation and vorticity. The circulation Γ is defined as the line integral of the velocity vector taken around a closed curve, enclosing a surface S within the region of fluid considered. Thus we have

$$\Gamma = \oint \mathbf{q} \cdot ds = \oint (u\,dx + v\,dy + w\,dz) \qquad (2.2.2)$$

According to the Stokes theorem,

$$\Gamma = \oint \mathbf{q} \cdot ds = \int_S \operatorname{curl} \mathbf{q} \cdot dS = 2 \int_S \omega \cdot n\,dS \qquad (2.2.3)$$

and therefore (2.2.2) may be written as

$$\Gamma = \iint 2\omega_x\,dy\,dz + \iint 2\omega_y\,dx\,dz + \iint 2\omega_z\,dx\,dy \qquad (2.2.4)$$

in which the components of rotation vector appear twice. They are said to be the components of the vorticity vector ζ such that $\{\zeta_x = 2\omega_x,\ \zeta_y = 2\omega_y,\ \zeta_z = 2\omega_z\}$. Thus it follows from (2.2.4) that $\Delta\Gamma = \zeta_n \Delta S$, where ζ_n is the component of the vorticity vector normal to the surface element ΔS. In other words, *the flux of vorticity through the surface is equal to the circulation along the curve enclosing the surface.*

For reference purposes only, we note that in a *frictionless fluid* an element *cannot acquire or lose rotation* (there are no shear forces to induce such a motion); a vortex tube always consists of the same fluid particles, regardless of its motion; and the circulation remains constant with time. These are the fundamental theorems of vorticity and were enunciated by Helmholtz (1858) and Lord Kelvin (see Sir William Thomson 1849). For a detailed discussion of these theorems the reader is referred to classic reference texts such as Batchelor (1967), Milne-Thomson (1960), or Lamb (1932).

In *real fluids*, vorticity may be generated, redistributed, diffused, and destroyed because the frictional forces are not conservative. In other words, vorticity is ultimately dissipated by viscosity to which it owes its creation. For example, the vorticity found in a vortex about four diameters downstream from a circular cylinder is about 70% of the vorticity shed from the separation point (Bloor and Gerrard 1966). Schmidt and Tilmann (1972) found a 50% reduction in circulation as the vortices move from 5-diameter to 12-diameter downstream positions. The remainder is partly diffused and partly canceled by the ingestion of fluid bearing oppositely signed vorticity. One should also bear in mind that the experiments yield only the *normal component* of vorticity. Thus the consequences of the stretching and twisting of vortex filaments as a consequence of three dimensionality in the wake of a body and hence the redistribution of vorticity into directions other than the normal are not accounted for. This relatively simple example points out not only some of the difficulties associated with the use of the inviscid-fluid assumption in attempting to model the motion of real fluids but also the fact that the most important region for a bluff body is not the far wake (commonly associated with the Kármán vortex street, 1911, 1912) but rather is that enclosing the body and the near wake where the vorticity is generated, diffused, and dissipated. Among the many types of interesting and industrially significant vortices, those created by the rolling-up of the vorticity generated on the wings of an aircraft are indeed extremely important as far as a following aircraft is concerned. The strength of a typical wake vortex (shed from a B-747) is about 400 m²/sec (about the strength of a typical tornado). Larger aircrafts, such as the A-380, have wingtip vortices with even larger circulations. Obviously the distance between the leader and the follower aircraft must be increased to avoid wake-vortex hazard. Submarines generate equally large vortices, which rise near the free surface and enhance their visibility through the scars they create on the ocean surface (Sarpkaya 1996a).

Medium-size aircrafts produce vortices, which have core radii of about 3 m, separated by a span of about 40 m. The mutual interaction of such vortices gives rise to sinusoidal instability and, farther downstream, to the formation of vortex rings. The instability structures (sinusoidal waves, linking of the vortices, inception of vortex rings, and puffs) are easily visible in the sky, from the near wake to a few miles downstream from the aircraft.

2.3 Velocity potential

Irrotational motion exists only when all components of the rotation vector are zero, i.e.,

$$\frac{\partial w}{\partial y} - \frac{\partial v}{\partial z} = 0, \quad \frac{\partial u}{\partial z} - \frac{\partial w}{\partial x} = 0, \quad \frac{\partial v}{\partial x} - \frac{\partial u}{\partial y} = 0 \qquad (2.3.1)$$

It is then possible to devise a continuous, differentiable, scalar function $\phi = \phi(x, y, z, t)$ such that its gradients satisfy (2.3.1) automatically. In fact, curl $q = 0$ is the necessary and sufficient condition for the existence of such a function. The difference of potential $\delta\phi$ along a portion of a streamline of length δs is $\delta\phi = v_s ds$ or $v_s = \partial\phi/\partial s$. Thus the velocity is given by the gradient of the potential known as the velocity potential. It is analogous to the force potential in a gravitational field, steady electric potential in a homogeneous conductor, and the steady temperature distribution function in a homogeneous thermal conductor.

In Cartesian and cylindrical polar coordinates, the velocity components are thus given by

$$u = \frac{\partial\phi}{\partial x}, \quad v = \frac{\partial\phi}{\partial y}, \quad w = \frac{\partial\phi}{\partial z}, \quad \text{i.e., } q = \text{grad } \phi \qquad (2.3.2)$$

and

$$v_r = \frac{\partial \phi}{\partial r}, \qquad v_\theta = \frac{\partial \phi}{r \partial \theta}, \qquad w = \frac{\partial \phi}{\partial z} \qquad (2.3.3)$$

respectively. Equations (2.3.2) and (2.3.3) satisfy (2.3.1) automatically, i.e., the potential flow is irrotational. It is also true that a potential exists only for an irrotational flow.

The introduction of ϕ into equation of continuity (2.1.4) results in a second-order linear differential equation, known as the Laplace equation:

$$\nabla^2 \phi = \frac{\partial^2 \phi}{\partial x^2} + \frac{\partial^2 \phi}{\partial y^2} + \frac{\partial^2 \phi}{\partial z^2} = 0 \qquad (2.3.4)$$

This equation was first introduced by Laplace in his celebrated book on celestial mechanics and is of *fundamental importance* in many branches of physics, mechanics, and mathematics. The solutions of Laplace's equation are known as *harmonic functions*. The real and imaginary parts of a complex function $F(z)$ are such harmonic functions. Their linear combinations are also solutions of the Laplace equation.

The direct determination of a harmonic function, which satisfies the given boundary conditions, is often a difficult problem. Intuition, heuristic reasoning, experience, and numerous methods (e.g., finite difference, finite element, and others) must be called on not only to obtain a solution but also to ascertain that the solution based on the inviscid-flow assumption is a reasonable approximation to the actual behavior of the fluid. No general guidelines can be given to determine when an idealized solution is a reasonable approximation and when it is not. Often, comparison with experiments and the experiences of others provide some guidance. In general, flows or regions of flow, which are rapidly accelerating or dominated by inertial forces, may be treated by potential-flow methods.

Flow about a sphere at Reynolds numbers of the order of unity cannot be treated by inviscid-flow methods because viscous forces dominate it even though the flow is essentially unseparated. On the other hand, efflux of jets, jet impingement, jet deflection, flow with relatively small amplitudes of oscillation about a large cylinder or streamlined body may be treated with potential-flow methods with due regard to diffraction and conditions of radiation.

Often the inviscid-fluid assumption for the unseparated flow of an actual fluid may be used as a first approximation to the outer flow (e.g., flow about a foil at moderate angles of attack). However, it must be emphasized that the potential-flow solution (in whole or in part, e.g., about an entire circular cylinder or only in the forebody of the cylinder) does not constitute even a first-order approximation. The state of the art is such that analytical results, in particular those based on the zero-vorticity assumption, must be compared with those obtained experimentally or with the DNSs of the viscous-flow counterpart.

2.4 Euler's equations and their integration

The assumption of zero shear enables one to reduce (2.1.1) to

$$\frac{\mathrm{d}u}{\mathrm{d}t} = X - \frac{\partial p}{\rho \partial x}, \quad \frac{\mathrm{d}v}{\mathrm{d}t} = Y - \frac{\partial p}{\rho \partial y}, \quad \frac{\mathrm{d}w}{\mathrm{d}t} = Z - \frac{\partial p}{\rho \partial z} \tag{2.4.1}$$

These are the celebrated Euler equations that were obtained by Leonhard Euler about 150 years before the evolution of the Navier–Stokes equations and about 60 years after Newton discovered a major part of mechanics. *Their beauty and originality lie in Euler's recognition that Newton's laws apply to every part of every system, whether discrete or continuous.* The inviscid-fluid assumption also implies that the pressure of the fluid is normal to the surface on which it acts. The use of the conditions of irrotationality [(2.2.1)] and the force potential enable one to reduce the three Euler equations into one equation:

$$\frac{1}{2}q^2 + \frac{\partial \phi}{\partial t} - \Omega + \frac{p}{\rho} = F(t) \tag{2.4.2}$$

where $q^2 = u^2 = v^2 + w^2$ and $F(t)$ is an arbitrary function of time only. The important point is that, at any instant, the left-hand side of (2.4.2) has the same value at all points of the region of irrotational motion and not merely about a streamline. Frequently $F(t)$ is absorbed into ϕ because this does not affect the physical quantities of interest. However, in some nonlinear problems this may cause difficulties (e.g., Whitham 1974) and must remain isolated.

For steady flow $\partial \phi / \partial t = 0$, which yields

$$\frac{1}{2}q^2 + \frac{p}{\rho} + gh = \text{constant} \tag{2.4.3}$$

This is of course the familiar Bernoulli equation and enables one to determine the pressure distribution once the potential function and hence the velocity distribution is obtained from the solution of Laplace's equation. It must be remembered that (2.4.3) is valid for steady flows with the provisions that either the flow is irrotational or the calculation is confined to one streamline or the velocity vector and the rotation vector coincide.

An essential feature of irrotational flow is that it represents the instantaneous response of the fluid, uninfluenced by its previous history, to the immediately prevailing boundary conditions, i.e., the properly posed boundary conditions determine it uniquely. For the case of flow generated by a solid body moving through homogeneous fluid, the appropriate boundary condition is that the normal components of velocity of the solid boundary must be impressed on the fluid adjacent to it. In a general form this is expressed as

$$\frac{\partial \phi}{\partial n} = V_n \quad \text{on the surface} \tag{2.4.4}$$

where n is the direction normal to the surface and V_n is the velocity of the surface normal to itself. If the boundary is rigid and fixed, such as a rocky sea floor, then V_n is zero and we simply have

$$\frac{\partial \phi}{\partial n} = 0 \tag{2.4.5}$$

In the case of a free surface, as on a water wave, a kinematic and a dynamic boundary condition are needed. The kinematic condition states that any particle, which lies at the free surface at any instant, will never leave it. This leads to (see, e.g., Lamb 1932)

$$\frac{\partial \phi}{\partial z} = \frac{\partial \eta}{\partial t} + u\frac{\partial \eta}{\partial x} + v\frac{\partial \eta}{\partial y} \quad \text{at } z = \eta \tag{2.4.6}$$

where $z = \eta(x, y, t)$ represents the free surface (for additional discussion of the free-surface conditions, see Chapter 4 and also Sarpkaya 1996a).

The dynamic free-surface condition requires that the pressure difference across the interface result in a force normal to the boundary that is due wholly to surface tension. This condition takes the form

$$p = p_a + \sigma(1/R_1 + 1/R_2) \tag{2.4.7}$$

where σ is the surface tension, R_1 and R_2 are the radii of curvature of the free surface in any two orthogonal directions, and $(p - p_a)$ is the pressure difference across the interface.

When the free surface is uncontaminated so that σ is taken as zero, the flow is irrotational such that Eq. (2.4.2) describes the pressure p within the fluid, and the (atmospheric) pressure just outside the liquid is constant, the free-surface condition reduces to (see Sarpkaya 1996a)

$$\frac{\partial \phi}{\partial t} + \frac{1}{2}q^2 + g\eta = F(t) \quad \text{at } z = \eta \tag{2.4.8}$$

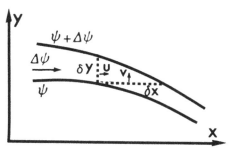

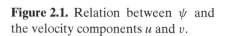

Figure 2.1. Relation between ψ and the velocity components u and v.

2.5 Stream function

Lagrange's stream function (first introduced by d'Alembert) is a scalar quantity that describes not only the geometry of a *two-dimensional flow* but also the components of the velocity vector at any point and the flow rate between any two streamlines.

For the type of flow and fluid under consideration, the flow rate between the two streamlines is independent of the path of integration. Thus, for a flow from left to right (see Fig. 2.1), one has

$$u = \frac{\partial \psi}{\partial y} \quad \text{and} \quad v = -\frac{\partial \psi}{\partial x} \tag{2.5.1}$$

Thus the partial derivatives of ψ with respect to any direction give the velocity components in the direction 90° clockwise to that direction. The definition of a stream function does not require that the motion be irrotational. In other words, ψ exists irrespective of whether the flow is rotational or irrotational, as long as it is continuous.

For an irrotational 2D flow, the condition of zero rotation yields

$$\omega_z = \frac{\partial v}{\partial x} - \frac{\partial u}{\partial y} = \frac{\partial^2 \psi}{\partial x^2} + \frac{\partial^2 \psi}{\partial y^2} = \nabla^2 \psi = 0 \tag{2.5.2}$$

Lagrange's stream function for an irrotational 2D flow is orthogonal to the potential function and as such may be regarded as the imaginary part of the complex function $F(z) = \phi + i\psi$. Also, ϕ and ψ in $F(z)$ may be interchanged because both satisfy Laplace's equation. Furthermore, each has to satisfy the condition leading to the existence of the other in order to represent a continuous, 2D, irrotational flow.

In polar coordinates the velocity components become

$$v_r = \frac{\partial \psi}{r \partial \theta} = \frac{\partial \phi}{\partial t} \tag{2.5.3a}$$

$$v_\theta = -\frac{\partial \psi}{\partial r} = \frac{\partial \phi}{r \partial \theta} \tag{2.5.3b}$$

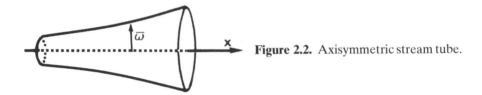

Figure 2.2. Axisymmetric stream tube.

Stokes' stream function, which is applicable only to *axisymmetric* flows, is based on the same concept, i.e., definition of a scalar function that yields the velocity components at any point, gives a measure of the flow rate, and thus automatically satisfies the equation of continuity. In cylindrical coordinates, the velocity components are given by (see Fig. 2.2)

$$u = \frac{\partial \psi}{\varpi \, \partial \varpi} = \frac{\partial \phi}{\partial x} \quad \text{and} \quad v_\varpi = -\frac{\partial \psi}{\varpi \, \partial x} = \frac{\partial \phi}{\partial \varpi} \tag{2.5.4}$$

that satisfy the equation of continuity given by

$$\frac{\partial (\varpi \, v_\varpi)}{\partial \varpi} + \frac{\partial (\varpi \, u)}{\partial x} = 0 \tag{2.5.5}$$

If the motion is assumed to be irrotational, then

$$\frac{\partial v_\varpi}{\partial x} - \frac{\partial u}{\partial \varpi} = \frac{\partial^2 \psi}{\partial x^2} + \frac{\partial^2 \psi}{\partial \varpi^2} - \frac{1}{\varpi} \frac{\partial \psi}{\partial \varpi} = 0 \tag{2.5.6}$$

In terms of ϕ, (2.5.6) is given by

$$\frac{\partial^2 \phi}{\partial x^2} + \frac{\partial^2 \phi}{\partial \varpi^2} + \frac{1}{\varpi} \frac{\partial \phi}{\partial \varpi} = 0 \tag{2.5.7}$$

The *sign difference* in the last terms of (2.5.6) and (2.5.7) should be noted.

It is possible to represent the velocity as a function of dual stream functions of three-dimensional (3D) flow fields that include, as special cases, both *Lagrangian* and *Stokesian stream functions*.

2.6 Basic inviscid flows

Even the drastic as well as convenient assumption of an inviscid flow does not lead to a direct solution of the equation of Laplace because the boundary conditions are often quite difficult to satisfy. Most of the practical useful solutions are indirect and require the combination of simpler known solutions. When the superposition of basic flow patterns is not sufficient to obtain a solution, one must resort to other exact or approximate methods such as conformal transformations, Fourier series, eigenfunctions, spherical or ellipsoidal harmonics, large-eddy simulations, distributed singularities, vortex-element methods, boundary-element methods, finite-element methods, and DNSs (see, e.g., Chatelain *et al.* 2008b, who used one *billion* vortex particles to simulate the evolution of aircraft wakes). The application of any one of these methods requires considerable work, computer power, and ingenuity, depending on the complexity of the problem and the accuracy desired. In any case one must, at the onset, ascertain that the time and effort required for attaining a solution is commensurate with the need for a solution and with the idealizations imposed on the real flow. For example, the use of the distributed sources and source panels to analyze the diffraction effects on a large circular cylinder is quite reasonable when separation is not likely to occur. On the other hand, the application of the same method to large bodies with sharp corners may be questionable because separation occurs at all relative displacements regardless of the magnitude of displacement. *For a viscous flow, the inviscid value of the added mass is not meaningful beyond a small fraction of time after the onset of flow* (the sharper the corners of the body, the shorter the time). In other words, the theoretical value of the added mass for bodies with sharp edges can be experienced only for a fleeting moment. The inviscid-flow values of the added mass (tabulated in books and in special reports) often give the impression that they are applicable to all unsteady *viscous* flows, at all stages of the flow, at all Reynolds numbers. Such an *incorrect* impression is so prevalent that one often encounters general force expressions or force decompositions for viscous flows using *the linear sum of a velocity-square-dependent drag force* plus an *ideal inertial force* (Lighthill 1986; Sarpkaya 2000, 2001b).

Tables 2.1 and 2.2 give the characteristics of basic two- and three-dimensional flows. Plots of the corresponding potential and stream functions may be found in many reference texts (see, e.g., Milne-Thomson 1960).

2.7 Force on a circular cylinder in unseparated inviscid flow

Because of its central importance in fluid mechanics, the inviscid flow about a circular cylinder and the resulting forces are described in some detail.

The proper combination of uniform flow U with a *doublet* represents the inviscid flow about a circular cylinder at rest. We have (see Table 2.1)

$$F(z) = U\left(z + \frac{c^2}{z}\right) \quad \text{and} \quad \psi = -\mu\frac{\sin\theta}{r} + Ur\sin\theta, \quad r \geq c \qquad (2.7.1)$$

where $\mu = Uc^2$ and c is the radius of the cylinder. The potential function becomes

$$\phi = U(r + c^2/r)\cos\theta \qquad (2.7.2)$$

Table 2.1. *Potential and stream functions for two-dimensional flows*

Uniform flow	$\phi = U_x,\ \psi = U_y,\ F(z) = U_z$
Point source	$\phi = (Q/2\pi)\mathrm{Ln}r,\ \psi = (Q/2\pi)\theta,$
	$F = (Q/2\pi)\mathrm{Ln}z$
Velocities	$u = \partial\phi/\partial x = \partial\psi/\partial y,\ \mathrm{v} = \partial\phi/\partial y = -\partial\psi/\partial x,$
	$\dfrac{dF(z)}{dz} = u - iv,\ v_r = \dfrac{\partial\psi}{r\,\partial\theta} = \dfrac{\partial\phi}{\partial r},$
	$v_\theta = -\dfrac{\partial\psi}{\partial r} = \dfrac{\partial\theta}{r\,\partial\theta}$
Doublet	$\phi = \mu x/r^2,\ \psi = \mu y/r^2,\quad F = \mu/z$
Vortex	$\phi(\Gamma/2\pi)\theta,\ \psi = (\Gamma/2\pi)\mathrm{Ln}r,$
	$F = -(i\Gamma/2\pi)\mathrm{Ln}z$
Uniform flow at angle α	$\phi = U(x\cos\alpha + \sin\alpha)\quad F = Uze^{i\alpha}$
	$\psi = U(y\cos\alpha - x\sin\alpha)$
Flow about a circular cylinder	$\phi = Ur\cos\theta + (Uc^2/r)\cos\theta$
	$\psi = Ur\sin\theta - (Uc^2/r)\sin\theta$
	$F = Uz + Uc^2/z$
Flow about a cylinder with circulation	$F = U(z + c^2/z) - (i\Gamma/2\pi)\mathrm{Ln}z$
Line source of uniform strength over length L	$\phi = (Q/4\pi L)\displaystyle\int_0^L \mathrm{Ln}[(x - \xi)^2 + y^2]d\xi$
	$\psi + (Q/4\pi L)\displaystyle\int_0^L [\tan^{-1}\{y/(x - \xi)\}]d\xi$

The velocity at the surface of the circular cylinder, necessarily tangent to the cylinder, is

$$v_\theta = -2U\sin\theta \qquad (2.7.3)$$

Then the pressure on the cylinder reduces to

$$p = \frac{1}{2}\rho U^2(1 - 4\sin^2\theta) \qquad (2.7.4)$$

where U is the maximum ambient velocity normal to the cylinder. The symmetry of the pressure distribution with respect to the x and y axes and hence the absence of both the drag and lift forces have been well known since d'Alembert and they are not paradoxes, simply consequences of *no viscosity* and *no separation*. In fact, one would hypothesize that there can be no

Table 2.2. *Potential and stream functions for axisymmetric flows*

Uniform flow	$\phi = Ux = Ur\cos\theta \quad \psi_s = 0.5Ur^2\sin^2\theta = 0.5U\omega^2$
	$u = \dfrac{\partial\phi}{\partial x}, \quad v_\omega = \dfrac{\partial\phi}{\partial\omega}, \quad \Delta Q = 2\pi\,\Delta\psi_s$
	$u = \dfrac{1}{\omega}\dfrac{\partial\Psi_s}{\partial\omega}, \quad v_\omega = -\dfrac{1}{\omega}\dfrac{\partial\Psi_s}{\partial x}$
Source (3D)	$\phi = -Q/(4\pi r), \quad \psi_s = -(Q/4\pi)\cos\theta$
Sink (3D)	$\phi = +Q/(4\pi r), \quad \psi_s = +(Q/4\pi)\cos\theta$
Doublet (3D)	$\phi = \mu\cos\theta/r^2, \quad \pi_s = -\mu\sin^2\theta/r$
Flow about a	$\phi = Ur[1 + c^3/(2r^3)]\cos\theta$
sphere of	$\phi = Ur[1 + G(r)]\cos\theta$
radius c	$\psi = 0.5Ur^2(1 - c^3/r^3)$
Translation of a sphere	$\phi = [Uc^3/(2r^2)]\cos\theta \quad \psi_s = -[Uc^3/(2r)]\sin^2\theta$
in a fluid	(direction of motion is in the x direction)
Axisymmetric	$\phi = [-Q/(4\pi L)]Ln[r_2 - x + L)/(r_1 - x)]$
line source of	$\psi = [-Q/(4\pi L)](r_1 - r_2)$
uniform distribution	Q = total strength of the line source

separation in inviscid flows. In some cases this leads to unreal flows (e.g., a streamline turning 180° around a sharp corner). For this reason, Herman von Helmholtz (1868) introduced the inviscid *discontinuous fluid motions* or *free-streamline theory*. This proved to be a powerful technique for the analysis of high-speed jets of considerable industrial significance (see, e.g., Birkhoff and Zarantonello 1957; Sarpkaya 1959, 1961).

If the ambient flow is a function of time, i.e., $U = U(t)$, then the pressure is given by Bernoulli's equation as

$$p = -0.5\rho q^2 - \rho\frac{\partial\phi}{\partial t} \tag{2.7.5}$$

Then the force exerted on the cylinder by the fluid in the x direction becomes

$$F = -\int_c p\,ds\,\cos\theta = \int_c 0.5\rho q^2\cos\theta\,ds + \int_c \rho\frac{\partial\phi}{\partial t}\cos\theta\,ds \tag{2.7.6}$$

Note that θ is measured in the counterclockwise direction. The second integral is clearly zero because of symmetry, as just seen. Thus, by inserting ϕ

from (2.7.2) into (2.7.6), one has

$$F = 2c^2 \rho \frac{\partial U}{\partial t} \int_0^{2\pi} \cos^2\theta \, d\theta = \pi\rho c^2 \frac{\partial U}{\partial t} + \pi\rho c^2 \frac{\partial U}{\partial t} \qquad (2.7.7)$$

It is trivial to show *through the use of a cylinder moving in a fluid otherwise at rest* that one half of the force in (2.7.7) is due to the pressure gradient to accelerate the flow and one half is due to the singularity (doublet) producing the circular cylinder. The genesis of both parts and the generalization of (2.7.7) to 3D bodies are discussed later.

2.8 Slow motion of a spherical pendulum in viscous flow

This section deals with forces acting on a body (a cylinder or sphere) performing *small-amplitude oscillations* (with amplitudes smaller than **one tenth** of the diameter) about a mean position in a viscous fluid otherwise at rest. *For larger amplitudes and for bodies with sharp corners, one has to resort to empirical theories and/or physical or numerical (DNS) experiments.* It must also be emphasized that the equations used are the linearized versions of the Navier–Strokes equations. In other words, the boundary-layer theory is not applicable *at any amplitude*.

The understanding of the forces acting on a body performing *sinusoidal oscillations in a viscous fluid* is extremely important. However, theoretical solutions (*limited to small amplitudes*), even for unseparated flow about the simplest bodies, such as a sphere or circular cylinder, require considerable mathematical, experimental, and numerical dexterity. For separated flows, the determination of the time-dependent forces (drag and lift) can be approximated only by empirical equations and verified by difficult experiments. The DNSs can be carried out at only relatively small Reynolds numbers (less than about 10^4) over a limited range of the governing parameters. This is the state of the art.

In preparation for a detailed discussion of the *inappropriately named* "*added mass*," we first introduce a remarkable solution by Stokes (1851) of the *unseparated* flow about a spherical **pendulum oscillating** *rectilinearly along a diameter with a velocity* $U = -A\omega \cos\omega t$ in a fluid otherwise at rest. The time-dependent force acting on the sphere (with the limitations of amplitude previously noted) is given by

$$F(t) = 3\pi\mu DU \left(1 + \frac{1}{2}\sqrt{\frac{\rho\omega D^2}{2\mu}}\right) + \frac{\pi\rho D^3}{6}\left(\frac{1}{2} + \frac{9}{2}\sqrt{\frac{2\mu}{\rho\omega D^2}}\right)\frac{dU}{dt} \qquad (2.8.1a)$$

that, for the case of flow *oscillating* about a sphere *at rest*, reduces to

$$\frac{F(t)}{\frac{1}{2}\rho\frac{\pi D^2}{4}U_m^2} = \frac{24}{Re}\left(1 + \frac{1}{2}\sqrt{\pi\frac{Re}{K}}\right)\cos\omega t + \frac{8\pi}{3K}\left(\frac{3}{2} + \frac{9}{2}\sqrt{\frac{K}{Re}}\right)\sin\omega t$$

(2.8.1b)

where the pressure gradient to accelerate the flow *about a sphere* increases the constant $1/2$ to $3/2$ in the second parenthesis of (2.8.1b). It must be noted that (2.8.1a) and (2.8.1b) are based on the assumption that *the sinusoidal motion has been well established.* They do not deal with the transit period in which an impulsively started sinusoidal flow reaches a periodic state.

The parameters governing the motion are the Reynolds number $Re = U_m D/v$ and the Keulegan–Carpenter (1956, 1958) number $K = U_m T/D$, with U_m as the amplitude of the velocity. Sarpkaya (1976a, 2005) was the first to show through the use of extensive experiments in a large U-shaped oscillating flow tunnel that $Re/K = \beta = D^2/vT = fD^2/v$ is a universal parameter for all oscillating flows (*laminar, turbulent, separated, or unseparated*), and it is not restricted to unseparated laminar flows at very small Reynolds numbers. These are discussed in more detail later.

Equation (2.8.1b) shows that both the *in-phase* and the *out-of-phase* components of the normalized force contain β, i.e., viscous and inertial forces. *The interdependency of the two components renders the creation of a physics-based force-prediction model, valid for all β, extremely difficult, if not impossible.*

One could decompose (2.8.1b) into as many as four components. However, in each case *the in-phase and the out-of-phase components of the force continue to depend on both Re and K.* Thus it is reasonable to think that the force exerted on a sphere (or any other bluff body) is not a simple juxtaposition of a velocity (as in Stokes) or velocity-square-dependent drag force and an acceleration-dependent inertial force. *Both depend on K and Re.* Consequently, *the inertial force will not be equal to its ideal value and the drag force will not be equal to its steady-state value at a given velocity.* Obviously this will complicate the analysis and design of smooth, roughened, and arbitrarily oriented structures subjected to wind and wave forces in all kinds of sea states and will require a certain degree of empiricism, based on experience gained from past successes and failures.

Unlike the preceding solution of Stokes, Basset (1888) expressed the effect of the smooth evolution of the unsteady viscous flow with a *"history term,"* which yields *exactly* the same result as the viscosity-dependent terms in Stokes' solution, but draws attention to the important fact that *unsteady flow is not a juxtaposition of instantaneous steady states* (i.e., it is not like the

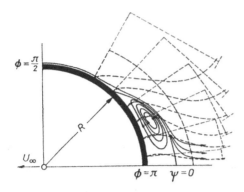

Figure 2.3. Boundary layer on the downstream side of a sphere accelerated impulsively after the onset of separation (from Boltze). The sphere has traversed a distance of 0.6 *R*.

frames of a digital motion picture, *it is uninterrupted in time except by separation*; Sarpkaya 1996b). Even though the solutions of Stokes and Basset are valid only for unseparated laminar flows at *very small Reynolds numbers and amplitudes,* the fundamental concepts gleaned from them are universally applicable to all unsteady flows about bluff bodies. It must be emphasized that separation causes large-scale pressure changes. Thus the theoretical value of the *"added inertia"* (mistakenly called the *"added mass"*) can be experienced only for a fleeting moment, especially for bodies with sharp edges. Figure 2.3 shows the instant of separation ($Ut/D = 0.196$) of a sphere started impulsively from rest (Boltze 1908).

2.9 Added mass or added inertia

About 200 years ago, Chevalier DuBuat, among many others, experimenting with spheres oscillating in water, and later Bessel in 1828, experimenting with spherical pendulums in air and water (*at very low Reynolds numbers and very small amplitudes*) found that it is necessary to attribute to the sphere an ***added mass*** Δm, making it behave as if it had a mass ($m + \Delta m$), larger than the actual mass of the sphere. In reality, ***no mass is added to or subtracted from the body***. The physical shape and mass of the body within the *incompressible control volume* remain invariant. *What is imparted to the fluid (or the body) is positive or negative accelerations or inertia (per unit mass) or changes in kinetic energy that are due to the motion of the body, which can be negative or positive.* In other words, the *increase* (*or* decrease) of *the kinetic energy* of the fluid within the control volume or the *quotient of the additional force* required to produce the accelerations throughout the fluid divided by the acceleration of the body manifest themselves as "added mass."

The mathematical study of the phenomenon was begun by Poisson, who disregarded the viscosity (and, of course, all consequences of separation, turbulence, roughness, change of direction, body rotation, density gradients, body proximity to others, etc.) and obtained $k = 0.5$ for the sphere, a result that was subsequently confirmed by Green, Plana, Stokes, and Lamb, all of whom used different methods (Dryden *et al.* 1956).

Bessel defined the *"added mass"* **in terms of the mass of the fluid displaced by the body,** i.e., $\Delta m/(displaced\ mass) = k$. His k values were 0.9 and 0.6 for air and water, respectively. Unfortunately, the *oversimplification* or the **misinterpretation** of his definition (based on a laminar, unseparated, linearized, V^1-*dependent equation*, Eq. (2.4.1), became a source of confusion and created the impression that the mass of the solid body has literally increased by some miraculous means and that there can be a "**positive**" or a "**negative**" mass. *Obviously, what is increased or decreased is the acceleration or deceleration of the fluid (**or the negative or positive inertia per unit mass**) or the increase or decrease of the kinetic energy of the fluid within the control volume, all of which can be negative or positive!* Thus it follows that what is mistakenly called "added mass" can be just as positive or negative as the inertia that cannot be invested with a bodily form!

It must be emphasized that the normalized *ideal* values of the added-mass coefficients (tabulated here and elsewhere; see Table 2.3) and the added-mass moment of inertia *in a viscous fluid* exist only for a **brief moment in time**, i.e., immediately after *the onset of the acceleration–deceleration or at the onset of the transformation of the existing steady state of the flow to an accelerated unsteady state, but **before the separation and viscous effects set in and/or before the amplitude or displacement of the motion exceeds a very small value** (about one tenth of the diameter of a cylinder or sphere).

It is also conceivable that a body, while in a new steady state or in a state of acceleration, may be subjected to an even larger acceleration or deceleration, making the dynamics of flow even more complex. Thereafter, viscous effects, separation, cavitation, body proximity, the kinetic energy of the fluid within the control volume, just to name a few, will irreversibly change and with them the kinetic energy, or the added inertia, or what is erroneously called the added mass. It must be clearly understood that the viscous effects manifest themselves as an integral part of the impulsively started flow as soon as the acceleration is imposed on the flow (examples are given in Chapter 3). In other words, the force acting on a cylinder subjected to impulsively started inertia is never free from viscous effects. Often confusion arises from the misunderstanding of the "added mass." It is not fully realized that "the added mass" is *added inertia or added kinetic energy* and carries with it all the consequences of no-slip, no-penetration, acceleration, and *viscous forces*.

Obviously, the impulsively imposed inertia dominates the flow for a very short time. Subsequently the flow separates, leading to usual vortex shedding, fluctuating lift, drag, and lateral forces. (Additional examples of impulsively started flow about circular cylinders and flat plates are given in

Table 2.3. *Added masses of various bodies**

	SHAPE	ADDED MASS PER UNIT LENGTH ←→ MOTION
(circle, c)	CIRCLE	$\rho\pi c^2$
(ellipse, b, a)	ELLIPSE	$\rho\pi b^2$
(ellipse, a, b)	ELLIPSE	$\rho\pi a^2$
(plate, $2w$)	PLATE	$\rho\pi w^2$

RECTANGLE (2b wide, 2a tall)

a/b			a/b		
∞	1.00 $\rho\pi a^2$		1	1.51 $\rho\pi a^2$	
10	1.14 "		0.5	1.70 "	
5	1.21 "		0.2	1.98 "	
2	1.36 "		0.1	2.23 "	(Wendel 1950)

DIAMOND (2b wide, 2a tall)

a/b		
2	0.85 "	
1	0.76 "	
0.5	0.67 "	
0.2	0.61 "	(Wendel 1950)

I-BEAM (2a, 2b, c)
a/c = 2.6
b/c = 3.6
           2.11 $\rho\pi a^2$ (Patton 1965)

REGULAR POLYGON (OR n SIDED), a

n = 3	0.654 $\rho\pi a^2$	
4	0.787 "	
5	0.823 "	
6	0.867 "	
∞	1.000 "	(Wendel 1950)

(*continued*)

Sarpkaya 1966). The important facts emerging from these experiments are these: Once the flow is set in motion, disturbed by separation and vortex shedding, the instantaneous forces acting on the cylinder cannot be predicted, except by 2D numerical simulations. In the course of the transition period there are no "Morison-" like equations because the evolution of the wake is still in a transient state (not a periodically definable motion). In any case, the impulsive inertia applied to a body (and the evolution of the

Table 2.3 *(continued)*

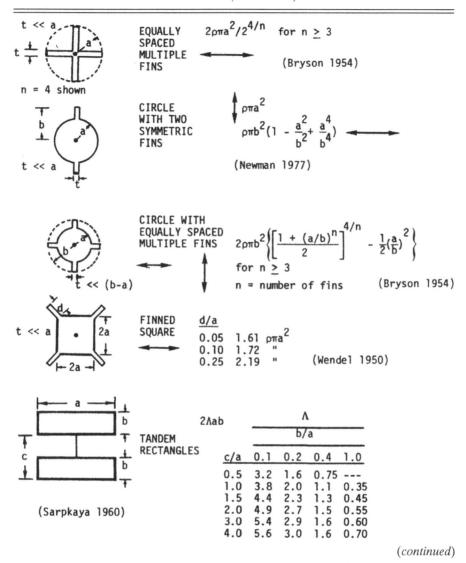

| | EQUALLY SPACED MULTIPLE FINS | $2\rho\pi a^2/2^{4/n}$ for $n \geq 3$ |
| | | (Bryson 1954) |

$n = 4$ shown

	CIRCLE WITH TWO SYMMETRIC FINS	$\rho\pi a^2$
		$\rho\pi b^2(1 - \dfrac{a^2}{b^2} + \dfrac{a^4}{b^4})$
		(Newman 1977)

	CIRCLE WITH EQUALLY SPACED MULTIPLE FINS	$2\rho\pi b^2\left\{\left[\dfrac{1 + (a/b)^n}{2}\right]^{4/n} - \dfrac{1}{2}(\dfrac{a}{b})^2\right\}$
		for $n \geq 3$
		n = number of fins (Bryson 1954)

	FINNED SQUARE	d/a		
		0.05	1.61	$\rho\pi a^2$
		0.10	1.72	"
		0.25	2.19	" (Wendel 1950)

TANDEM RECTANGLES	$2\Lambda ab$		$\dfrac{\Lambda}{b/a}$		
	c/a	0.1	0.2	0.4	1.0
	0.5	3.2	1.6	0.75	---
	1.0	3.8	2.0	1.1	0.35
	1.5	4.4	2.3	1.3	0.45
	2.0	4.9	2.7	1.5	0.55
	3.0	5.4	2.9	1.6	0.60
	4.0	5.6	3.0	1.6	0.70

(Sarpkaya 1960)

(continued)

flow about the body) depends on the character of the motion of the body. One should not expect to formulate a force–time relationship to quantify the early stages of the kinematics and dynamics of **impulsively started** flows. *As noted earlier, the unsteady flow of a viscous fluid is* **not a juxtaposition of instantaneous steady states.** *For* very *slow unseparated periodic motions* of a smooth body (say a sphere or circular cylinder, similar to those experimented with by Bessel and DuBuat or theorized by Stokes), an approximate "added mass" may be defined.

For cylinders and all other bodies subjected to waves and vagaries of the currents and gusts, one can speak of *only* the *drag* and *inertia* coefficients

Table 2.3 *(continued)*

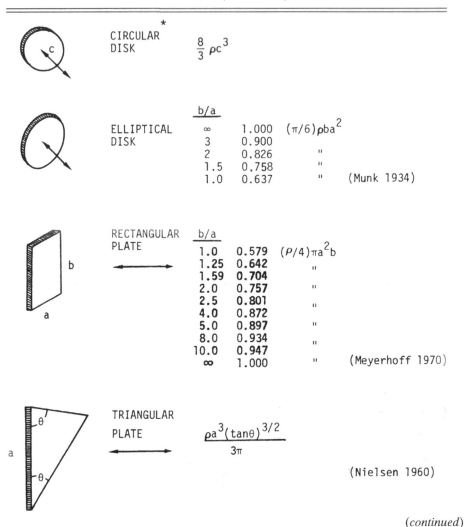

CIRCULAR DISK*		$\frac{8}{3}\rho c^3$		

	b/a			
ELLIPTICAL DISK	∞	1.000	$(\pi/6)\rho ba^2$	
	3	0.900		
	2	0.826	"	
	1.5	0.758	"	
	1.0	0.637	"	(Munk 1934)

RECTANGULAR PLATE	b/a			
	1.0	0.579	$(\rho/4)\pi a^2 b$	
	1.25	0.642	"	
	1.59	0.704		
	2.0	0.757	"	
	2.5	0.801	"	
	4.0	0.872	"	
	5.0	0.897	"	
	8.0	0.934	"	
	10.0	0.947	"	
	∞	1.000	"	(Meyerhoff 1970)

TRIANGULAR PLATE	$\dfrac{\rho a^3 (\tan\theta)^{3/2}}{3\pi}$	
		(Nielsen 1960)

(continued)

(obtained experimentally) and of the Morison equation[*] to use the *cycle-averaged values of the force coefficients (C_d and C_m). Then one cannot use an ideal added-mass coefficient (picked up from the tables) in dealing with separated steady or unsteady flows, if one is going to use the empirical Morison equation.* This has been a source of confusion over the years simply because the meaning of **added inertia** has not been fully understood.

Numerical predictions (say using DNS) at low Re will not (unfortunately) eradicate the so-called Morison equation for centuries to come because the laws of the physics of turbulence have not yet been discovered. The ideal values of the added mass provide only an *upper limit* (albeit often grossly

---

[*] $F = \frac{1}{2}C_d\rho D|U|U + \rho C_m \frac{\pi D^2}{4}\frac{dU}{dt}$ *(for a stationary structure in unidirectional flow).*

Table 2.3 *(continued)*

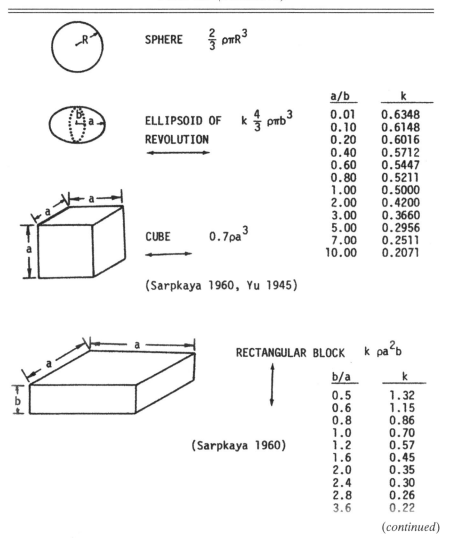

SPHERE $\frac{2}{3}\rho\pi R^3$

ELLIPSOID OF $k\frac{4}{3}\rho\pi b^3$
REVOLUTION

a/b	k
0.01	0.6348
0.10	0.6148
0.20	0.6016
0.40	0.5712
0.60	0.5447
0.80	0.5211
1.00	0.5000
2.00	0.4200
3.00	0.3660
5.00	0.2956
7.00	0.2511
10.00	0.2071

CUBE $0.7\rho a^3$

(Sarpkaya 1960, Yu 1945)

RECTANGULAR BLOCK $k\,\rho a^2 b$

b/a	k
0.5	1.32
0.6	1.15
0.8	0.86
1.0	0.70
1.2	0.57
1.6	0.45
2.0	0.35
2.4	0.30
2.8	0.26
3.6	0.22

(Sarpkaya 1960)

(continued)

large and grossly inaccurate!) for the *inertial component* of the total force in Morison's empirical equation. They are the added-mass values tabulated in books. *They are accurate only within a small fraction of time (say about 1 sec) following the impulsive start of the motion. Immediately thereafter, separation and vortex shedding take over, and there is no way (other than the recording of the force) to determine how the vortices, separation, and the lift and drag forces will evolve. When the flow about the body becomes reasonably periodic, one can use a Morison-like equation. In other words, there are two major stages to an impulsively started flow about a cylinder: The initial instants before the separation begins and the period during which a number of vortices shed and the alternative vortex shedding evolves. When the oscillation reaches a quasi-steady state, the forces and the vortex shedding about*

Table 2.3 *(continued)*

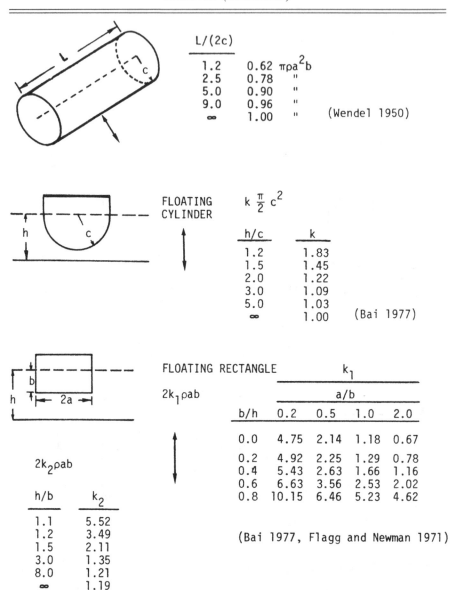

L/(2c)		
1.2	0.62	$\pi\rho a^2 b$
2.5	0.78	"
5.0	0.90	"
9.0	0.96	"
∞	1.00	" (Wendel 1950)

FLOATING CYLINDER $k \frac{\pi}{2} c^2$

h/c	k
1.2	1.83
1.5	1.45
2.0	1.22
3.0	1.09
5.0	1.03
∞	1.00 (Bai 1977)

FLOATING RECTANGLE

$2k_1 \rho ab$

	k_1			
		a/b		
b/h	0.2	0.5	1.0	2.0
0.0	4.75	2.14	1.18	0.67
0.2	4.92	2.25	1.29	0.78
0.4	5.43	2.63	1.66	1.16
0.6	6.63	3.56	2.53	2.02
0.8	10.15	6.46	5.23	4.62

$2k_2 \rho ab$

h/b	k_2
1.1	5.52
1.2	3.49
1.5	2.11
3.0	1.35
8.0	1.21
∞	1.19

(Bai 1977, Flagg and Newman 1971)

* All bodies shown above are assumed to be very thin.

the cylinder become more or less repetitive, never becoming exactly the same because of a large number of secondary influences.

For example, the forced oscillations of the water column in a moonpool (a vertical well passing through the air–sea interface in offshore platforms) can be calculated only at the moments of zero velocity (maximum and minimum displacements), ignoring the shedding of the distorted ring vortices. At all other times, *the added inertia and damping coefficients* can be determined only experimentally.

The added-inertia concept is not confined to marine hydrodynamics. For example, while taking off, hovering, or landing, not only aircraft but also birds and insects give rise to and make use of the added inertia provided by the rapid flapping of the wings (or the rapid rise or descent of the aircraft, stronger vortices, larger lift, etc.; see, e.g., Sarpkaya and Kline 1982).

There are 21 independent components of the added-inertia matrix that include the translations and rotations along and about all three axes. In the case of a body with three orthogonal planes of symmetry, the 21 coefficients reduce to 6. For a sphere, 3 of these are equal to 1/2, and the other 3 are equal to zero.

The *instantaneous values* of the drag, lift, and any one component of the added mass depend on many parameters. Only a few of them are given here: the size, shape, orientation, the path of motion of the body, flexibility of the body, density of fluid, stratification of the surrounding fluid, dissolved gasses in water, cavitation, separation of flow (vortex shedding, etc.), proximity of other bodies (walls, or a large number of other rigid or deformable bodies), the flexibility, deformability, and surface roughness (size, shape, and distribution of rigid as well as soft excrescences), or free-surface proximity, waves, currents, and internal waves, the prevailing wave and current motion near the free surface, moving from one medium to another (air–water interface), density stratification, sloshing of liquids (if any) inside the body, porosity of the body (the number, size, shape and distribution of holes, hole sizes and their shapes, and pitches) and many more. Evidently, the quantification of the motion of a body in a relatively dense medium (say in water) or in an air–water interface is very complex and poses extremely difficult theoretical, numerical and experimental problems.

The tabulated values of the added-inertia coefficients here (see Table 2.3) and elsewhere *simply represent the theoretical values arrived at assuming unidirectional, inviscid, **unseparated**, impulsively started flow. For a viscous flow, the inviscid value of the added mass is not meaningful beyond a small fraction of time* (the sharper the corners of the body, the shorter the time). The inviscid-flow values of the added mass (tabulated in books and in special reports) often give the impression that they are applicable to all unsteady *viscous* flows, at all stages of the flow, at all Reynolds numbers. **Such an incorrect impression is so prevalent that one often encounters general force expressions or force decompositions for viscous flows using *the linear sum of a velocity-square-dependent drag force* plus an acceleration-dependent ideal inertial force** (Lighthill 1986; Leonard and Roshko 2001). This is in spite of the fact that Stokes (1851) showed over 160 years ago that *viscosity and the state of the wake* do affect the added inertia (even in simpler flows) and that if the ideal inertial force is subtracted from the total force it still leaves some inertial force in the drag force. In other words, the force acting on a body *cannot* be reduced to the sum of the *ideal inertial force* plus a *velocity-square-dependent drag force*. Both forces modify each other. These are discussed in more detail in Chapter 3. Here, a relatively simple example is given to demonstrate the preceding facts and pave the way to more complicated time-dependent flows.

2.10 An example of the role of the added inertia

Let us consider the rise of a small air bubble (*assumed to be nonexpandable*) near the bottom of a bottle. Let us ignore the existence of all other bubbles, free-surface effects, and rigid boundaries. When the bubble is released from rest, it first accelerates and then reaches a constant velocity. The question is the determination of the initial acceleration of the bubble after its release. This relatively complicated problem can be simplified if we consider the fact that the flow about the bubble is not separated (*there is no boundary layer at the liquid–gas interface*), and the density of the gas in the bubble is much smaller than that of the liquid. The upward forces acting on the body are the buoyant force $\rho_w g V_b$, where ρ_w is the density of the surrounding liquid, g is the gravitational acceleration, and V_b and ρ_a are the volume and the density of the gas in the bubble, respectively. The downward force is the weight of the gas in the bubble. Then we have

$$\rho_w g V_b - \rho_a g V_b = \rho_a V_b \frac{dv}{dt} \qquad (2.10.1)$$

Simplifying, one has $dv/dt = (\rho_w/\rho_a) - 1$. Clearly this is a very large acceleration because ρ_w is much larger than ρ_a. The only reasonable explanation is that the bubble must be accelerating a greater mass (or imparting large changes in kinetic energy to its surroundings), which was not accounted for in the original formulation. Obviously the mass in question is the added mass or the added inertia (the second term in the parentheses). Thus one has

$$\rho_w g V_b - \rho_a g V_b = (\rho_a V_b + A_{ic} \rho_w V_b) \frac{dv}{dt} \qquad (2.10.2)$$

in which A_{ic} is the *added-inertia coefficient*. In general, it can be determined theoretically, experimentally, or both. Combining (2.10.2) with (2.10.3), we have

$$\frac{dv}{dt} = \frac{g(\rho_w - \rho_a)}{\rho_a + A_{ic} \rho_w} \qquad (2.10.3)$$

In this particular case $A_{ic} = \frac{1}{2}$ (well known as the added-mass coefficient for a sphere).

Clearly we may ignore ρ_a, noting that the density of the gas in the bubble is much smaller than the density of the surrounding fluid. Then the initial acceleration of the bubble reduces to about $2g$. *This is the difference the added inertia or added mass might make in the hydrodynamics of bodies.* One must hasten to note that the added mass is not always one half of the displaced mass of the body and its effect is not always as profound as in the preceding example.

2.11 Forces on bodies in separated unsteady flow

In general, a body has a mass M_b of its own and experiences a resistance (in *viscous* fluids) that is primarily due to *separation of the flow* or the resulting pressure forces. This is called the *form drag*. Assuming that the instantaneous values of the drag and inertial forces can be added (*this is a quantum jump in hydrodynamics* but *we do not know any better!*), the total force acting on a *cylinder moving* unidirectionally with a time-dependent velocity $U(t)$ becomes

$$F = \frac{1}{2}C_d(t)\rho A_p|U|U + [M_b + \rho V_b C_a'(t)]\left(\frac{\partial U}{\partial t} + U\frac{\partial U}{\partial x}\right) \qquad (2.11.1)$$

in which C_a' is the time-dependent added-mass coefficient. An experimentally determined average value of C_a' may be used (at present, there are no other alternatives, except for the unseparated flows discussed earlier).

For a body at rest in a unidirectional time-dependent flow, the force exerted on the body becomes

$$F = \frac{1}{2}C_d\rho A_p|U|U + \rho V_b[1 + C_a']\left(\frac{\partial U}{\partial t} + U\frac{\partial U}{\partial x}\right) \qquad (2.11.2)$$

in which A_p is the projected area of the body on a plane normal to the flow and V_b is the volume of the body. In the so-called "Morison equation" (to be discussed later) the term $\frac{dU}{dt}$ is replaced with $\frac{\partial U}{\partial t}$. This is acceptable only if the convective acceleration terms $\left(\frac{U\partial U}{\partial x} + \frac{W\partial U}{\partial z}\right)$ are small enough relative to the local acceleration $\frac{\partial U}{\partial t}$. In general, C_a' and C_d depend on time, body shape, representative Reynolds number, relative displacement of the fluid, and the parameters characterizing the history of the motion. **In practice**, they are time averaged and plotted as functions of Re and K, enabling the representation of the in-phase and out-of-phase components of the in-line force in an *approximate* equation, as previously noted. In other words, the inertia coefficient used in Morison's equation is $C_m = (1 + C_a)$, **where C_a is the cycle-averaged C'_a.** Likewise, C_d will henceforth represent its *cycle-averaged value.*

It is apparent from the foregoing that our inability to calculate the time-dependent forces on bluff bodies leads us to consider each time-dependent flow as unique, and the most appropriate means of analysis and evaluation of the force-transfer coefficients must be discovered through intuition, heuristic reasoning, and above all through observations, measurements, and numerical simulations.

As will be noted later, the separation of the flow gives rise to a time-dependent transverse force or lift force regardless of whether the ambient flow is steady or time dependent. If the relative displacement of fluid in one direction is not large enough, separation does not necessarily lead to an alternating vortex shedding (for K less than about 4–5). The separated flow about circular cylinders is discussed in more detail in Chapter 3.

The discussion of the fluid forces acting on a cylinder will not be complete without the consideration of the relative motion of the body in a fluid stream whose velocity may vary with both time and distance in the direction of cylinder motion, i.e., $U = U(x, t)$. Let us assume that the body is subjected to a displacement x, velocity $\dot{x}$, and acceleration $\ddot{x}$ in the direction of the incident stream and the diffraction effects are negligible. Then the force acting on the cylinder becomes

$$F = \frac{1}{2}\rho C_d A_p |U - \dot{x}|(U - \dot{x}) + \rho V_b(1 + C'_a)\left[\frac{\partial U}{\partial t} + (U - \dot{x})\frac{\partial U}{\partial x}\right] - \rho V_b C'_a \ddot{x}$$

(2.11.3)

The first term on the right-hand side of (2.11.3) represents the form drag; the second term, the inertial force that is due to the local and convective accelerations of the fluid about the body; and the third term, the inertial force that is due to the motion of the body, as it would be in a fluid otherwise at rest. The negative sign in front of the last term is due to the fact that the inertial fluid force acting on the accelerating body is in the opposite direction to the drag and inertial forces resulting from the motion of the fluid about the body.

Equation (2.11.3) may also be written as

$$F = \frac{1}{2}\rho C_d A_p |U - \dot{x}|(U - \dot{x}) + \rho V_b C_m \left[\frac{\partial U}{\partial t} - \ddot{x}\frac{C'_a}{C_m} + (U - \dot{x})\frac{\partial U}{\partial x}\right]$$

(2.11.4)

where the inertia coefficient $C_m = 1 + C'_a$. The simplicity of this equation should not obscure the following facts. First, the drag and inertia coefficients are not time invariant and depend on the Reynolds number, relative motion of the fluid, history of the motion, relative roughness, etc. The variation of the velocity at a given point with time and at a given time with space, the omnidirectionality of the ambient flow, the motion of the body in directions other than those of the ambient flow, interference with the wakes of other neighboring bodies, wall- and free-surface proximity effects, just to mention a few of the practically important conditions, play significant roles in the determination of the drag and inertia coefficients. Second, the form drag is not always necessarily expressible in terms of the square of the relative velocity. In fact, it may be necessary to devise more suitable expressions for the form drag depending on the relative magnitudes of the time-dependent velocities $U(t)$ and $\dot{x}(t)$.

2.12 Kinetic energy and its relation to added mass

The kinetic energy of a fluid region is given by

$$T = \frac{1}{2}\rho \iiint (u^2 + v^2 + w^2)dxdydz \qquad (2.12.1)$$

that, for an ideal-fluid flow with a single-valued potential, may be written as

$$T = \frac{1}{2}\rho \int \phi \frac{\partial \phi}{\partial n} dS \qquad (2.12.2)$$

where n is the outward normal and dS is the elemental area. The integral is evaluated over the boundary. The velocity components from (2.5.3) are often used to determine the ideal value of the fluid force, which must be added to the force necessary to accelerate the actual mass of the body in *vacuum*.

Let

$$\phi = U_i\phi_i = U_1\phi_1 + U_2\phi_2 + \cdots + U_6\phi_6\phi \qquad (2.12.3)$$

in which U_1, U_2, and U_3 represent the translational components of the velocity of the body; ϕ_1, ϕ_2, and ϕ_3 are the corresponding potentials for *unit velocity*; U_4, U_5, and U_6 are the angular velocities; and ϕ_4, ϕ_5, and ϕ_6 are the corresponding potentials for unit angular velocity (see Fig. 2.4). In other words, each ϕ represents the velocity potential that is due to a body motion with unit velocity in the mode indicated by the index i.

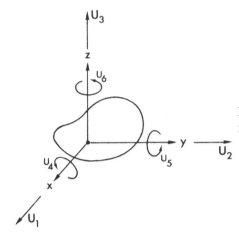

Figure 2.4. Rotational and translational motions of a body.

Substituting Eq. (2.12.3) into Eq. (2.12.2), one has

$$\frac{2T_f}{U_i U_j} = A_{ij} = \rho \int \phi_i \frac{\partial \phi_j}{\partial n} dS \qquad (2.12.4)$$

or

$$C_{ij}^a = \frac{A_{ij}}{\rho V_b} = \frac{1}{V_b} \int \phi_i \frac{\partial \phi_j}{\partial n} dS \qquad (2.12.5)$$

where V_b represents a suitable reference volume and dS is the elemental surface area.

Of the 36 added-mass coefficients, 9 are for translation, 9 for rotation, and 18 from the interaction between translation and rotation. Because of the fact that $C_{ij}^a = C_{ji}^a$ for inviscid fluids, one has 6, 6, and 9 components, respectively, or 21 in all.

The coefficients C_{ij}^a are independent of time and the translational and rotational velocities and accelerations. They depend on only the shape of the body, the proximity of other bodies or free surface, and the choice of the coordinates. Only the sum of the six elements represented by C_{ii}^a is independent of the choice of coordinates.

The foregoing could have been presented with a greater degree of mathematical sophistication. However, the type of flows considered herein and the motion of real fluids in general are such that the fundamental assumptions of inviscid fluid and no separation leading to (2.12.4) are rarely satisfied. The special cases for which the added-mass coefficients obtained through the use of the velocity potential may be used are as follows:

1. Initial instants of the motion when the body is set in motion impulsively from rest;

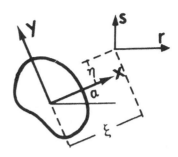

Figure 2.5. Coordinates (x, y) and (r, s).

2. When the body is subjected to relatively small-amplitude oscillations (so as to preclude the effects of separation) at a relatively high frequency;

3. Initial instants of the change of velocity of the body when the steady velocity of the body is changed impulsively to a new steady velocity.

The technique of vibrating the body in water is commonly used (Yu 1945; Stelson and Mavis 1955; Sarpkaya 1960; Chandrasekaran *et al.* 1972; Fickel 1973; Ciesluk and Colonell 1974; Brooks 1976) to determine the added-mass coefficient and to compare it with that obtained from the potential theory. The **ideal values of the added-mass coefficients** for various shapes of bodies may be found in J. L. Taylor (1930), McLachlan (1932), Wendel (1950), Weinblum (1952), Bryson (1954), Ackermann and Arbhabhirama (1964), McConnell and Young (1965), Patton (1965), Sedov (1965), Landweber (1967), Myers *et al.* (1969), Waugh and Ellis (1969), Meyerhoff (1970), Moretti and Lowery (1975), Chen *et al.* (1976), Au-Yang (1977), Gibson and Wang (1977), and Newman (1977).

For 2D bodies, the added-mass matrix with respect to another Cartesian coordinate system (r, s) is given by (Sedov 1965)

$$A_{rr} = A_{xx} \cos^2 \alpha + A_{yy} \sin^2 \alpha + A_{xy} \sin 2\alpha$$
$$A_{ss} = A_{xx} \sin^2 \alpha + A_{yy} \cos^2 \alpha - A_{xy} \sin 2\alpha$$
$$A_{rs} = \frac{1}{2}(A_{yy} - A_{xx}) \sin 2\alpha + A_{xy} \cos 2\alpha$$
$$A_{r\beta} = (A_{xx}\eta - A_{xy}\xi + A_{x\theta}) \cos \alpha + (A_{xy}\eta - A_{yy}\xi + A_{y\theta}) \sin \alpha$$
$$A_{s\beta} = -(A_{xx}\eta - A_{xy}\xi + A_{x\theta}) \sin \alpha + (A_{xy}\eta + A_{yy}\xi + A_{y\theta}) \cos \alpha$$
$$A_{\beta\beta} = A_{xx}\eta^2 + A_{xy}\xi^2 - 2A_{xy}\xi\eta + 2(A_{x\theta}\eta - A_{y\theta}\xi) + A_{\theta\theta}$$

in which α represents the angle through which the r–s axes are rotated with respect to the x–y axes; ξ and η are the origin of the r–s axes with respect to the x–y axes; θ is the rotation about an axis through the origin of the x–y coordinates; and β is the rotation about the origin of the r–s axes (see Fig. 2.5). Evidently the added-mass expressions with respect to the r–s

Table 2.4. *Added-mass coefficients for finite cylinders and plates (Blevins 1977)*

Cylinder (length $= L$, radius $= c$)			Plate (length $= L$, width $= w$)		
Added mass$/(\pi \rho L c^2)$			Added mass$/(\pi \rho L w^2/4)$		
$L/(2c)$	Strip theory	Experiment	L/w	Strip theory	Exact solution
1.2	1.0	0.62	1.0	1.0	0.5790
2.5	1.0	0.78	2.0	1.0	0.7568
5.0	1.0	0.90	4.0	1.0	0.8718
9.0	1.0	0.96	10.0	1.0	0.9469

system are considerably simplified if the x–y axes are axes of symmetry so that

$$A_{xy} = A_{x\theta} = A_{y\theta} = 0 \quad \text{or} \quad A_{12} = A_{16} = A_{26} = 0$$

In general, the analytical or numerical evaluation of the added-mass coefficients for bodies with relatively complex geometries is somewhat difficult and time consuming. Under such circumstances it is advantageous to use either experimental methods or relatively simple and approximate numerical methods such as strip theory. This method consists of summing the added masses of individual 2D strips, provided that (i) the flow over a narrow slice or strip of the body is essentially 2D, (ii) the interaction between the adjacent strips is negligible, (iii) the end effects are relatively small, and/or (iv) the body is sufficiently slender. There are no simple rules to determine as to when and to what degree of approximation will these conditions be satisfied. Evidently, the more slender the body (the flow is assumed to be normal to the long axis of the body), the more suitable the strip theory. For example, a submarine, a ship hull, or a long cigar-shaped body are more suitable for the strip theory than a cube or a sphere. Table 2.4 shows a comparison of the added-mass coefficients obtained through the use of the strip theory with those obtained experimentally and theoretically. It should be noted that the strip theory cannot take into account the 3D flow effects *even if the flow is assumed to be unseparated*. Consequently the predictions of the strip theory differ significantly from those obtained experimentally and theoretically as the body becomes increasingly stubby.

In unseparated flow, the added mass always decreases the natural frequency of the body from that which would be measured in vacuum. The relationship between the two frequencies may be written as

$$f(\text{in a fluid})/f(\text{in vacuum}) = (1 + \Delta m/m_b)^{-1/2} \qquad (2.12.6)$$

where Δm is the added mass and m_b is the mass of the body. For a pipe of mass m_b per unit length (in air) the preceding equation may be written as

$$f(\text{in water})/f(\text{in vacuum}) = [1 + (\Delta m + m_w)/(/m_b + m_w)]^{-1/2} \quad (2.12.7)$$

in which Δm represents the added mass and m_w is the mass of water inside the pipe, all for unit length. Apparently both the added mass and the fluid inside the pipe play significant roles in determining the vibrational characteristics of structures in the ocean environment. Furthermore, water surrounding the structure increases the so-called fluid damping of the vibrations. This effect is discussed in detail in Chapters 6 and 7.

For the separated motion of real fluids, the added-mass coefficients depend, in general, on the parameters characterizing the history of the motion, time, and Reynolds number (suitably defined) in addition to the parameters such as the type and direction of motion, proximity effects, etc. There is often no single mathematical or experimental procedure to separate the inertial component of the force from that attributable to the form drag. This brings us to the fundamental question of "how does one express the time-dependent force acting on a body undergoing an arbitrary time-dependent motion?" For inviscid fluids, this is a relatively simple matter, and the appropriate force and moment expressions have been given by Taylor (1928a), Birkhoff (1950), Landweber (1956), Landweber and Yih (1956), Cummins (1957), Milne-Thomson (1960), and Sarpkaya (1963). For real fluids, however, the preceding question is not answerable. In fact, the determination of resistance in time-dependent flows presents an enormously complex texture of conditions and threatens to remain a perpetual problem. Only in the case of a few and relatively manageable time-dependent motions can one devise approximate force equations. Even then one must be aware of the fact that the approximate equation may be valid for only a limited range of the governing parameters. Ordinarily, one performs a dimensional analysis of the pertinent parameters (assuming that one knows beforehand what the most pertinent parameters are), carries out a series of experiments (with due consideration of scale effects, transition to turbulence, etc.), and attempts to determine the range of validity of the proposed equation. In doing so, the coefficients obtained with steady flows or from potential theory serve as limiting values or simple benchmarks. In all practical applications, the awareness of the limitations of the equations and coefficients used is just as important as the equations and coefficients themselves.

In the foregoing, we have not touched on the added mass of bodies moving at or near the liquid–air interface, the added mass of fluids in closed systems (see e.g., Sarpkaya 1962), and the added mass of bodies in an unbounded fluid that is due to the expansion or collapse of nearby cavities.

For a body undergoing small sinusoidal oscillations on or near a free surface (see, e.g., Frank 1967), the added-mass coefficients have frequency-dependent in-phase and out-of-phase components. For low-frequency oscillations in the vertical direction at the liquid–air interface, one may assume $\partial \phi / \partial n = 0$ at the free surface. For large-frequency oscillations, however, the free-surface condition is given by $\phi = 0$. Extensive research has been devoted to the calculation of the in-phase and out-of-phase components of the resulting force for both low- and high-frequency oscillations of various shapes of bodies in surge, sway, yaw, heave, roll, and pitch. For a detailed discussion of the subject, the reader is referred to Havelock (1963), Ogilvie (1964), Kim (1966), Vugts (1968), Flagg and Newman (1971), Wehausen (1971), Bai (1977), Takaki (1977), Faltinsen (1990), and the references therein.

This chapter shows once again that hydrodynamics can provide a rational description of only certain flows, even if the details are not analyzed completely and the most practical problems require a strong interaction among theory, laboratory, numerical experiments, and physical insight.

3

Separation and Time-Dependent Flows

3.1 Introduction and key concepts

Fluid mechanics is fortunate to have fundamental equations and pioneering contributions that are primarily due to Sir Isaac Newton (1643–1727), Gottfried W. Leibniz (1646–1716), Leonhard Euler (1707–1783), John Bernoulli (1667–1748), Daniel Bernoulli (1700–1782), Jean le Rond d'Alembert (1717–1783), Joseph-Louis Lagrange (1736–1813), Pierre Simon de Laplace (1749–1827), Barré de Saint Venant (1797–1886), James A. Froude 1818–1894), Sir William Thomson (Lord Kelvin) (1824–1907), Sir George G. Stokes (1819–1903), Herman L. F. von Helmholtz (1821–1894), Ernst Mach (1838–1916), Claude Navier (1785–1836), John William Strutt (Lord Rayleigh) (1842–1919), Arnold J. W. Sommerfeld (1868–1951), Ludwig Prandtl (1875–1953), G. I. Taylor (1886–1975), among many others.

Although our understanding of the fundamental concepts of fluid mechanics is sound, the applications are in a continual state of flux. In the years to come, either the laws of physics of turbulence will be discovered or delegated to *heuristic reasoning,* meticulous observations and measurements, physical and numerical simulations, and rigorous analyses.

Batchelor (1967) stated, "One of the most important problems of fluid mechanics is to determine the properties of the flow due to moving bodies of simple shape, over the entire range of values of *Re*, and more especially for the large values of *Re* corresponding to bodies of ordinary size moving through air and water." The advances in fluid mechanics over the past 50 years have only enhanced the importance of the problem.

Separated flows in general and *time-dependent flows* in particular arise in many engineering situations, and the prediction of the fluid–structure interaction (forces and dynamic response) presents monumental mathematical, numerical, and experimental challenges in both laminar and turbulent flows (Sarpkaya 1991a; Bouak and Lemay 1998). Thus it is appropriate to review some of these unsteady flows with special emphasis on marine hydrodynamics: unsteady separation; special time-dependent flows such as impulsively and nonimpulsively started flow about cylinders (Sarpkaya 1991c); sinusoidally oscillating flows (henceforth SOFs); excursion of the separation points on a circular cylinder in oscillating flow; drag, inertia, and lift coefficients; Morison's equation; and oscillating body (flow) in a uniform current. It is hoped that the material covered herein will enhance the understanding of many exciting phenomena in marine hydrodynamics and improve and inspire new applications and research.

The unsteady flows may be classified as slowly varying flows, flows with significant but manageable time dependence, and flows with significant as well as arbitrary time dependence. Defining x_0 to be a reference length (e.g., diameter), U_0 to be a reference velocity, and $(dU/dt)_0$ to be a reference acceleration, we note that $U_0(dU/dx)_0/(dU/dt)_0$, $U_0^2/x_0(dU/dt)_0$, and $(x_0^2/\nu)(1/U_0)(dU/dt)_0$ must all be considered in assessing the significance of the local acceleration to the convective acceleration and the rates of diffusion (the last two parameters, one of which may be replaced with Reynolds number U_0x_0/ν). The first parameter leads to U_mT/D, first derived by Schlichting (1932). It is now known as the Keulegan–Carpenter (1956, 1958) number K because in the mid-1950s Schlichting's well-known book on boundary-layer theory had not yet been published and his original paper (written in German in 1932) did not attract the attention of the then-budding offshore industry. Otherwise, and more appropriate, the Keulegan–Carpenter number would have been justly called the Schlichting number. In any case, *K represents only a length ratio* $(2\pi A/D)$. The second parameter also leads to K. In other words, one *would not expect a correlation* between a kinematic quantity like K and a dynamic quantity like the drag coefficient (or the inertia coefficient). This fact had not been recognized until 1976. The third parameter leads to $D^2/\nu T$, *which* is identical to Re/K. *As noted earlier, it* was *first discovered by* Sarpkaya (1976e) and injected into *the marine hydrodynamics* (for details see, Sarpkaya 2005).

Extensive data were obtained by Sarpkaya at high Reynolds numbers in a *sinusoidally oscillating* flow, generated in a large U-shaped water tunnel (Fig. 3.1).

The discovery of $\beta(= Re/K = D^2/\nu T)$ and the use of a large U-shaped water tunnel brought order into the chaotic field data (shown in Fig. 3.2a and 3.2b) and to the interpretation of the experiments.

Historically, the *universality* of $D^2/\nu T = Re/K$ *for all Re and K values* (not just for SOFs at *Re* less than about 5) *was not discovered for 124 years, until* Sarpkaya *universalized it in 1976*. The reason for this is that it was assumed to be a parameter naturally resulting from Stokes' analysis of pendulum experiments (du Boit and others in 1880s) at *very small amplitudes* and Reynolds numbers. Stokes analysis (1851) (based on his linearized equations for *unseparated laminar flow* about a cylinder) was carried out

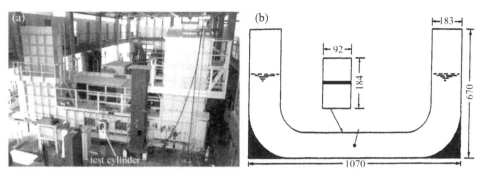

Figure 3.1. (a) A photograph and (b) a schematic drawing of a U-shaped water tunnel. The tall square structure in the foreground (left) is the filtration system. The working section is 148 cm high, 92 cm wide, and 10.7 m long. The two 6.7-m vertical sections are each 183 cm wide by 92 cm long.

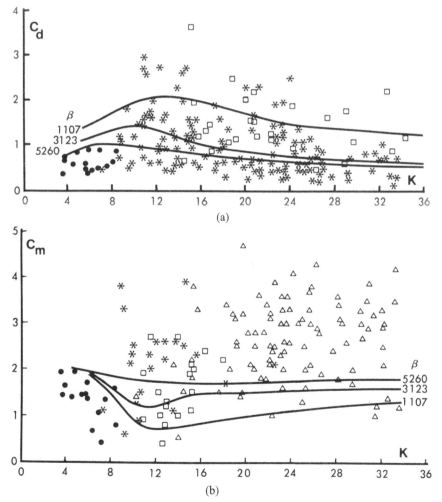

Figure 3.2. (a) Drag coefficients and (b) inertia coefficients from field data, as compiled by Wiegel *et al.* (1957). Solid curves from Sarpkaya (1976e).

for the purpose of proving the hypothesis that the kinematic viscosity (his *friction index*) was a fluid property dependent only on temperature. Stokes repeatedly stated that his *linearized* solutions can be used only for oscillations of very small amplitudes (note that he could not solve the non-linearized equations). Furthermore, concepts regarding *Reynolds number, separation, turbulence, etc., did not exist during his lifetime.*

3.2 Consequences of separation

Separation is a fundamental unsolved phenomenon in modern fluid dynamics, and the numerical and experimental study of its occurrence in both steady and unsteady ambient flow about bluff bodies is of paramount importance (Gad-el-Hak and Bushnell 1991). A definition of separation that is meaningful for all kinds of *unsteady flows* has not yet been established. It is now a well-known fact that the use of *vanishing surface shear stress* or *flow reversal* as the criterion for separation is unacceptable. In unsteady flow, the surface shear can change sign with flow reversal, but without breakaway. Conversely, the boundary-layer concept may break down before any flow reversal (Sears and Telionis 1975). If the vanishing of the wall shear is not the appropriate criterion for separation, then what is? The many criteria that have been proposed required the use of laminar boundary-layer equations (some at an infinite Reynolds number!) and *a priori* knowledge of certain flow features, such as the free-stream velocity or base pressure, which must be identified either experimentally or theoretically.

The so-called MRS (for Moore, Rott, and Sears) criterion (the proposition that the unsteady separation is the simultaneous vanishing of the shear and the velocity at a point within the boundary layer, and away from the solid surface, in a coordinate convected with the separation velocity) requires *a priori* knowledge of the speed of separation (Rott 1956; Sears 1956; Moore 1957; Chang 1970). The outer flow is usually unknown and affected in an unknown manner by the boundary layer itself (Williams 1977). It is proposed in calculations of laminar boundary layers, using boundary-layer equations, that if at some location the calculated flow is found to be on the point of separation, then the corresponding real boundary layer will actually separate at or near that location, even though the boundary-layer approximation may have failed there. A failure of the boundary-layer approximation is not readily distinguishable from a limit point reached when the imposed pressure gradient is no longer adequate to describe the flow, in view of the well-known fact that the boundary-layer development is very dependent on the imposed pressure gradient.

Historically, it was the occurrence and perplexing consequences of separation that led Prandtl in 1904 to his brilliant discovery of the *boundary-layer* or "*transition layer*" concept. It is rather ironic that, even though the boundary-layer theory has revolutionized fluid dynamics, the phenomenon, which gave impetus to its inception, is associated with its failure. Aside from its intrinsic interest, the subject is of practical importance in many fields of engineering. In fact, the forces exerted on bodies subjected to vortex-excited oscillations (Sarpkaya 2004) and the springing and ringing response of compliant systems require a clear understanding of the flow in the range of Keulegan–Carpenter numbers from 10^{-3} to 100 and $\beta (= fD^2/\nu)$ from about 1 to 10^8.

In unsteady flow, the mobile separation points on a bluff body (when they are not fixed by sharp edges) may undergo large excursions. The formation of a wake gives rise not only to a form drag, as would be the case if the motion were steady, but also to significant changes in the inertial forces. *The velocity-dependent form drag is not the same as that for the steady flow of a viscous fluid, and the acceleration-dependent inertial resistance is not the same as that for an unseparated unsteady flow of an inviscid fluid.* In other words, *the drag and the inertial forces are interdependent as well as time dependent.*

The geometry of the wake varies with time, and a solution to a specific problem, if one exists, is not unique in terms of only the instantaneous Reynolds number and some suitable acceleration parameter, because the characteristics of the instantaneous state depend also on *how that particular state has been reached.* These experimental facts render the treatment of boundary layers on bluff bodies subjected to periodic wake return extremely difficult, particularly when the state of the boundary layer changes during a given cycle. As noted by Simpson (1989), in his incisive review of turbulent boundary-layer separation, "Imposed and self-induced unsteadiness strongly influence detached flows. Detachment from sharply diverging walls leads to large transitory stall in diffusers and flapping shear layers around bluff bodies." This behavior makes such flows difficult to interpret and or to calculate (Henderson 1995).

Recently, Haller (2004) and Kilic *et al.* (2005) used the method of averaging to improve separation criteria for 2D unsteady flows with a well-defined asymptotic mean velocity. However, they reached the obvious conclusions that "the application of averaging methods to moving unsteady separation remains an open question. In moving separation, the location of the separation point changes in time, and hence will not be captured by averaging over infinite times."

Experiments by Sarpkaya (2006b) have shown that (Fig. 3.3) the separation is 3D, far from being a single eruption of a double-sided single shear layer or the departure of dye filaments from the surface of a self-contained bubble. The increase of the three-dimensionality of the flow, evolution of various sizes of structures, changes of the gradients of velocity, acceleration, and pressure along the moving separation, secondary separations within the primary separation zone, occasional eruption of multiple shear layers and

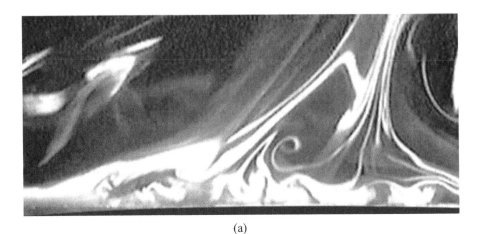

(a)

Figure 3.3. (a) A sample of separation ($K = 3.95$) composed of a number of shear layers (magnified about 20 times). The size of the photographed region is about 3 mm by 1.5 mm (Sarpkaya 2006b).

the enhancement of small-scale structures (see Figs. 3.3a through 3.3c) render the numerical simulation of separation in time-dependent flows a challenge for the distant future (Sonneville 1976; West and Apelt 1993; Marxen *et al.* 2003).

In the analysis of many steady and unsteady flows about bluff bodies at sufficiently large Reynolds numbers, the requirement for separation-point specification constitutes the weakest link. Even in *steady* 2D flow, separation points can be predicted only approximately for laminar flows and hardly at all for turbulent flows. Clearly the central obstacle toward a complete description of the bluff-body problem is the insufficient understanding of the near-wake mechanics.

Faced with an impossible task, man's ingenuity has turned to finding a compromise between experimentally determined facts and mathematical expediency. The improvement of experimental techniques in acquiring facts and the development of high-speed computers, which made possible the numerical solution of the governing nonlinear differential equations, have stimulated extensive research on time-dependent flows. The beauty and the vigor of potential-flow theory have been applied directly to the prediction of the characteristics of a class of fluid flows in which either the boundary-layer separation does not occur or the stream surface constitutes an interface between a liquid and a gas. For real fluids, the applicability of potential-flow results is subject to the usual qualifications that are due to the effect of viscosity, however small the viscosity may be.

Evidently, the coefficients obtained for unseparated unsteady flows are not applicable to occurrences in which the duration of flow in one direction is long enough and the body form blunt enough for separation to occur. Thus it is necessary to determine the relationships among various resistance

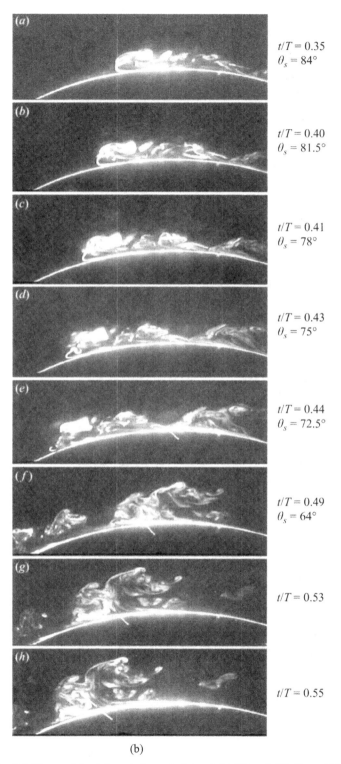

(b)

Figure 3.3. (b) The ambient flow is from left to right ($K = 3.45$, $Re = 23500$). The separation front is advancing from right to left. The secondary separations (identified by white lines) are seen in frames (*e–h*). (Sarpkaya 2006b).

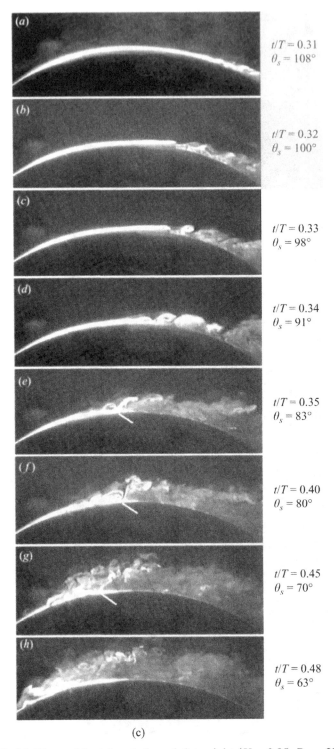

(c)

Figure 3.3. (c). The ambient flow is from left to right ($K = 3.95$, $Re = 27\,000$). The separation front is advancing from right to left. Frames (*e–g*) show the eruptions in shear layers (white arrows). (Sarpkaya 2006b).

components in terms of the unsteadiness of the ambient flow, the geometry of the body, the roughness of the object, the degree of the upstream turbulence, the speed of the separation, and the past history of the flow.

The history of the motion deserves considerable attention, particularly in periodic flows or oscillating bodies, because the wake that forms behind the body becomes the fluid at the leading edge of the body at the start of the next step in the flow cycle. Motions of this type are of great practical significance in the determination of wave forces on partly or fully submerged rigid and elastic bodies, in the investigation of wind- and gust-induced oscillations of towers (Gurley and Kareem 1993; Zhou and Kareem 2001) and launch vehicles, and in the study of a large class of problems that spans the two just cited.

Observations as well as numerical experiments show that the wake of a bluff body comprises an alternating vortex street. The character of the vortices immediately behind the cylinder and in the wake farther downstream depends, for a steady ambient flow about a smooth cylinder, on the Reynolds number and the intensity and length scale of the turbulence present in the ambient flow. For a time-dependent flow, the instantaneous state as well as the past history of the flow plays significant roles, and it is not possible to give a general set of normalized parameters. Often heuristic reasoning and laboratory and numerical experiments will have to supplement the information gathered from relatively idealized solutions in order to correctly identify the most important governing parameters. The separation point may be mobile as on a circular cylinder or fixed at an edge, e.g., an edge of an inclined flat plate or of the base of a triangular section.

3.3 Body and separation

When the separation point is fixed, the drag and pressure coefficients vary *considerably less* than those for the circular cylinder, pointing out the important fact that the mobility of the separation points strongly affects the character of the force coefficients. However, they are *not completely independent* of the Reynolds number, especially in the lower ranges where various transitions occur in the wake. It is not correct to assume that the problems associated with the downstream of fixed separation points are less complex than those associated with the downstream of the mobile separation points (see, e.g., Huse and Muren 1987 and Miau *et al.* 2006, who used the wavelet analysis of the signals obtained by a micro-electrical-mechanical-system film-sensor-array), flush with the cylinder surface. At $\theta = 85°$ (near separation), the unsteady flow exhibited two time scales: one was associated with vortex shedding, and the other (at least one order of magnitude longer) with the excursion of the flow separation in a circumferential region of the extent of about $5°$.

When the separation point is mobile, its motion is coupled with the shedding of vortices (Sarpkaya 1991c). The vorticity feeding the shear layer fluctuates with time because the separation point moves into higher- or lower-velocity regions. The separation occurs beyond the point of maximum velocity (about 10° downstream of this), i.e., the pressure gradient is positive at the mobile separation point whereas it is negative up to the fixed separation point. In addition, and to further complicate the matter, the flow upstream of the separation points is affected by the occurrence, the motion, and the character of the separation zones or by the circumstances leading to the formation, growth, and motion of the wake.

The front stagnation point oscillates about its mean position in such a manner that its motion is 180° out of phase (Dwyer and McCroskey 1973; Cao *et al.* 2007) with that of the separation point. The streamlines emanating from the separation points may be convex (at a fixed separation point or at a mobile separation point at the front of a cylinder) or concave (when separation occurs on the back of the cylinder).

It is evident from the previous discussion that there is a fundamental interaction between the body and the separated flow, particularly in the region enclosing the body and its near wake. The dynamics of this interaction is of major importance in determining the time-dependent fluid resistance and the characteristics of flow-induced vibrations (see, e.g., Yamaguchi 1971, Sarpkaya 1977a, 1979b, 1987). Unfortunately it is not yet possible to predict theoretically the behavior of the flow when separation leads to a large-scale wake comprising alternating vortices. This is particularly true for a time-dependent flow about a cylinder (e.g., wave motion, 2D sinusoidal flow with or without a mean flow, etc.). In such cases, the separation points undergo large excursions and the roles played by the *forebody* (part of the body upstream of the separation points) and the *afterbody* (downstream of the separation points) interchanges periodically, as seen in Fig. 3.4. Furthermore, the boundary-layer equations (if they can ever be integrated through the separation points interacting with the wake) can deal only with flows with relatively small Reynolds numbers.

When the vortices, vortex-feeding layers, and even the boundary layer over the forebody become turbulent, significant changes take place in the forces acting on the body. This is vividly illustrated through measurements and motion pictures. For example, the expository experimental study by Roshko and Fiszdon (1969) has shown that, when the Reynolds number lies between about 1 and 50, the entire flow is steady and laminar. In the range of Reynolds numbers from about 50 to 200, the flow retains its laminar character but the near wake becomes unstable and oscillates periodically. At still higher Reynolds numbers but below about 1500, turbulence sets in and spreads downstream (Sarpkaya 1993b). In the region between 1500 and

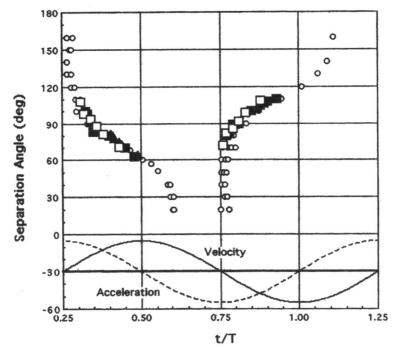

Figure 3.4. Instantaneous separation angles within one period for $K = 3.56$, $Re = 23500$ (filled triangles); $K = 3.95$, $Re = 27000$ (filled squares); and $K = 4.45$, $Re = 30300$ (open squares). Measurements with the differential probes and the flush-mounted hot-film sensors for $K = 3.95$. $Re = 27000$ are shown (open circles). No distinction was made between them because of their remarkably close agreement.

2×10^5, the transition and turbulence gradually move upstream along the free-shear layers and the far wake becomes increasingly irregular. When the transition coincides with the separation point at a Reynolds number of about 5×10^5 (depending on the intensity of the free-stream turbulence and the peculiarities of the test apparatus), the flow undergoes first a laminar separation, followed by a reattachment to the cylinder, and then a turbulent separation to form a narrower wake. This results in a large fall in both the lift and the drag coefficient.

Because of the more diverse interest in the steady drag force, the phenomenon leading to the sharp decrease in drag became known as the drag crisis. The fact that the lift force as well as other characteristics of the flow also undergoes dramatic changes in the same Reynolds number range suggests that the phenomenon be called *resistance crisis*. It is very important to remember that the Reynolds number at which the resistance crisis occurs depends very much on the character of the flow. It is not correct to assume that the resistance crisis will occur at the same Reynolds number for a time-dependent flow. For a uniformly accelerating flow about a cylinder, the

Table 3.1. *Incompressible flow regimes and their consequences*

	A Subcritical	B Critical	C Supercritical	D Post-supercritical
Boundary layer	Laminar	Transition	Turbulent	Turbulent
Separation	About 82 deg.	Transition	120–130 deg.	About 120 deg.
Shear layer near separation	Laminar		Laminar separation, bubble turbulent reattachment	Turbulent
Strouhal number	$S = 0.212 - \dfrac{2.7}{Re}$	Transition	0.35–0.45	About 0.29
Wake	$Re < 60$ laminar; $60 < Re < 5000$ vortex street $Re > 5000$ turbulent	Not periodic		
Approximate Re range	$< 2 \times 10^5$	2×10^5 to 5×10^5	$5 \times 10^5 - 3 \times 10^6$	$> 3 \times 10^6$

resistance crisis may occur at higher Reynolds numbers. On the other hand, for a decelerating flow or for a periodic flow about a cylinder, the resistance crisis may occur at smaller Reynolds numbers. In general, it is worth remembering that, when the flow is sensitive to the variations in a particular parameter, it becomes equally sensitive to others whose influence may have been otherwise negligible (e.g., roughness, free-stream turbulence, vibrations, end conditions, uniformity of the flow).

In the absence of roughness and excessive free-stream turbulence, the transition in drag coefficient between $Re = 3 \times 10^6$ and $Re = 2 \times 10^7$ is interpreted (e.g., Roshko 1961; Tani 1964; Batchelor 1967, p. 342; Jones *et al.* 1969) to be due to transition of the separated boundary layer to a turbulent state, the formation of a separation bubble, reattachment of the rapidly spreading turbulent free-shear layer, and finally, separation of the turbulent boundary layer at a position farther downstream from the first point of laminar separation. The reduction of the wake size as a consequence of the retreat of the separation points then results in a smaller form drag (see Table 3.1 and Fig. 3.5) where these different ranges are distinguished).

The subsequent increase in the drag coefficient between $Re = 10^6$ and $Re = 10^7$ is then interpreted to be a consequence of the transition to a turbulent state of the attached portion of the boundary layer. At higher Reynolds numbers the trend is not clear and additional experiments, however difficult to perform, are needed. The best conjecture is that *dramatic changes are not likely to occur* in the boundary layers at Reynolds numbers several

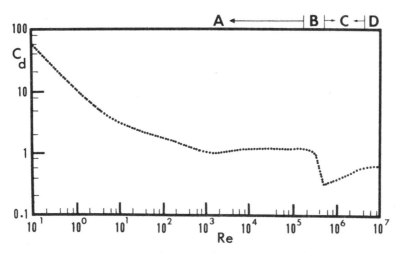

Figure 3.5. Drag coefficient for circular cylinders as a function of Reynolds number (Schlichting 1968).

orders of magnitude larger than 10^7. The scale of turbulence may decrease, and the diffusion and cancellation of vorticity may increase. These in turn lead to higher Strouhal numbers and lower drag coefficients. Be that as it may, there is, at present, a large Reynolds number gap between the highest Reynolds numbers attained in the experiments and the Reynolds numbers to be encountered by the designers of marine structures.

The experimental observations previously cited do not consider the physical movement of the body, free-stream turbulence, and the roughness effects (see, e.g., Guven 1975). Experience has shown that elastic structures near linear-resonance conditions can develop flow-induced oscillations by extracting energy from the flow about them. The oscillations of the structure modify the flow and give rise to nonlinear interaction. This is in addition to any nonlinearity, which can arise from the restoring force (variable support stiffness) and/or from response-dependent structural damping. The understanding of these nonlinear interactions is of paramount importance. It is well worth remembering the facts noted in this brief review of the basic concepts as well as the approximations that must be introduced into the design of an offshore structure while one is reading the rest of this text.

The free-stream turbulence and roughness effects are discussed in Section 3.7. For a more detailed discussion of separation and flow about cylinders the reader is referred to Morkovin (1964), Lienhard (1966), Mair and Maull (1971), Wille (1974), Sarpkaya (1979b, 1987), Sarpkaya and Butterworth (1992), Sarpkaya (1992), Cassel *et al.* (1996), Chesnakas and Simpson (1997), Walker (2003), Zdravkovich (1997, 2003), Kilic *et al.* (2005), Sarpkaya (2006b), and Allam and Zhou (2007).

3.4 Strouhal number

The net result of the alternate vortex shedding from a circular cylinder is an oscillating side thrust in a direction away from the last detached vortex. This side thrust or lift force (sometimes referred to as the lateral force or transverse force or out-of-plane force) exists practically at all Reynolds numbers regardless of whether the body is allowed to respond dynamically or not. In other words, the alternate shedding of vortices gives rise to a transverse pressure gradient.

In 1878, Cének (Vincent) Strouhal, (a Czech scientist) discovered a relationship among the vortex-shedding frequency f_{st}, cylinder diameter D, and the velocity of the ambient flow U in connection with his work on a special method of creation of sound by using a vibrating string. This relationship, denoted by $St(= f_{st}D/U)$, is known as the Strouhal number. It may also be written as D/UT, where T is the period of oscillation. The inverse of this number is UT/D, or $U_m T/D = 2\pi A/D = K = St^{-1}$ (for a SOF). In other words, K, *as the inverse of St*, is not a new number. However, Strouhal's St, Schlichting's U_m/fD, and Keulegan–Carpenter's K were created at different times, by different means, toward different purposes. It must be noted for later use that K is *only for a SOF about a vertical cylinder with no current or shear*. Otherwise, modified (approximate) versions of K must be used, as discussed in detail later.

Unsteady hydrodynamic loads arising from the transverse pressure gradient acting on the afterbody can excite dynamic response. Body natural frequencies near the exciting frequency raise the spectrum of load and response enhancement. Thus the spectral content of the forcing functions is important to dynamic structural response analysis. The shedding process may be random (broadband) over a portion of the Reynolds number range for which a statistical response analysis is required based on the spectral content. Also required is a measure of the *correlation length* along the cylinder. Evidently, the characterization of the vortex-shedding process by a simple frequency is a practical simplification. Power spectral density analyses of unsteady cylinder loads reveal that, in certain Reynolds number regimes, the shedding process is practically periodic and can be characterized by a single Strouhal frequency.

At subcritical Reynolds numbers, the energy-containing frequencies are confined to a narrow band, and the Strouhal number is about 0.2 for smooth cylinders (see Fig. 3.6). It must be emphasized that only an average Strouhal number may be defined for Reynolds numbers larger than about 20000.

In the critical Reynolds number regime, a broadband power spectral density is usually observed for a rigidly held cylinder. At higher Reynolds numbers, the Strouhal frequency rises to about 0.3 and the shedding process is

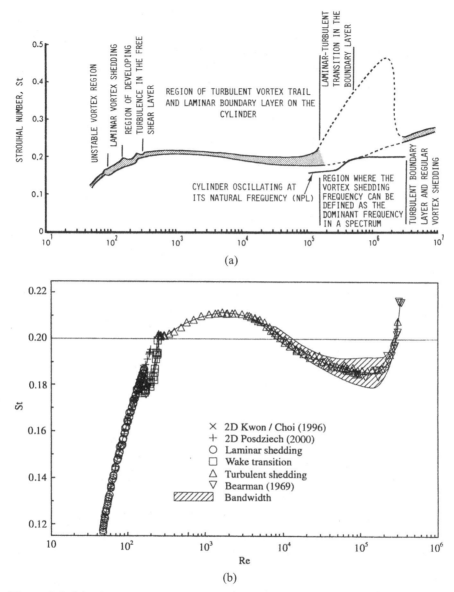

Figure 3.6. (a) The Strouhal–Reynolds number relationship for circular cylinders (Lienhard 1966). (b) Strouhal number versus Reynolds number.

quasi-periodic. The spectral content of the exciting forces (Schewe 1983a, 1983b; Zdravkovich 1997) is particularly important for bodies, which may undergo in-line and/or transverse oscillations because the vortex-shedding frequency locks on to the frequency of the transverse oscillations of the cylinder when the vortex-shedding frequency is in the neighborhood of the natural frequency of the cylinder. This often raises questions as to what happens when the flow is in the critical regime. The question is a complex one because the so-called critical regime does not necessarily occur at Reynolds numbers corresponding to those for a stationary cylinder. The motion of the cylinder changes the boundary-layer characteristics, the occurrence of the laminar separation bubble (assumed to be responsible for the transition in the critical and supercritical regions), and the Strouhal frequency. Thus calculations based on stationary cylinder values in the critical regime may prove to be false. Experiments conducted at the *National Physical Laboratory* (NPL) (1969) have shown [see the inset in Fig. 3.6(a) and the NPL data therein] that, when the cylinder is free to oscillate, the *sharp rise in the Strouhal number does not occur and remains at a value nearly equal to that found at subcritical Reynolds numbers.*

Various attempts have been made to devise a universal Strouhal number, which would remain constant for differently shaped 2D and axisymmetric bodies (Simmons 1977). The universal Strouhal numbers introduced by this and many other authors employ the wake width as the characteristic length and velocity outside the shear layers as the characteristic velocity. Their value is not so much in their ability to predict but rather to uncover the intricate relationship, say, among the flow velocity, vortex-shedding frequency, wake width, base pressure, formation length, time mean of the rate of shedding of vorticity, etc. Time has shown that isolating a small number of simple scales to describe the wake development is not possible.

The constancy of the Strouhal number over a broad range of Reynolds numbers does not imply that the base pressure remains constant and that a single 2D vortex emanates from a separation line. In reality, there is not only a phase shift between various sections along the vortex, separated by a correlation length (*the equivalent length over which the velocity fluctuations at similar points in the wake may be described as perfectly correlated*), but also variations in both the intensity and the frequency of vortex segments.

The variation of the base pressure with Reynolds number, in the range where the Strouhal number remains practically constant (see Fig. 3.7), may be related to the variation of the mean vorticity flux or to the variation of the correlation length with the Reynolds number, turbulence, length-to-diameter ratio, and surface roughness (Humphreys 1960; Phillips 1985). Table 3.2, as compiled by King (1977a), gives an approximate idea about the typical values of the correlation length. The net effect of the spanwise variations of the vortex tube is that the transverse force (lift) coefficient obtained from a pressure integration is not necessarily identical with that obtained from the direct measurements of the lift force. *Partial spanwise correlation leads to variations in both the frequency and the amplitude of the lift force*, the variation of the latter being more pronounced than that of the former.

The lack of correlation exists not only spanwise but also chordwise (Gerrard 1965; Chaplin and Shaw 1971; Wilkinson *et al.* 1974), and the chordwise

Table 3.2. *Correlation lengths*

Reynolds number	Range correlation length	Source
$40 < Re < 150$	$15D$–$20D$	Gerlach and Dodge (1970)
$150 < Re < 10^8$	$2D$–$3D$	Gerlach and Dodge (1970)
$1.1 \times 10^4 < Re < 4.5 \times 10^4$	$3D$–$6D$	El Baroudi (1960)
$Re < 10^5$	$0.5D$	Gerlach and Dodge (1970)
$Re = 2 \times 10^5$	$1.56D$	Humphreys (1960)

correlation is related to the spanwise correlation. Comparison between various results suggests that with increasing Reynolds number over the range 10^4–10^5, the chordwise correlation for square (Kumar *et al.* 2008) and circular cross-section cylinders is maintained or improved. This leads to an increase in fluctuating lift coefficient. The reasons for these variations are not quite clear. The end effects (Humphreys 1960; Stansby 1974; Etzold and Fiedler 1976), wall boundary layers, free-stream turbulence (McGregor 1957; Surry 1969), and nonuniformity of the flow are mentioned often as possible reasons. The complexity of the 3D nature of the flow about a cylinder is clearly demonstrated with measurements by Tournier and Py (1978). It would not be correct to assume that the mobility of the separation points is primarily responsible for the imperfect coherence. *Even bodies such as 90° wedges, square cylinders, with fixed separation lines, do not exhibit perfect correlation* (Kumar *et al.* 2008). However, the variation of the base-pressure

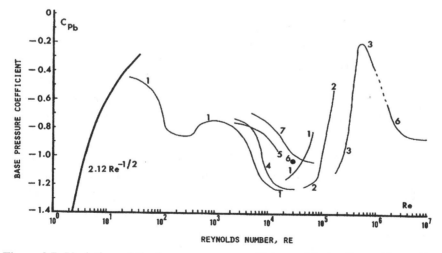

Figure 3.7. Variation of the base-pressure coefficient with the Reynolds number for circular cylinders: 1, Thom (1928); 2, Fage and Falkner (1931); 3, Flachsbart (1932); 4, Schiller and Linke (1933); 5, Roshko (1953); 6, Fage (1931); 7, Gerrard (1965).

coefficient for bodies with mobile separation lines is greater than that for bodies with fixed separation lines.

3.5 Near wake and principal difficulties of analysis

As noted earlier, the strength of the vortices plays an important role, particularly in the near wake. Laboratory and numerical experiments have shown that (Fage and Johansen 1928; Abernathy and Kronauer 1962; Sarpkaya 1963; Bloor and Gerrard 1966; Mair and Maull 1971; Schmidt and Tilmann 1972) the net circulation of a rolled-up vortex of the street is 40% to 60% smaller than that generated in the boundary layer during a shedding cycle. Prandtl determined that the initial vorticity decreases to about half where the first vortex centers appear. Vorticity is ultimately dissipated by the viscosity to which it owes its generation. Nevertheless, one may think of loss of circulation through cancellation of oppositely signed vorticity. Primarily, there are three mechanisms whereby oppositely signed vorticities are brought close together: Vorticity generated on the forebody is carried by the shear layers near that generated on the afterbody; vorticity of the deformed and cut sheet is carried across the near wake by the entrainment of the irrotational fluid; and finally, vorticity is swept across the entire wake (Zdravkovich 1969). The percentage quoted in the literature (Berger and Wille 1972) for the total loss of vorticity often implies that the vortices, once having acquired a certain circulation, retain that circulation throughout the rest of their motion. The fact that circulation decreases continuously with time or distance is demonstrated clearly by the experiments of Schmidt and Tilmann (1972) and Bloor and Gerrard (1966). This is very similar to the decay of the aircraft wake vortices.

The amount of vorticity generated in the boundary layers and the amount dissipated are of prime importance not only for the flow past stationary bluff bodies but also for those undergoing resonant oscillations. In fact, the entire bluff-body problem may be reduced to the determination of the vorticity distribution throughout the flow field (e.g., by DNS; see Wissink and Rodi 2003). The determination of the vortex strengths is difficult and sensitive to the theoretical and experimental means employed. It is evident from the foregoing and from a more detailed study of the references cited that the description of the near wake of a bluff body is in need of further computational (say DNS or large-eddy simulation, LES), theoretical, and experimental work.

Much of what is known about the consequences of separation has come from laboratory experiments. It has not yet been possible to develop a numerical model with which experiments may be conducted to explain the observed or inferred relationships between various parameters and to guide and complement the laboratory experiments. The principal difficulties are as follows:

1. Separation Points. They represent a mobile boundary between two regions of vastly different scales. This in turn leads to complex physical nonuniformities in relatively narrow regions, which cannot be handled within the framework of the boundary-layer theory (Williams 1977). LESs, finite-difference methods, finite-element methods (FEMs), boundary-element methods (BEMs) and DNSs are expected to shed more light on separation (one of the most complex phenomena in fluid dynamics).

2. The discretization of the continuous process of vorticity generation by line vortices in the vicinity of a mobile or fixed singular point (discrete-vortex model) strongly affects the existing nonuniformities and promotes earlier separation. Attempts to preserve the prevailing flow conditions, say by limiting the influence of the nascent vortices while satisfying a relatively simple separation criterion, lead to hydro-dynamical inconsistencies and nondisposable parameters (Sarpkaya and Shoaff 1979b, 1979c).

3. Reynolds Number. Finite-difference schemes for bluff-body flows are limited to relatively small Reynolds numbers whether the scale of the flow is assumed to be governed by a constant viscosity or by a constant-eddy viscosity. The large recirculation region of the flow often comprises turbulent vortices even when the boundary layer is laminar. The transition to turbulence moves upstream in the shear layers as the Reynolds number is increased from about 10^3 to 5×10^4. At $Re = 5 \times 10^4$, it reaches the shoulder of the cylinder (Bloor 1964). It does not move appreciably further upstream before the critical Reynolds number is reached. Thus the distribution, turbulent diffusion, and decay of vorticity and the interaction between the wake and the boundary layers cannot be subjected to numerical simulation without recourse to some heuristic turbulence models and inspired insight. The representation of the wake by clouds of point vortices or discretized spiraling sheets (see, e.g., Abernathy and Kronauer 1962; Fink and Soh 1974; Clements 1977; Sarpkaya and Shoaff 1979b; Sarpkaya 1989, 1991a, 1994) is not immune to scaling problems. In fact, not a particular Reynolds number but only a particular flow regime may be specified, depending on the separation criteria used. In recent years, the discrete-vortex model has lost its prominence. Its luster has been transferred to DNS, LES, BEM, FEM, and others, as discussed in great detail in Sarpkaya (1989).

4. Three-Dimensionality. As noted earlier, even a uniform flow about a stationary cylinder exhibits chordwise and spanwise variations. These 3D effects may play a major role in the stretching of vortex filaments and in the redistribution of vorticity in all directions. The numerical models are not in a position to account for such complex effects. Even if the DNSs and particle methods at Reynolds numbers as high as 10^5 were possible (Chatelain *et al.* 2008a) in the years to come, the large parameter space in any application precludes a purely numerical solution. One may hope to assess the effects of three-dimensionality by means of 2D numerical experiments.

3.6 Lift or transverse force

The lift force data are far less plentiful and far less consistent than the drag force data (see Fig. 3.8). The scatter in these plots is attributed to various causes. Humphreys (1960) experimentally demonstrated the effect of the end gaps. He obtained relatively larger lift coefficients when the cylinder ends were sealed at the wall. In general, many researchers noted the sensitivity of the lift force to the stream turbulence. The degree of rigidity of the mounting of the cylinders may also play an important role. As noted earlier, even the smallest transverse oscillations of a cylinder (say, $A/D = 0.05$) distinctly regularize the vortex shedding and considerably increase the spanwise correlation (see, e.g., Blevins 1990). As noted by Norberg (2001), dramatic variations in C_L occur at turbulent shedding conditions. It is also apparent from the rms values of the lift coefficients (Norberg 2001) that striking changes occur in C_L at Re close to 250 on either side of the data. The numerical simulations show even greater scatter than the rms values of the measurements in the range of Re from about 400 to about 10^4. Clearly the vortex shedding is such a phenomenon that it will never yield scatter-free data at any Reynolds number and the reasons for it cannot be fully quantified.

3.7 Free-stream turbulence and roughness effects

Fage and Warsap (1930) were the first to investigate the effects of grid-generated turbulence, tripping wires, and surface roughness on the flow past circular cylinders (see also Goldstein 1938 and Schlichting 1968 for partial accounts of this study). Fage and Warsap increased surface roughness both by covering their cylinders with different grades of abrasive paper and by installing trip wires in the boundary layer. They changed the free-stream turbulence level by placing a coarse rope mesh at different distances upstream from the cylinders. In all cases ($10^4 < Re < 2.5 \times 10^5$), the transitions shifted

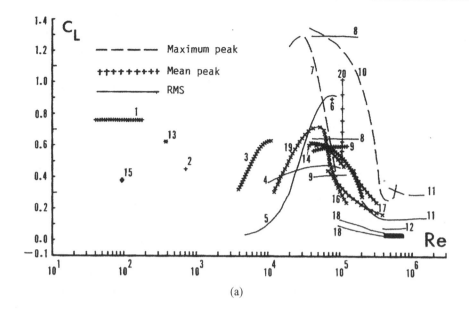

(a)

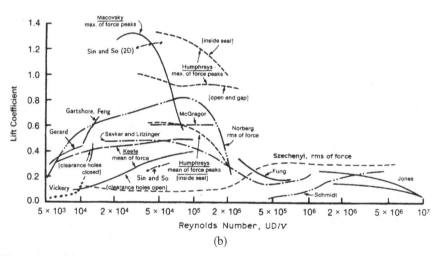

(b)

Figure 3.8. (a) Lift coefficient as a function of the Reynolds number for circular cylinders: 1, Phillips (1956); 2, Schwabe (1935); 3, Bishop and Hassan (1963); 4, Keefe (1962); 5, Gerrard (1961); 6, Bingham *et al.* (1952); 7, Macovsky (1958); 8, Vickery and Watkins (1962); 9, McGregor (1957); 10, Humphreys (1960); 11, Fung (1960); 12, Schmidt (1965); 13, Jordan and Fromm (1972); 14, Macovsky (1958); 15, Dawson and Marcus (1970); 16, Weaver (1961); 17, Goldman (1958); 18, Bublitz (1971); 19, Warren (1962); 20, Schmidt (1965). (b) Measurements of vortex-induced lift coefficient on a stationary cylinder (Yamaguchi, 1971; Szechenyi, 1975; Jones *et al.* 1969; Gartshore, 1984; Sin and So, 1987; Savkar and Litzinger 1982) (after Blevins 1990).

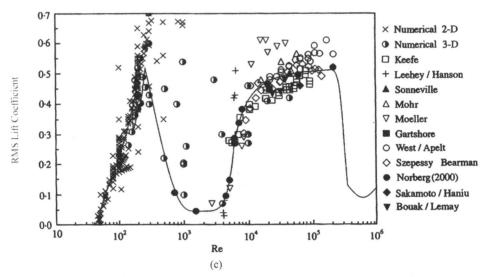

Figure 3.8. (c) The rms lift coefficient versus Reynolds number (after Norberg 2001).

strongly to lower *Re* with increasing disturbance. Although a variation in upstream turbulence did not appear to alter the severity of the dip in the drag coefficients, increases in cylinder roughness very definitely did. In fact, the dip in C_d almost vanished on an extremely rough cylinder. Fage and Warsap attributed the increase in the drag coefficient with roughness in the supercritical Reynolds number range to the retardation of the boundary-layer flow by roughness and, hence, earlier separation. They also noted that "when the surface is very rough the flow around the relatively large excrescences and so around the cylinder, is unaffected by a change in a large value of the Reynolds number."

Achenbach (1968) made measurements of pressure and skin friction over a range of Reynolds numbers up to $Re = 3 \times 10^6$. His measurements showed, among other things, that beyond some large value of *Re* the drag coefficient becomes independent of *Re*, in conformity with Fage and Warsap's suggestion, and that, in the range of Reynolds number *independence*, the drag coefficient shows a definite dependence on the relative roughness k/D (see Fig. 3.9). Achenbach also found that the larger roughness results in higher skin friction and hence in greater retardation of the boundary layer. These in turn result in earlier separation and larger absolute values of the base-pressure coefficient. One should not infer from this that flows over rough cylinders with identical separation points will result in identical pressure distributions and drag coefficients. Such a correspondence between separation position and drag coefficient does not exist because the pressure distribution is affected not only by the location of separation but also by the boundary layer ahead of separation. For example, Achenbach

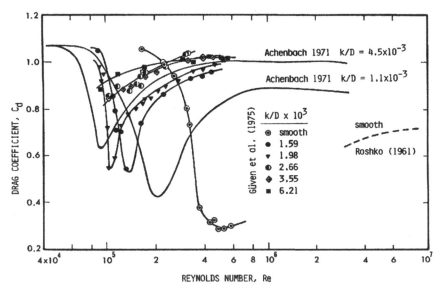

Figure 3.9. Drag coefficient for rough cylinders as a function of Reynolds number.

obtained an identical separation angle of $\theta_S = 110°$ for $k/D = 1.1 \times 10^{-3}$ at $Re = 4.3 \times 10^5$ and for $k/D = 4.5 \times 10^{-3}$ at $Re = 3 \times 10^6$, but the pressure distribution and the drag coefficients for the two cases were considerably different. The foregoing is an oversimplification of the complex interaction among the wake, separation points, and the flow over the forebody of the cylinder. One cannot intelligently isolate the flow over a given region or at a singular point so as to explain the observations or measurements over the remainder without getting involved in a circular argument. There are considerable differences among the results of Miller (1977), Guven (1975), Achenbach (1971), and Fage and Warsap (1930).

Batham (1973) reported experiments on the effects of surface roughness of the continuously distributed type ($k/D = 2.17 \times 10^{-3}$) and of free-stream turbulence on the mean and fluctuating pressure distributions on circular cylinders at $Re = 1.11 \times 10^5$ and $Re = 2.35 \times 10^5$. Szechenyi (1975) carried out an extensive investigation in which he measured steady drag coefficients and unsteady lift coefficients of rough-walled cylinders over a range of Reynolds numbers up to $Re = 6.5 \times 10^6$. Both Batham (1973) and Szechenyi (1975) were interested in *simulating the pressure distributions at high Reynolds numbers through the use of roughness*. Szechenyi suggested that the drag coefficient in the supercritical regime is a function of the roughness Reynolds number Vk/ν only for $k/D = 1.16 \times 10^{-4}$ to 2×10^{-3}. Guven *et al.* (1975) argued that this observation is at variance with their own measurements as well as with previous findings.

Guven (1975) has undertaken a comprehensive investigation of the effect of surface roughness on cylinders and found that (a) roughness has a strong influence on circular cylinders even at very large Reynolds numbers; (b) beyond some large values of Re that depend on the surface roughness, the pressure distribution becomes independent of Re and is determined by the characteristics of the surface roughness; (c) free-stream turbulence appears to have no influence on the pressure distribution at such large Reynolds numbers; (d) the effects of external ribs are generally similar to those of distributed roughness even though there are strong local influences of ribs; (e) at large Re, a cylinder with $k/D = 10^{-5}$ behaves essentially like a smooth cylinder even though the boundary layer may be fully rough, because the additional boundary resistance that is due to such small surface roughness is very small; (f) for distributed roughness, the major effects of roughness on the pressure distribution are observed in the range of k/D up to 2.5×10^{-3}. For large values of k/D, additional effects of the pressure distribution are relatively small.

For a highly irregular and relatively rough surface subjected to a marine environment, the Reynolds number and the roughness Reynolds number are quite large. Under these circumstances, the drag and inertia coefficients approach nearly constant values, *independent of Re and* $Re_k = Vk/v$. This still does not eliminate the need for the apparent diameter of the cylinder in calculating the projected area and the effective volume of the structures (for use in calculating the inertial force). If a marine-roughened pipe is available, average diameters (at suitable elevations) may be directly measured, including the average height of the protrusions. If, on the other hand, a structure is to be built in a location for which no prior information exists regarding the growth of marine roughness, it is advisable either to place pipes at various depths at the location to gather such pertinent information as a function of time or to take strong antifouling measures. All told, for structures immersed in an environment with plenty of sunlight and warm currents, the combination of hard roughness elements and soft excrescences renders the determination of the effective diameter and of the effective roughness size rather difficult. The designer must therefore rely on past experience and sound engineering judgment (Sarpkaya 1978b, 1991b, 1991d).

Only for guidance, we note that there is not a single effective diameter or mean roughness, and, for that matter, a single set of lift, drag, and inertia coefficients for a given segment of a pile. The reason for this is not only the change of the wave/current velocities along the pile but also the change of the roughness elements with time and space. To compound the difficulties associated with roughness, the type and shape of the fouling elements change with depth and the particular ocean. For example, in the middle Atlantic, sea roach and blue-green algae cover the topmost region. In the Mid-littoral, one may find barnacles, anemones, squirts, tubiculous amphipods, and oysters. In the Sub-littoral one may find red algae, hydroids, sponges, and kelp (Gosner 1978). Thus each section of a structure presents complexities of its own in the determination of forces, moments, and natural frequencies and calls for engineering judgment.

In the foregoing we have dealt mostly with the drag coefficient. As noted earlier, vortex shedding gives rise to a transverse force. Szechenyi (1975) found that the lift coefficient C_L is a function of the roughness Reynolds number. For $Re_k > 10^3$, he suggested $C_L(\text{rms}) = 0.25$ to 0.30 at a Strouhal number of $f_0 D/V = 0.25$ to 0.27. This Strouhal number is quite close to that obtained by Roshko (1961) at post-supercritical Reynolds numbers with smooth cylinders. Roshko attributed the reestablishment of the regular vortex shedding to a return to the generation of 2D vortices. This, of course, does not explain why the vortices regularized.

As will be discussed later, similar results are obtained with sand-roughened cylinders in 2D sinusoidal flows (Sarpkaya, 1991c). The importance of this observation is that roughness may enhance the possibility of the occurrence of hydroelastic oscillations partly by regularizing the vortex shedding, partly by increasing the coherence length (Shih and Hove 1977), and partly by increasing the total lift force. Finally, there does not seem to be much information on the effect of roughness on the lift and drag coefficients for yawed cylinders in uniform flow (Glenny 1966). Such information will be useful not only for marine engineering but also in aerodynamics in connection with the determination of in-plane and out-of plane forces acting on artificially roughened slender bodies at large angles of attack.

For additional information on the effects of roughness in boundary layers, the reader is referred to papers by Simpson (1973), Guven (1975), Furuya *et al.* (1976), Miller (1977), Granville (1978), Gosner (1978), Knight and McDonald (1979), and Leonardi *et al.* (2007).

Not all of the differences between the drag coefficient versus *Re* curves for rough cylinders are attributable to the test conditions (e.g., intensity and scale of turbulence in the ambient flow, nonuniformity of the velocity distribution, end-gap effects, length-to-diameter ratio, blockage effect, wall proximity, method of measurement, flexibility of the mounting of the cylinder, care taken in gluing the edges joining the sandpaper, etc.). The difficulty of uniquely specifying "roughness" is partly responsible for the differences between the steady-flow data reported by various workers, particularly in the drag-crisis region. It has been known for quite some time (see Schlichting 1968a) that the effective surface roughness may be larger or smaller than the nominal relative roughness based on the geometric size of the roughness element, depending on the shape and arrangement of the roughness elements. In other words, the shape, size and physical distribution, and packing of the roughness elements in a given flow demand some justification for the one-parameter characterization of the effects of roughness in terms of k/D. There exists very little systematic investigation of the influence of roughness density for 3D roughness. Dvorak (1969) collated the data on roughness density and presented them as functions of a parameter λ_r, where λ_r = (total surface area)/(plain roughness area). He concluded that the peak skin-friction drag is associated with values of λ_r near 5.

Experiments with relatively small rigid roughness elements glued on a large smooth cylinder do not reflect all the complexities encountered in expressing the roughness of a pipe subjected to marine environment. The structure may be covered with rigid (scale, barnacles, mussels, etc.) and soft (seaweeds, anemones) excrescences. The thickness of the accumulated growth may considerably increase the effective diameter or the characteristic size of the structure. Furthermore, the size (and the *shape*) of the accumulated growth may change with time and along the structure depending on the prevailing temperature, currents, ecological effects of the structure on the existing marine life, etc. In the North Sea, for example, the thickness of the accumulated roughness reached about 8 inches in two years with a rate of growth of about 1 inch per month. As a consequence, the diameter of certain members increased 4–12 inches. Thus it is not a simple matter to decide what fraction of the protrusions constitutes an average roughness or equivalent sand roughness and what fraction constitutes an increase in the radius of a circular cylinder. One needs to cut sample pipe sections and find a mean radius and a mean roughness height (averaged over several sections along the pipe). One needs to get accustomed to the idea that fluid mechanics in general and marine hydrodynamics in particular represent an enormously complex texture of conditions even for a simple roughness element.

3.8 Impulsively started flows

3.8.1 *Introductory comments*

Unsteady flow past bluff bodies has attracted a great deal of attention because a number of problems of practical importance are unsteady. Among the numerous theoretical, numerical, and experimental investigations, "impulsively started" steady flow about a cylinder has occupied a prominent place partly because of its intrinsic interest toward the understanding of the evolution of separation, vortex formation, and growth; partly because of its practical importance in various aerodynamic applications (e.g., the impulsive flow analogy, flow about missiles, dynamic stall, launching of structures in the ocean); and partly because it provides the most fundamental case for the comparison and validation of various numerical methods and codes. However, neither impulsive start nor impulsive stop is physically realizable. The flow must be accelerated from rest to a constant velocity or decelerated from a constant velocity to rest, or to another velocity, in a pre-scribed manner. This fact gives rise to a series of new questions: (1) What is the effect of the initial acceleration, prior to the establishment of a steady uniform flow, on the characteristics of the resulting time-dependent flow? (2) Are there critical values of the governing parameters above or below which the flow may be regarded as almost impulsively started? (3) How does the rate of accumulation of vorticity, as well as its cross-wake trans-fer, depend on the initial history of the motion? These questions may be explored through the use of a constant, rather than arbitrary, acceleration of the ambient flow over a specified period.

3.8.2 *Representative impulsively started flows*

Such flows are common examples of nonsteady boundary layers and have some practical importance, particularly when an accident sets the fluid impulsively in motion about the bodies immersed in it (e.g., a loss-of-coolant accident in boiling-water nuclear reactors).

Impulsively started flow is one of those unsteady flow situations for which analytical and numerical solutions exist at least for small times and relatively low Reynolds numbers. At the early stages of the motion the vorticity does not have enough time to diffuse. Thus the boundary layers are very thin and the flow is essentially irrotational. In other words, this is the time period during which the *added-inertia* concept is valid. The fluid force acting on the body is *primarily inertial*, and the inertia coefficient C_m should be equal to 2 (cylinder at rest, the fluid accelerating; otherwise $C_m = 1$), and one may suppose that the value of C_d should be identical with that applicable to a constant velocity. For bodies without sharp corners (e.g., a circular cylin-der), the separation does not occur immediately. Furthermore, it does not necessarily initiate at the downstream stagnation point (as in the case of a circular or elliptic cylinder).

For 2D cylinders, it can be shown that the separation begins after a time t_s at a place where the absolute value of dU/dx is largest. The relationship between t_s and dU/dt is (see Schlichting 1968)

$$1 + \left(1 + \frac{4}{3\pi}\right) \frac{dU}{dx} t_s = 0 \qquad (3.8.1)$$

For a circular cylinder started impulsively from rest to a constant velocity, the distance covered until separation begins is $s = 0.351c$, c being the radius of the cylinder. The separation begins at the rear stagnation point. For a uniformly accelerating circular cylinder the same distance is $s = 0.351c$ and obviously greater than that for the case of impulsive motion.

For axisymmetric bodies, t_s is given by the expression (Schlichting 1968)

$$1 + t_s \left[\frac{dU}{dx} \left(1 + \frac{4}{3\pi} \right) + 0.15 \frac{U}{r} \frac{dr}{dx} \right] = 0 \qquad (3.8.2)$$

For a sphere impulsively set in motion, $s = 0.392c$. The distance covered by the sphere until separation begins is larger, as in the case of the cylinder, when the sphere is accelerated uniformly from rest. Evidently the rate of acceleration as well as the history of acceleration is important in the calculation of the relative distance covered prior to the occurrence of separation.

For a circular cylinder undergoing sinusoidal oscillations, the flow may be assumed to accelerate uniformly, at least during the early stages of acceleration. Then the relative distance or preferably π times that distance, denoted by K, is $K = 2\pi s/2c = 0.52\pi = 1.63$. *For bodies with sharp corners, separation starts immediately and the inertial force imparted to the fluid is not necessarily equal to that given by the potential theory for the unseparated flow (the early evolution of the inertial force depends on the rate of diffusion of vorticity and Re).*

The role played by separation on the inertial force imparted on the body is an intriguing one and must be understood clearly. Consider an impulsive change superimposed on an already established flow pattern. Just prior to the impulsive change, the drag coefficient is given by its steady-state value at the prevailing Reynolds number. Sears, as reported by Rott (1964), showed that the initial motion following the impulsive change of the conditions consists of the superposition of the velocity pattern existing just before the change and the inviscid-flow velocity pattern that is due to the impulsive boundary values. In other words, *at the initial instants of the impulsive change* the drag coefficient is equal to its steady-state value and $C_m = 1 + C_a$ (C_a being equal to that given by the *potential theory*, e.g., $C_a = 1$ and $C_m = 2$ for a circular cylinder). As time progresses, *neither C_d nor C_a remains the same. They change with the evolution of the flow, ever dominated by the past history and ever affected by the gross features of the current state.* Furthermore, it should be emphasized once again that the inertial force per unit mass is acceleration.

Theoretical and numerical investigations of the impulsively started motion of a circular cylinder in a fluid otherwise at rest are confined mostly *to early times and very small* Reynolds numbers. Such a motion was first considered by Blasius (1908), and his work was later extended by Goldstein and Rosenhead (1936), Görtler (1944, 1948), Schuh (1953), Watson (1955), Wundt (1955), Bouard and Coutanceau (1980), Braza *et al.* (1986), and Sarpkaya (1986a, 1989). It was found, as noted earlier, that, after a certain lapse of time, the boundary layer separates from the surface of the cylinder, the time and location of separation depending on the Reynolds number and the bluffness of the body. The separation points then move rapidly around the cylinder until at large times they coincide with the average positions of the points of laminar separation for steady flow.

Experiments at relatively low Reynolds numbers have been reported by Schwabe (1935), Taneda and Honji (1971), Taneda (1972), and Coutanceau and Bouard (1977). There are very few experimental data for impulsively started flow at sufficiently high Reynolds numbers, i.e., for Reynolds numbers in the supercritical and post-supercritical regimes. This is partly because of the experimental difficulties encountered in establishing a vibration-free impulsively started steady flow and partly because of the instrumentation required to measure the transient quantities involved. Figure 3.10 shows the evolution of the force acting on a cylinder in a representative impulsively started flow at relatively high Reynolds numbers. Additional works are presented in Sarpkaya (1966, 1979b, 1990b, 1991c), Bouard and Coutanceau (1980), Sarpkaya and Kline (1982), and Sarpkaya and Ihrig (1986) for impulsively as well as nonimpulsively started accelerating flows about cylinders.

It is seen from Figs. 3.10a and 3.10b that the drag coefficient in the initial stages ($Vt/c \leq 4$) of an impulsively started flow can exceed its steady value by as much as 30%, partly because the vorticity is slow to diffuse and therefore accumulates rapidly in the close vicinity of the cylinder. The growth of the vortices is so rapid that they become much larger than their quasi-steady-state size and soon reach unstable proportions. This leads to the observed large drag coefficient (see also Roos and Willmarth 1971 for a similar observation with spheres). One of the vortices and then the other sheds from its shear layer through a cutoff mechanism (Gerrard 1965; Bloor and Gerrard 1966). Shortly thereafter, the drag coefficient decreases sharply and the lift coefficient begins to increase. In the following text, Fig. 3.11 shows a comparison of the measured and calculated drag coefficients; Fig. 3.12 shows the measured lift coefficient; Fig. 3.13 shows the evolution of the wake, predicted with the discrete-vortex model; and Fig. 3.14 shows the evolution of the lift and the wake in the later stages of the impulsively started flow. The details of the discrete-vortex model are given in Sarpkaya (1989, 1994).

The impulsive flow has long been regarded as analogous to the evolution of separated flow about slender bodies moving at high angles of attack in the subsonic to moderately supersonic velocity range. The approximate flow similarity between the development of the cross flow with distance along an inclined body of uniform diameter and the development with time of the flow on a cylinder in impulsive flow is known as the "cross-flow analogy." It was first proposed by Allen and Perkins (1951) and subsequently used by many other researchers to calculate the in-plane normal force and the out-of-plane force (side force normal to the plane of flight). A detailed

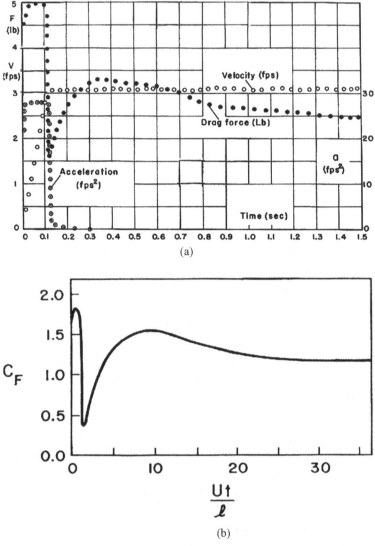

Figure 3.10. (a) Variation of velocity, acceleration, and drag force during a representative test run with a $D = 2.75$-inch circular cylinder. (b) Force coefficient $C_F (= 2F/\rho D V^2)$ for an impulsively started steady flow (synthesized from Fig. 3.10a).

discussion of the analogy, extensive measurements for various nose shapes and body combinations, and the most pertinent references may be found in Thomson and Morrison (1969, 1971); Bostock (1972); Thomson (1972); Lamont (1973); Lamont and Hunt (1976); Wardlaw (1974); Ericsson and Reding (1979); and Sarpkaya *et al.* (1982). It must be noted that *the analogy is far from perfect.* A blunt-nosed cylinder at high angles of attack generates a stationary asymmetric vortex array, which is similar to the Kármán vortex street. Thus the approximate space–time equivalence is possible because

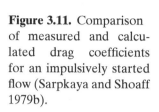

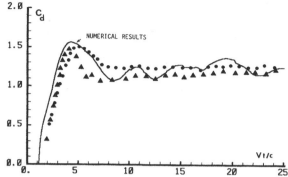

Figure 3.11. Comparison of measured and calculated drag coefficients for an impulsively started flow (Sarpkaya and Shoaff 1979b).

the vortices have an axial degree of freedom for their liftoff to make up the asymmetric array.

3.9 Sinusoidally oscillating flow

3.9.1 *Introduction*

Small-amplitude sinusoidal oscillations have been used extensively to determine the "added mass" of bodies of various shapes (see, e.g., Sarpkaya 1960; Stelson and Mavis 1965; Chandrasekaran *et al.* 1972; Fickel 1973; Ciesluk and Colonell 1974; Brooks 1976; Skop *et al.* 1976). We are concerned here primarily with the relatively large-amplitude oscillations of flow around the body for which the separation of flow and turbulence plays a very important role.

We begin with the discussion of the simplest possible oscillating flow (SOF) to show that much remains to be resolved. One should be aware early on that the SOF about a circular cylinder represents an enormously complex texture of conditions in all states of K, Re, cylinder orientation,

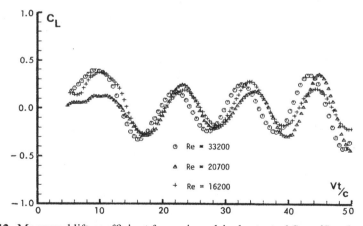

Figure 3.12. Measured lift coefficient for an impulsively started flow (Sarpkaya 1978b).

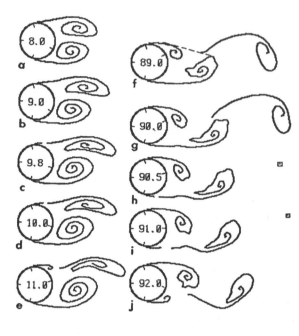

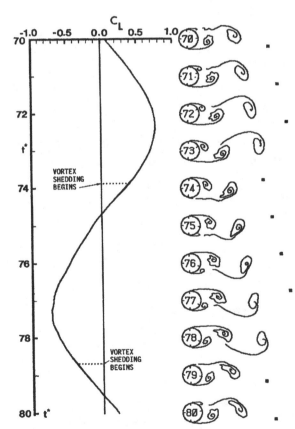

Figure 3.13. Evolution of wake in an impulsively started flow (Sarpkaya and Shoaff 1979b).

Figure 3.14. Evolution of lift and wake in the later stages of an impulsively started flow (Sarpkaya and Shoaff 1979b).

surface roughness, end conditions, and ambient turbulence, just to name a few.

There are several fundamental differences between the unidirectional flow and the SOF and between the SOF and the wavy flow over a cylinder. When a cylinder is subjected to a sinusoidal flow normal to its axis, the flow does not only accelerate from and decelerate to zero but changes direction as well during each cycle. This produces a reversal of the wake from the downstream to the upstream side whenever the velocity changes sign. The separation points undergo large excursions. The boundary layer over the cylinder may change from fully laminar to partially or fully turbulent states and the Reynolds number may range from subcritical to post-supercritical over a given cycle. The vortices that have been or are being formed or shed during the first half of the flow period are also reversed around the cylinder during the wake reversal, giving rise to a transverse force (as a consequence of their mere convection) with or without additional vortex shedding. This is particularly pronounced for amplitudes of flow oscillation for which the number of newly formed vortices during the period of flow reversal is not much greater than that of the vortices that have survived dissipation (laminar and/or turbulent) and convected around the cylinder during the wake reversal. Particularly significant are the changes in the lift, drag, and inertial forces when the reversely convected vortices are not symmetric.

The wavy flows are of course relatively more complex (Chapter 4). Aside from the effects of the free surface, the orbital motion of the fluid particles gives rise to 3D flow effects. The rotation of the wake about a horizontal cylinder and the exponential decay of the representative wave velocity along a vertical cylinder further complicate the matters. In fact, it is for these reasons that a number of investigators preferred to separate the additional effects brought about by the waviness of the flow from those resulting from the periodic reversal of the flow in a simply sinusoidal rectilinear motion.

SOF about cylinders has been investigated by a large number of researchers (Keulegan and Carpenter 1958; Rance 1969; Sarpkaya 1976b, 1976c, 1976d, 1976e, 1977a; and the references cited therein). As in the case of other time-dependent flows, the most serious difficulty with SOF lies in the description and the quantification of the time-dependent force itself. Some insight may be gleaned into the nature and decomposition of this force from a remarkable paper by Stokes (1851) on the motion of pendulums. Stokes has shown (as noted in Chapter 2) that the force acting on a sphere oscillating in a liquid with the velocity $U = -A_0 \cos \omega t$ is given by

$$F(t) = \frac{\rho \pi D^3}{6} \left(\frac{1}{2} + \frac{9}{2} \sqrt{\frac{2\mu}{\rho \omega D^2}} \right) \frac{dU}{dt} + 3\pi \mu D U \left(1 + \frac{1}{2} \sqrt{\frac{\rho \omega D^2}{2\mu}} \right) \qquad (3.9.1)$$

This force is composed of two parts: an inertial force and a drag force, *linearly dependent on acceleration and velocity*, respectively. Both components depend on viscosity. Furthermore, the flow is *assumed to be unseparated and laminar*. The decomposition of the time-dependent force into the two said components is somewhat arbitrary. The same force may be decomposed into three or four parts, and each part may be given a separate meaning. For example, we may write

$$F(t) = \frac{1}{2}\left(\frac{\rho\pi D^3}{6}\right)\frac{dU}{dt} + 3\pi\mu DU$$

$$+ \frac{9}{2}\left(\frac{\rho\pi D^3}{6}\right)\sqrt{\frac{2\mu}{\rho\omega D^2}}\frac{dU}{dt} + \frac{1}{2}\pi\mu DU\sqrt{\frac{\rho\omega D^2}{2\mu}} \qquad (3.9.2)$$

in which the first term on the right-hand side represents the added mass (its ideal value) times acceleration; the second term, the linear viscous resistance to the steady motion of a sphere at very small Reynolds numbers (say $Re < 1$); the third term, either the effect of history or the motion on the inertial force or simply the viscous effects in SOF on the acceleration-dependent forces; and the last term, the history effect on the linear drag. Also, one may combine the last two terms and regard them as history-dependent modifications to the ideal values of the inertia and drag forces.

As yet a theoretical analysis of the problem for separated flow is difficult, and much of the desired information must be obtained both numerically and experimentally. In this respect, the experimental studies of Morison and his co-workers (1950) on the forces on piles that are due to the action of progressive waves have provided a useful and somewhat heuristic approximation. The forces are divided into two parts, one that is due to the drag, as in the case of flow of constant velocity, and the other that is due to acceleration or deceleration of the fluid. This concept necessitates the introduction of a drag coefficient C_d and an inertia coefficient C_m in the expression for force. In particular, if F is the force per unit length experienced by a cylinder, then U and dU/dt represent, respectively, the undisturbed velocity and the acceleration of the fluid. It is assumed that the wave slope and the associated pressure gradient are roughly constant across the diameter of the body and the wave scattering is negligible (discussed in Chapter 5).

In the Stokes sphere problem in which the Reynolds number is very small, drag is proportional to the first power of velocity. In Morison's equation, drag is proportional to the square of the velocity because the flow is separated and the drag is primarily due to pressure rather than to the skin friction. It is evident that Morison's equation is a heuristic extension to separated time-dependent flows of the solution obtained by Stokes. It is also evident that *the validity of the equation and the limits of its application will have to be determined experimentally.*

The fact that the drag and inertia coefficients in the Morison form of the resistance equation depend on both the Reynolds number ($Re = U_m D/v$, where $U_m = A\omega$) and the relative amplitude A/D or $K = 2\pi A/D$ may be demonstrated by writing Morison's force and Stokes force for a SOF ($U = -U_m \cos \omega t$) about a sphere at rest as

$$F = -\frac{1}{2}\rho \frac{\pi D^2}{4} C_d U_m^2 |\cos \omega t| \cos \omega t + \rho \frac{\pi D^3}{6} C_m U_m \omega \sin \omega t \qquad (3.9.3a)$$

or

$$\frac{F}{\frac{1}{2}\rho \frac{\pi D^2}{4} U_m^2} = -C_d |\cos \omega t| \cos \omega t + \frac{8\pi}{3} \frac{1}{K} C_m \sin \omega t \qquad (3.9.3b)$$

and, from (3.13), written in current symbols and in terms of a new parameter $\beta = Re/K = D^2/vT$ (see Sarpkaya 1976e, 2000, 2005), (3.9.3b) reduces to

$$\frac{F}{\frac{1}{2}\rho \frac{\pi D^2}{4} U_m^2} = -\frac{24}{Re}\left(1 + \frac{1}{2}\sqrt{\pi \frac{Re}{K}}\right)\cos \omega t + \frac{8\pi}{3K}\left(\frac{3}{2} + \frac{9}{2}\sqrt{\frac{K}{Re}}\right)\sin \omega t$$
$$(3.9.4)$$

in which one part of 3/2 in the second set of parentheses is due to the pressure gradient to accelerate the flow about a sphere. Equation (3.9.4) yields

$$C_d = \frac{24}{Re}\left(1 + \frac{1}{2}\sqrt{\pi \frac{Re}{K}}\right) = \frac{24}{Re}\left(1 + \frac{1}{2}\sqrt{\pi \beta}\right) \qquad (3.9.5)$$

and

$$C_m = \frac{3}{2} + \frac{9}{2}\sqrt{\frac{1}{\pi} \frac{K}{Re}} = \frac{3}{2} + \frac{9}{2}\sqrt{\frac{1}{\pi \beta}} \qquad (3.9.6)$$

where $\beta = Re/K = D^2/vT$ and $T = 2\pi/\omega$.

The drag and inertia coefficients for the Stokes force depend on both K and Re. However, there is a unique relationship between C_d and C_m, dependent only on Re, i.e.,

$$\left(\frac{C_d}{24/Re} - 1\right)\left(C_m - \frac{3}{2}\right) = \frac{9}{4} \qquad (3.9.7)$$

The parameter $24/Re$ in (3.9.7) is the steady-flow drag coefficient for a sphere in the Stokes regime and the constant $(3/2)$ is the ideal value of C_m for a sphere. Thus, in unseparated Stokes flow *the oscillations increase both the drag and the inertia coefficient above their corresponding steady-state values.* The fact that this is not always so for separated flows will become apparent later. Experiments show that *only for small values of K and β that C_m exceeds its ideal potential-flow value.*

Aside from historical and scientific reasons (i.e., *to confirm that viscosity is a fluid property and the* Navier–Stokes *equations are here to stay!*), the foregoing is of little direct importance to the designer of offshore structures, but it is very instructive as are all *basic investigations,* however limited their scope or range of applicability may be.

On the basis of irrotational flow around a cylinder, C_m should be equal to 2 (cylinder at rest, the fluid accelerating; otherwise $C_m = 1$), and one may suppose that the value of C_d should be identical to that applicable to a constant velocity. However, numerous experiments have shown that this is not the case and that C_d and C_m show considerable variations from those just cited. Even though *no one has suggested a better alternative* (Sarpkaya 2001b), the use of the Morison's equation gave rise to a great deal of discussion on what values of the two coefficients should be used. Furthermore, the importance of the effect of viscosity, roughness, rotation of the velocity vector, upstream turbulence, spanwise coherence, free surface, yaw, and the effect of neighboring elements remained in doubt because experimental evidence published over the past 50 years has been quite inconclusive. The problem has further been compounded by the difficulty of accurately measuring the velocity and accelerations to be used in the Morison equation. In general, the nature of the equation rather than the lack of precision of measurements or the difficulty of calculating the kinematic characteristics of the flow from the existing wave theories has been criticized.

The drag and inertia coefficients obtained from a large number of field tests, as compiled by Wiegel (1964), show extensive scatter whether they are plotted as a function of the Reynolds number or the so-called Keulegan–Carpenter number K ($K = U_m T/D$). The reasons for the observed scatter of the coefficients remained unknown for a long time. The scatter was attributed to several reasons or combinations thereof such as the irregularity of the ocean waves, free-surface effects, 3D nature of the flow, inadequacy of the averaged resistance coefficients to represent the actual variation of the nonlinear force (particularly when there are only one or two vortices in the wake), omission of some other important parameter that has not been incorporated into the analysis, the effect of ocean currents on separation, vortex formation, and hence on the forces acting on the cylinders, etc.

3.9.2 *Fourier-averaged drag and inertia coefficients*

The first systematic evaluation of the Fourier-averaged drag and inertia coefficients was made by Keulegan and Carpenter (1958) at relatively low Reynolds numbers through measurements on submerged horizontal cylinders and plates placed in the node of a standing wave, applying theoretically derived rather than measured values of the velocities and accelerations.

Keulegan and Carpenter expressed the force in terms of a Fourier series, assuming the force to be an odd-harmonic function of $\theta = 2\pi t/T$, i.e., $F(\theta) = -F(\theta + \pi)$, as

$$2F/(\rho D U_m^2) = 2[A_1 \sin\theta + A_3 \sin 3\theta + A_5 \sin 5\theta$$
$$+ \cdots + B_1 \cos\theta + B_3 \cos 3\theta + B_5 \cos 5\theta + \cdots] \quad (3.9.8)$$

Keulegan and Carpenter were able to reconcile (3.9.8) with the equation proposed by Morison, O'Brien, Johnson, and Schaaf (1950) (henceforth MOJS) by writing (3.9.8) in the following form:

$$\frac{2F}{\rho D U_m^2} = \frac{\pi^2}{K} C_m \sin\theta + 2[A_3 \sin 3\theta + A_5 \sin 5\theta + \cdots]$$
$$- C_d |\cos\theta| \cos\theta + 2[B_3' \cos 3\theta + B_5' \cos 5\theta \cdots] \quad (3.9.9)$$

in which U is assumed to be given by $U = -U_{rn} \cos\theta$. Evidently (3.9.9) reduces to the equation proposed by MOJS, i.e., to

$$\frac{2F}{\rho D U_m^2} = \frac{\pi^2}{K} C_m \sin\theta - C_d |\cos\theta| \cos\theta \quad (3.9.10)$$

provided that *the coefficients C_m and C_d are independent of θ*, i.e., each term has the same constant value (dependent on K and Re) and A_n and B_n are zero for n equal to or greater than 3. These are, indeed, monumental assumptions in view of the fact that there is nothing better.

The Fourier averages of C_d and C_m are obtained by multiplying both sides of Eq. (3.9.10) once with $\cos\theta$ and once with $\sin\theta$ and integrating between the limits $\theta = 0$ and $\theta = 271$. This procedure yields

$$C_d = -\frac{3}{4} \int_0^{2\pi} \frac{F \cos\theta}{\rho D U_m^2} d\theta \quad (3.9.11a)$$

$$C_m = \frac{2 U_m T}{\pi^3 D} \int_0^{2\pi} \frac{F \sin\theta}{\rho D U_m^2} d\theta \quad (3.9.11b)$$

The drag and inertia coefficients may also be evaluated through the use of the method of least squares. This method consists of the minimization of the error between the measured and calculated forces. It is not shown here because the Fourier analysis and the method of least squares yield identical C_m values and the C_d values differ only slightly.

One can show through the use of (3.9.11a) and (3.9.11b) that the rate of change of force with time is zero at the time of maximum acceleration and is proportional to C_m/KT at the time of maximum velocity. Thus the determination of C_m, in particular through the use of force at the time of maximum acceleration, depends on the particular values of C_m, K, and T, and may not be quite accurate. In general, it is recommended that either the Fourier-averaged or the least-squares-averaged force-transfer coefficients be used.

Equation (3.9.11) also shows that the maximum in-line force does not occur at the time of maximum velocity, but rather *it leads it*. The maximum force coefficient $C_F(spp)$ may be calculated from (3.9.11) to yield

$$C_F(spp) = C_d + \frac{\pi^4 C_m^2}{4C_d K^2} \quad \text{for} \quad K > \frac{\pi^2 C_m}{2C_d} \tag{3.9.12a}$$

and

$$C_F(ssp) = \frac{\pi^2 C_m}{K} \quad \text{for} \quad K < \frac{\pi^2 C_m}{2C_d} \tag{3.9.12b}$$

In general, the dividing value of K may be assumed to be about $K = 10$. For relatively small values of K, the force is said to be *inertia dominated*. For relatively large values of K, the force is *drag dominated*. The ratio of the maximum inertial force to maximum drag force is $R = \pi^2 C_m / K C_d$.

For purposes of orientation, it is useful to indicate the range of K values over which the various components of force become predominant. These are

K smaller than about 10 ... inertia increasingly important
K larger than about 15 ... drag increasingly important
K larger than about 5 ... lift force important

A more detailed discussion of the various loading regimes will be taken up later.

3.9.3 *Experimental studies on C_d and C_m*

It is recognized that the coefficients previously cited are not constant throughout the cycle and are either time-invariant averages or peak values at a particular moment in the cycle. A simple dimensional analysis of the flow under consideration shows that the time-dependent coefficients for a uniformly roughened cylinder may be written as

$$\frac{2F}{\rho L D U_m^2} = f\left(\frac{U_m T}{D}, \frac{U_m D}{\nu}, \frac{k}{D}, \frac{t}{T}\right) \tag{3.9.13}$$

in which F represents the in-line or the transverse force. Equation (3.9.13) combined with (3.9.10), taking for now the latter to be valid, yields

$$C_d = f_1(K, Re, k/D, \ t/T) \tag{3.9.14}$$

There is no simple way to deal with (3.9.14) even for the most manageable time-dependent flows. Another and perhaps the only other alternative is to eliminate time as an independent variable and consider suitable time-invariant averages as given by Eqs. (3.9.11a) and (3.9.11b). Thus, for periodically oscillating flows, the Reynolds number is not necessarily the most suitable parameter. The primary reason for this is that U_m appears in both K and Re. This is not convenient for experiments in a U-shaped water tunnel (Sarpkaya 1976d) because the period T cannot be varied enough to make the effort worthwhile. Thus, replacing Re with $Re/K = D^2/UT$ (or by fD^2/U), in (3.35), one has

$$C_i(\text{a coefficient}) = f_i(K, \beta, k/D) \tag{3.9.15}$$

in which $\beta = D^2/UT$. Its universality was not known until 1976. It was assumed that it had nothing to do with separated flows at any Reynolds number. The facts regarding the universalization of the parameter β to fully non-linear unsteady separated flows about bluff bodies (nearly 125 years after its first appearance in a *linearized* analysis of *unseparated* viscous flow) with very slow oscillations about a cylindrical rod and sphere by Stokes (1851) is discussed in detail in Sarpkaya (2005). *There was then no Keulegan–Carpenter number, no Reynolds number, no Stokes number, and thus no Re/K.* Obviously one could not have known it without the use of a large U-shaped oscillating flow tunnel.

From the standpoint of dimensional analysis, either the Reynolds number or β could be used as an independent variable. For a series of experiments conducted with a cylinder of diameter D in water of uniform and constant temperature, β is constant if the period T is kept constant. Then the variation of a force coefficient with K may be plotted for constant values of β. Subsequently one can easily recover the Reynolds number from $\beta K = Re$ and connect the points, on each $\beta = $ constant curve, representing a given Reynolds number.

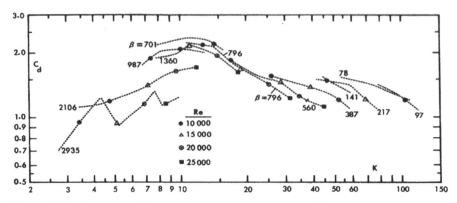

Figure 3.15. A replot of Keulegan–Carpenter's drag coefficients as functions of K for 12 values of β (Sarpkaya 1976e).

From the standpoint of laminar boundary-layer theory, β represents the ratio of the rate of diffusion through a distance δ (i.e., v/δ^2, where δ is the boundary-layer thickness) to the rate of diffusion through a distance D (i.e., v/D^2). This ratio is also equal to $(D/\delta)^2$ and, when it is large, gradients of velocity in the direction of flow are small compared with the gradients normal to the boundary, a situation to which the boundary-layer theory is applicable (Rosenhead 1963).

Let us now reexamine the Keulegan–Carpenter data (1958), partly to illustrate the use and significance of β as one of the governing parameters and partly to take up the question of the effect of Reynolds number on the force coefficients.

The data given by Keulegan and Carpenter may be represented by 12 different values of β. The drag and inertia coefficients are plotted in Figs. 3.15 and 3.16 and connected with straight-line segments. The identification of the individual data points in terms of the cylinder diameter, as was done by Keulegan and Carpenter and by Sarpkaya (1975) irrespective of the

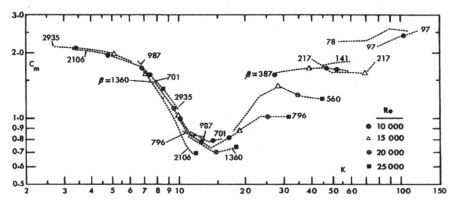

Figure 3.16. A replot of Keulegan–Carpenter's inertia coefficients as functions of K for 12 values of β (Sarpkaya 1976e).

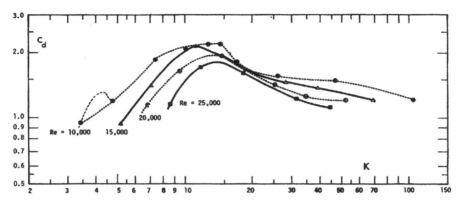

Figure 3.17. A replot of Keulegan–Carpenter's drag coefficients as functions of K for four selected values of the Reynolds number (Sarpkaya 1976e).

β values, gives the impression of a scatter in the data and invites one to draw a mean curve through all data points. Such a temptation is further increased by the fact that the data for each β span over only a small range of K values. Thus the drawing of such a mean curve obscures the dependence of C_d and/or C_m on β and hence on Re.

These figures show, within the range of Re and K values encountered in Keulegan–Carpenter data, that (a) C_d depends on both K and Re and decreases with increasing Re for a given K; and that (b) C_m depends on both K and Re for K larger than approximately 15 and *decreases* with increasing Re. A similar analysis of Sarpkaya's data (1975) also shows that C_d and C_m depend on both K and Re and that C_m *increases* with increasing Re. Notwithstanding this difference in the variation of C_m, between the two sets of data, Figs. 3.17 and 3.18 put to rest the long-standing controversy regarding the dependence or lack of dependence of C_d and C_m on Re and show the importance of β as one of the controlling parameters in interpreting the data, in interpolating the K values for a given Re, and in providing guidelines for further experiments as far as the range of K and β is concerned.

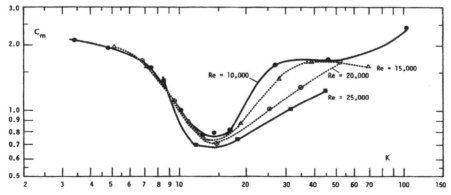

Figure 3.18. A replot of Keulegan–Carpenter's inertia coefficients as functions of K for four selected values of the Reynolds number (Sarpkaya 1976e).

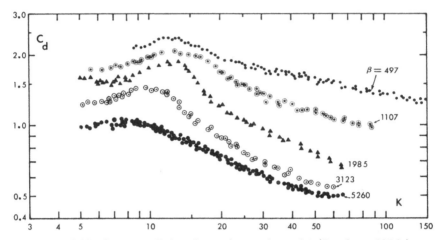

Figure 3.19. C_d versus K data for various values of β (Sarpkaya 1976e).

As noted earlier, there has been growing awareness of the fact that the coefficients obtained at relatively low Reynolds numbers may not be applicable at higher Reynolds numbers, that the transverse forces acting on the elements of offshore structures may be as much or more important (e.g., for the dynamic response of the structural elements) than the in-line forces given by the MOJS formula, and that the initial or growing marine roughness may significantly alter the forces acting on the structure. In view of the foregoing considerations, Sarpkaya (1976e, 1976d) conducted a series of experiments with smooth and sand-roughened cylinders in the U-shaped vertical water tunnel (Fig. 3.1).

In these experiments the drag and inertia coefficients have been evaluated through the use of Eqs. (3.9.11a) and (3.9.11b), i.e., through the use of the Fourier analysis. Figures 3.19 and 3.20 show C_d versus K and C_m versus K for five representative values of β. Mean lines drawn through the data shown in Figs. 3.19 and 3.20 are presented in Figs. 3.21 and 3.22 together

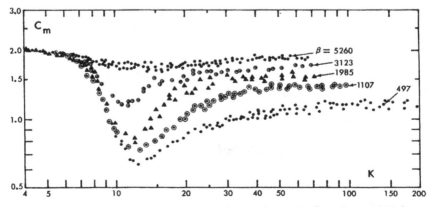

Figure 3.20. C_m versus K data for various values of β (Sarpkaya 1976e).

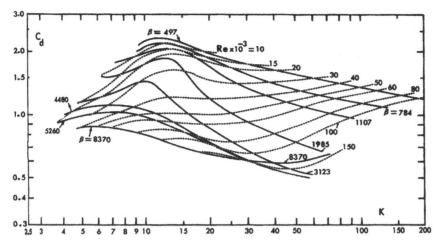

Figure 3.21. C_d versus K plots for various values of β (Sarpkaya 1976e).

with the constant Reynolds number lines obtained through the use of $K = Re/\beta$. Evidently there is a remarkable correlation between the force coefficients, Reynolds number, and the Keulegan–Carpenter number. In the *inertia-dominated* regime (K less than about 5), the extraction of the drag force from the total force is extremely difficult. This is particularly true for the drag coefficients for the *smaller cylinders, smaller β values, and smaller K values (less than about 5)*. In Figs. 3.19 and 3.21, *no drag coefficient is plotted* in the regions where such uncertainty might exist. Subsequently a series of special experiments has been performed with small cylinders and very sensitive force gauges (Chapter 7) to delineate the characteristics of the drag and damping coefficients in the inertia-dominated regime.

The entire data are shown in Figs. 3.23 and 3.24 as functions of Re for constant values of K. These figures clearly show that C_d decreases with

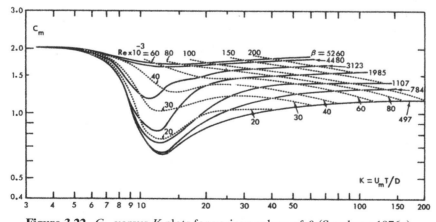

Figure 3.22. C_m versus K plots for various values of β (Sarpkaya 1976e).

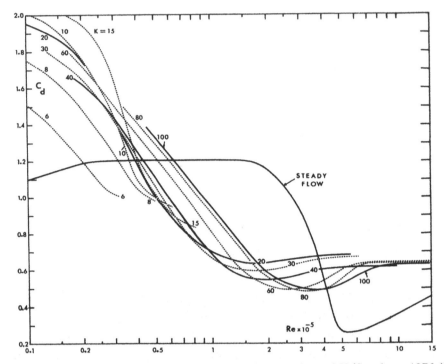

Figure 3.23. C_d versus Reynolds number for various values of K (Sarpkaya 1976e).

increasing *Re* to a value of about 0.5 (depending on *K*) and then gradually rises to a constant value (post-supercritical value) within the range of Reynolds numbers encountered. The inertia coefficient increases with increasing *Re*, reaches a maximum, and then gradually approaches a value of

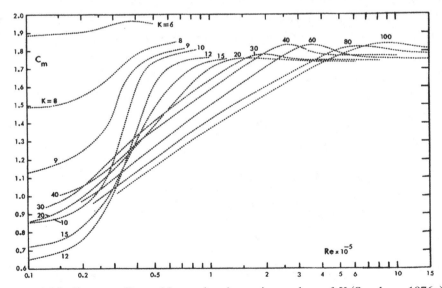

Figure 3.24. C_m versus Reynolds number for various values of K (Sarpkaya 1976e).

about 1.85. The Keulegan–Carpenter data show an opposite trend. It is now believed that the Keulegan–Carpenter data for C_d are not quite reliable for K values larger than about 15. Figure 3.23 shows that the drag coefficient for a cylinder in SOF is not always larger than that for steady flow at the same Reynolds number. For $K = 100$, for example, C_d for the oscillating flow is larger than its steady-state value for Re smaller than about 60 000 and larger than about 400 000. In the range of Reynolds numbers between the two previously cited, the drag coefficient for the oscillating flow is considerably lower than that for steady flow. The reason for this is surely the earlier transition in the boundary layers.

It is a well-known fact that the occurrence of drag crisis in steady cross flow about a cylinder depends on the characteristics of turbulence in the ambient flow, blockage and length-to-diameter ratio of the cylinder, the vibration amplitude and frequency of the test body, surface roughness, and on other peculiarities of the wind tunnel or water tunnel in which the experiments are performed. In fact, it is for this reason that the minimum value of the drag coefficient and the base pressure in steady flow are widely scattered. Because the formation of the separation bubbles is largely responsible for the low values of C_d, one could state that the formation and the extent of the separation bubbles are very sensitive to the factors previously cited. *A priori*, one would expect the same to occur in SOF about a cylinder. During a given cycle, the flow at both sides of the cylinder contains a number of vortices and large-scale turbulence. Thus it is natural to assume that they would give rise to earlier transition. In fact, Figs. 3.23 and 3.24 show that the drag and inertia coefficients reach their asymptotic values at smaller Reynolds numbers.

The relationship between C_d and C_m has been of some interest. A plot of C_m versus C_d (Fig. 3.25) shows that there is not a unique relationship among them, independent of K and Re. In general, for a given K, C_d increases as C_m decreases. The more fundamental issues concern the occurrence and motion of the separation points, the distribution of the instantaneous pressure around the cylinder, and the relationships among the vortex shedding, transverse force, and the drag and inertia components of the in-line force (see, e.g., Grass and Kemp 1979; Matten *et al.* 1979; Sarpkaya and Butterworth 1992; and Sarpkaya 2006b).

It is appropriate here to note through the comparison of Figs. 3.2a and 3.2b (compiled by Wiegel *et al.* 1957) with Figs. 3.19–3.21 that the ocean experiments may be better understood and the reasons for the large scatter may be better explained through the controlled laboratory experiments, not the other way around. This is particularly important for the cases in which the effect of Re is relatively weak (say, for Re smaller than about 15000). It is only then that one learns not only the mechanics (if not the physics)

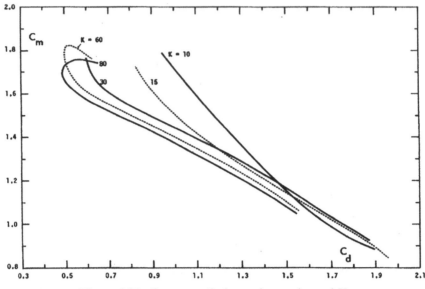

Figure 3.25. C_m versus C_d for various values of K.

of the wave–structure interaction but also the spectacular ocean–structure interaction.

The foregoing discussion shows that the time-averaged force coefficients reflect *only in a very crude way* the state of an extremely complex time-dependent flow. The key to the understanding of the instantaneous behavior of the lift, drag, and inertia forces *is the understanding of the formation, growth, and motion of vortices.* At present, this is possible only in a continuum and only experimentally. Once the continuum is bounded and its boundaries are discretized, one faces numerous numerical approximations and limitations. The infinity reduces to a short distance (judged by the maker), the number of particles become never large enough, and the devices that perform the arithmetic never fast enough. These facts render the assessments of accuracy of the predictions rather expensive. In short, man has not yet mastered the art of the very dramatic crossing of a boundary into a digitized world. There are, to be sure, numerous computational fluid-dynamic techniques, which, in turn, require discretization techniques to let the flow move through the boundaries into incalculable uncertainties. The majority of the numerical solutions have been devoted to steady flows, and are thus of limited value to marine hydrodynamics. Only a few examples are cited here: Breuer's (1998, 2000) work on the numerical and modeling influences on LESs for the flow past a circular cylinder ($Re = 3900$); numerical studies of flow over a circular cylinder at $Re = 3900$ by Kravchenko and Moin (2000); low Reynolds number flow around an oscillating circular cylinder at small Keulegan–Carpenter numbers by Dutsch *et al.* (1998); Nehari *et al.*'s (2004) 3D analysis of the unidirectional oscillatory flow around a circular cylinder at low K and β; Lu *et al.*'s (1997a, 1997b) application of LES to an oscillating flow past a circular cylinder; and a number of other contributions (using LES, DNS, and the discrete-vortex model) by Sarpkaya (1975), Stansby (1977, 1979), Graham (1979), Sawaragi and Nakamura (1979), Sarpkaya and Shoaff (1979b, 1979c), Sarpkaya *et al.* (1992), Tutar and Holdo (2000), and Wissink and Rodi (2003), all of which require several assumptions and some fine-tuning.

3.9.4 *Transverse force and the Strouhal number*

Vortex shedding and the resulting alternating force in steady flow have been studied extensively and the existing data on lift coefficients were presented in Fig. 3.8. In spite of the considerable interest, however, the transverse force in SOFs received very little attention. Recently it became clear from observations of the oscillations of long piles and strumming of cables that the lift forces are important not only because of their magnitude but also because of their alternating nature, which under certain circumstances may lead to the phenomenon known as the lock-in or vortex synchronization, or vortex-induced vibrations (VIVs). This phenomenon may cause failure that is due to fatigue and increased in-line force. Obviously the total instantaneous force acting on the structure is increased by the lift force and modified by the oscillations of the body, mode shape of the oscillation, strum-suppression devices, the ovalling of the cross section, etc.

The transverse force exerted on smooth and sand-roughened cylinders in SOFs has been measured by Sarpkaya (1975, 1976b, 1976c, 1976d, 1976e) for a wide range of Reynolds numbers, The lift coefficient $[C_L(\text{max})]$ reaches its maximum value (see Fig. 3.26) in the neighborhood of $K = 12$ and decreases sharply with increasing K, mostly because of the decrease of coherence along the cylinder.

As noted earlier, the lift force is a consequence of the pressure gradient across the wake. The alternating pressure gradient increases with increasing asymmetry of the strength and position of the vortices. Thus one may conclude that the flow for K values in the neighborhood of 12 must exhibit maximum asymmetry. Part of this asymmetry is due to the shedding of one or two new vortices and partly due to the convection of the vortices shed in the previous half-cycle back toward the cylinder. Neither the convected vortices nor the ones shed newly are necessarily symmetrical and result in a maximum wake asymmetry. The interaction of the convected vortices with the flow about the cylinder is such as to enhance the asymmetry by forcing the shedding of a new vortex from the side of the cylinder nearest to the convected vortex. This is easily understood if one considers the high velocities induced on the cylinder by the convected vortex and its image inside

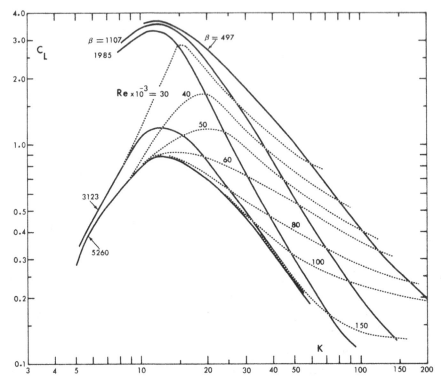

Figure 3.26. Lift coefficient versus K for various values of the Reynolds number and the frequency parameter.

the cylinder. Figure 3.27 shows that, for Re smaller than about 20000, C_L depends primarily on K. In the Reynolds number range from about 20000 to 100000, C_L depends, to varying degrees, on both Re and K. Above a Reynolds number of about 100000, the dependence of C_L on Re and K is quite negligible and certainly obscured by the scatter in the data. However, the magnitude of the lift force relative to the in-line force is not negligible.

The minimum value of K at which lift or the asymmetry in the vortices develops is, by the very nature of the vortices, extremely sensitive to the experimental conditions. Experiments by Sarpkaya (1976e) have shown that there is a 90% percent chance that the asymmetry will appear at $K = 5$. At $K = 4$, there is only a 5% percent chance that the asymmetry will appear for very short periods of time. The determination of the onset of asymmetry is of special importance not only in connection with the ocean structures but also with bodies of revolution flying at high angles of attack (see, e.g., Lamont and Hunt 1976). It has previously been noted in connection with the discussion of impulsively started flows and uniformly accelerating flows (see Subsections 3.8.1 and 3.8.2) that the asymmetry sets in when s/D is equal to about 3 for the impulsive motion and to about 4.8 for the uniformly accelerating motion. Assuming $K = 5$ for the onset of asymmetry in SOF, one finds

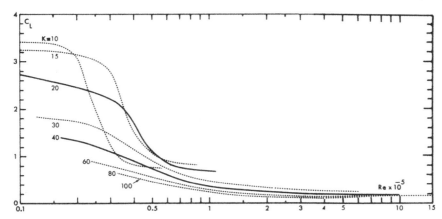

Figure 3.27. Lift coefficient versus Reynolds number for various values of K (Sarpkaya 1976e).

that $s/D = A/D = 5/(2\pi) = 0.8$. Strictly speaking, a direct comparison of the SOF with unidirectional flows is not justified. Furthermore, the effect of flow deceleration is very profound in promoting instability and earlier separation. Nevertheless, a simple-minded comparison of the relative distances covered by the fluid in various flow situations prior to the onset of asymmetry shows that asymmetry occurs much sooner in oscillating flows, as would be expected.

Aside from its magnitude, the most important feature of the transverse force is its frequency of oscillation. This frequency varies with K, Re, and time in a given cycle and also from cycle to cycle. This leads to multiple lift coefficients at a given K and Re (Sarpkaya 1976e; Maull and Milliner 1978). As noted by Maull and Milliner, this multiplicity raises doubts about any elementary analysis of the lift signal because it must be nonstationary in a statistical sense and therefore the rms value, for instance, will be a function of the length of the record taken. Experiments show that (see, e.g., Maull and Milliner 1978) the occurrence of peaks in the rms lift is closely associated with the progressive occurrence of higher-order components in the frequency domain. With increasing K the rms force reaches its first maximum at approximately $K = 13$, where the dominant frequency is twice the particle frequency. The next maximum occurs at $K = 18$ and is associated with a dominant frequency component at three times the particle frequency. At higher K, the lift coefficient decreases and the said frequency component is associated with higher and higher multiples of the particle frequency. This is evidenced by Fig. 3.28, where the frequency ratio $f_r = f_v/f_w$ (the ratio of the maximum frequency in a cycle, defined here as the reciprocal of the shortest interval between two maxima, to f_w) is plotted for a smooth cylinder. A point on each line represents the maximum value of K for a given Re and f_r. In other words, a line such as $f_r < 4$ means that the alternating force

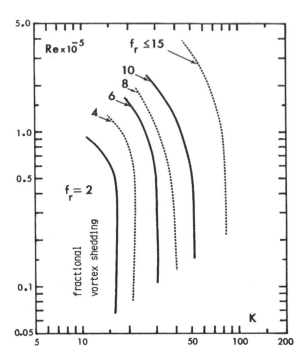

Figure 3.28. Relative frequency of vortex shedding as a function of K and Re.

does not contain frequencies larger than $f_r = 4$ for K and Re values in the region to the left of the line. Intermediate values of f_r such as $f_r = 3, 5$, etc., are not shown to keep the figure relatively simple.

Several facts are of special importance and are discussed in detail. First, Fig. 3.28 begins with $K = 5$. As noted earlier, there is occasional vortex shedding for K values between 4 and 5. Second, each $f_r = N$ line does not represent an absolute line of demarcation between the frequencies $N - 1$ and $N + 1$. Occasionally, a frequency of $N + 1$ will occur on the $N - 1$ side of the N line, and vice versa. Third, the frequency of vortex shedding is not a pure multiple of the flow-oscillation frequency. At first this would appear anomalous, but a closer examination of the behavior of the vortices shows that a fractional value of f_r is perfectly understandable. Evidently, f_r, as an integer, is a measure of the number of vortices actually shed during a cycle. However, those vortices that do not break away from their shear layers before the flow is reversed are partially developed and result in incomplete shedding. Thus, the fractional part of f_r indicates an incomplete shedding. This is particularly true for f_r in the neighborhood of 2 or 3. Finally, it should be noted that the frequencies previously cited correspond to the vortices shed at or near the maximum velocity in the cycle. As noted earlier, the vortex-shedding frequency is not constant throughout the cycle. Furthermore, the variation of the shedding frequency is not a simple harmonic function as it would be if the vortex-shedding process directly responded to the instantaneous velocity.

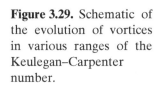

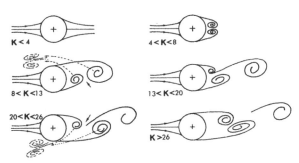

Figure 3.29. Schematic of the evolution of vortices in various ranges of the Keulegan–Carpenter number.

The fractional shedding of vortices for K values in the region of 5 to 15 causes incalculable changes in the flow pattern and in the in-line and transverse forces. In fact, no separated, time-dependent flow is more complex than the one in which there are only one or two vortices. For small values of K, two vortices begin to develop at the start of the cycle in one direction, but the vortices do not acquire identical strengths for various reasons. As the flow reverses, the larger of the vortices is swept past the cylinder but the weaker one dissipates partly because of turbulent diffusion. The consequences of this single shedding are that the in-line force becomes asymmetrical and the vortex that is swept away plays an important role in the formation of the new vortices when the flow reverses its direction. The dominant vortex establishes, by its sense of rotation (and the opposite sense of rotation of its image inside the cylinder), a preferred location for the generation of a new dominant vortex. The new vortex and the one convected downstream may form a pair and increase the transverse pressure gradient and thus give rise to significant lift forces, as noted earlier. For certain values of K, the convected vortex during the flow reversal may be rapidly thrown out of the flow field because of the large local transport velocities induced at it by the newly formed vortices (see Fig. 3.29). The foregoing discussion points out the complexity of the growth and motion of vortices in SOF and the reasons for the difficulty in devising a universal expression for the time-dependent force (such as Morison's equation) that would be equally applicable to SOFs at all Re and K values. This matter will be taken up later in discussing the merits and shortcomings of the Morison equation.

As noted in connection with steady flows, it is customary to express the vortex-shedding frequency in terms of the Strouhal number. In SOF there is not a unique Strouhal number. For sake of simplicity, one may define a Strouhal number in terms of the maximum velocity of the flow as $St = f_v D/U_m = f_r/K$. Sarpkaya's experiments (1976e) have shown that St so defined remains reasonably constant at 0.22 for f_r larger than about 3. Figure 3.28 shows that St depends on both Re and K. For very large values of Re (i.e., in the post-supercritical region) St rises to about 0.3. The average value of *St*, based on all vortices shed during a given cycle, was found to be between 0.14 and 0.16.

3.9.5 *Roughness effects on C_d, C_m, C_L, and St in SOF*

Of the scores of papers dealing with fluid loading on offshore structures, only a few have treated the effect of roughness on the force-transfer coefficients. Yet it is a fact that the structures in the marine environment become gradually covered with rigid as well as soft excrescences. Thus the fluid loading and the structural response, because of identical ambient flow conditions, may be significantly different from those experienced when the structure was clean, partly because of the "roughness effect" of the excrescences on the flow and partly because of the increase of the "effective diameter" and the effective mass (natural frequency and damping) of the elements of the structure.

In the absence of any data appropriate to SOFs or wavy flows, it has been assumed that the drag coefficients obtained from tests in steady flow over artificially or marine-roughened cylinders are applicable to wave-force calculations, at least when the loading is predominantly drag. *It is not generally appreciated that the consequences of all nearly steady flows are not always identical to those of steady flows.* Even for large amplitudes of oscillation, there is only a finite vortex street comprising vortices of nearly equal strength because of the nearly steady nature of the flow. As the flow reverses, the situation is not the same as that of a uniform flow (with or without free-stream turbulence) approaching a roughened cylinder but rather that of a finite vortex street approaching a rough-walled cylinder. Such a flow cannot be regarded as identical to steady flow with some turbulence of fairly uniform intensity and scale as the data presented herein show.

The salient features of the influence of roughness on the cross flow around a cylinder in steady *flow* have been discussed previously. It has been pointed out that the supercritical value of the drag coefficient depends on both the character of the flow and the surface condition of the cylinder; the drag coefficient in the post-supercritical region returns more or less to its steady subcritical value (Coceal *et al.* 2007); the larger the effective roughness, the larger the retardation of the boundary layer; and the disturbances generated by the roughness elements cause an incalculable change in the critical region of the flow. It appears from the foregoing that roughness should play an equally significant role on the characteristics of a SOF about a circular cylinder.

Sarpkaya (1976b, 1976e, 1990b) carried out a series of experiments with sand-roughened cylinders in SOF. Part of his results are shown in Figs. 3.30 and 3.31 for two values of *K* as a function of *Re*. Each curve on each plot corresponds to a particular relative roughness. Also shown on each figure are the corresponding drag and inertia coefficients for the smooth cylinder at the

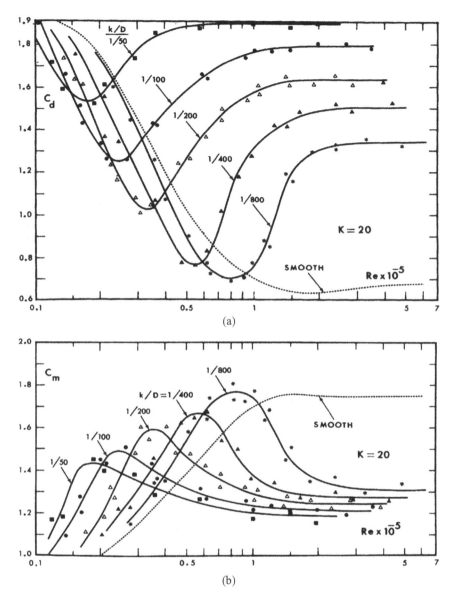

Figure 3.30. (a) C_d versus Re for rough cylinders, $K = 20$ (Sarpkaya 1976e). (b) C_m versus Re for rough cylinders, $K = 20$ (Sarpkaya 1976e).

corresponding K value. The $k/D =$ constant curves on each plot undergo changes similar to those found for steady flow about rough cylinders (see Fig. 3.9). For a given relative roughness, the drag coefficient does not significantly differ from the smooth-cylinder value at very low Reynolds numbers. As the Reynolds number increases, C_d for the rough cylinder decreases rapidly, goes through the region of drag crisis at a Reynolds number considerably lower than that for the smooth cylinder, and then rises sharply to a nearly constant post-supercritical value. The larger the relative roughness, the larger the magnitude of the minimum C_d and the smaller the Reynolds

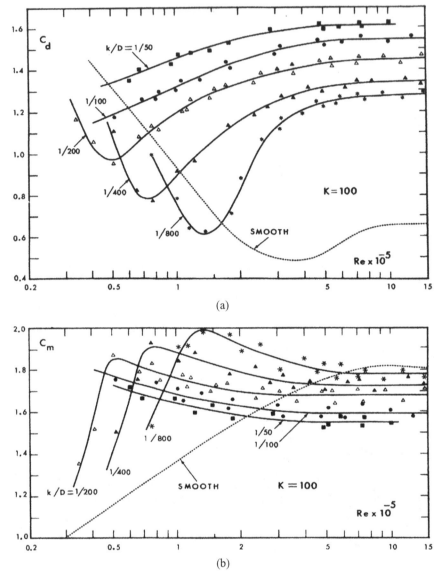

Figure 3.31. (a) C_d versus Re for rough cylinders, $K = 100$ (Sarpkaya 1976e). (b) C_m versus Re for rough cylinders, $K = 100$ (Sarpkaya 1976e).

number at which that minimum occurs. However, there appears to be a minimum Reynolds number below which the results for rough cylinders do not significantly differ from those corresponding to smooth cylinders. In other words, the *Reynolds number must be sufficiently high for the roughness to play a role on the drag and flow characteristics of the cylinder.*

The figures for the drag coefficient also exhibit a few other interesting features. First, even a relative roughness as small as 1/800 can give rise to post-supercritical drag coefficients that are considerably higher than those for the smooth cylinder. *Further increases in roughness have a smaller effect than the initial change from a smooth to the first rough cylinder.* Second, the asymptotic values of the drag coefficient (within the range of K and Re values encountered) for roughened cylinders can reach values that are considerably larger than those obtained with steady flows over cylinders of similar roughness ratio. In other words, it is not safe to assume that the post-supercritical drag coefficients in SOF flows will be identical to those found in steady flows and will not exceed a value of about unity.

In steady flow about a cylinder, roughness precipitates the occurrence of drag crisis and gives rise to a minimum drag coefficient that is larger than that obtained with a smooth cylinder. This is partly because of the transition to turbulence of the free-shear layers at relatively lower Reynolds numbers (because of disturbances brought about by the roughness elements) and partly because of the retardation of the boundary-layer flow by roughness (higher skin friction) and hence earlier separation. In SOF about a cylinder, roughness appears to play an even more complex role because of the time dependence of the boundary layer and the position of the separation points. In particular, the magnitude of the transverse force (to be discussed later) strongly suggests that the combined effect of uniformly distributed roughness and time dependence (even in the drag-dominated region of K values) is *to increase the strength of the vortices and the spanwise coherence* relative to that in steady flow at the same Reynolds number about the same cylinder.

The flow visualization studies of Grass and Kemp (1979) of the oscillatory flow past smooth and rough cylinders have shown that the angle of separation follows a well-defined time-varying path through each half cycle of oscillation as in the case of Sarpkaya's (2006b) data (see Figs. 3.3b, 3.3c, and 3.4). In the case of rough cylinders, the separation angle initially reduces much more rapidly and remains considerably smaller than that for the smooth cylinder. As noted by Grass and Kemp, this observation is consistent with the larger drag coefficient measured by Sarpkaya (1976d) under closely similar flow and relative roughness conditions.

The Reynolds number at which the drag crisis occurs gives rise to an "inertia crisis." For a given relative roughness, C_m rises rapidly to a maximum at a Reynolds number that corresponds to that at which C_d drops to a minimum. At relatively higher Reynolds numbers, C_m decreases somewhat and then attains nearly constant values, which are lower than those corresponding to the smooth cylinders. It is also apparent from the inertia coefficient curves that the smaller the relative roughnesses, the larger the maximum inertia coefficients. For a relatively smaller roughness such as $k/D = 1/800$, the terminal value of C_m is not entirely unexpected. It has been previously noted that whenever there is a rise in the drag coefficient, there also is a decrease in the inertia coefficient. It is apparent from Eqs. (3.9.11a) and (3.9.11b) that C_d increases and C_m decreases as the phase difference between the maximum force and the maximum velocity decreases. Thus the noted countervariation of C_d and C_m is a consequence of the use of Morison's equation with time-invariant coefficients and not a consequence of a fluid-mechanical phenomenon.

Sarpkaya's rough cylinder data for $K = 100$ are plotted in Fig. 3.31 as functions of the *roughness Reynolds number* ($Re_k = U_m k/v$) for representative values of k/D. Similar plots may be prepared for other values of K through the use of the data given by Sarpkaya (1976d).

It is rather remarkable that C_d and C_m become practically independent of k/D for Re_k larger than about 300. In other words, for sufficiently large values of the roughness Reynolds number, the drag and inertia coefficients for a uniformly roughened cylinder in a given SOF are determined by the height of the excrescences (above the mean diameter) rather than by the diameter of the cylinder (fully rough regime). The importance and the consequences of this result for post-supercritical Reynolds number simulation for steady flow over roughened cylinders have been discussed by Szechenyi (1975). It must be emphasized that the designer needs to know the apparent diameter of the pipe in order to predict accurately the forces acting on the roughened pipe (drag increases with D and inertia, with D^2). Furthermore, increased marine growth results in higher values of measured response at low frequencies. Consequently there is a strong need for the acquisition of marine-fouling data for various oceans and construction sites. It is only through such information that one can determine the effective diameter of a pipe as a function of water depth and time (say in years) at the construction site.

The entire data for the lift coefficients for the *roughened* cylinders are presented in Fig. 3.32 as functions of K for $k/D = 1/200$. It is not too surprising that *the smooth-cylinder data at relatively low values of β form more or less the upper limit of the rough-cylinder data.*

The Strouhal number St for rough cylinders remains essentially constant at a value of about 0.22. To be sure, there are variations from one cylinder to another and from a given combination of Re and K to another. Nevertheless, the Strouhal number is fairly constant for all roughness, relative amplitudes, and Reynolds numbers (larger than about 20000). This fact is of special importance in determining the in-line and transverse vibrational responses of the elements of a structure to wave-induced forces and *in devising approximate models for the lift force.* One must, however, bear in mind the fact that the spanwise coherence along a vertical cylinder in the ocean environment may be reduced by the variation of the velocity vector with time and depth and that the lift coefficients previously cited *represent the maximum possible values.*

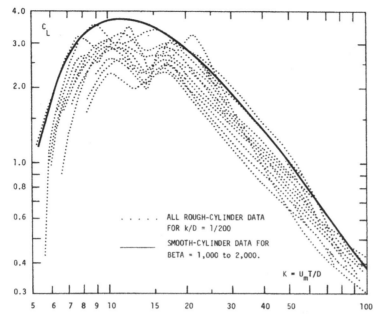

Figure 3.32. Lift coefficients for rough cylinders as functions of K for various relative roughnesses (Sarpkaya 1976e).

3.9.6 *A critical assessment of the Morison equation*

Morison's equation was proposed as an approximate solution to a complex problem. Its critical assessment means different things to different people. To an offshore engineer it usually connotes the examination of the applicability of the said equation to the prediction of the wave- and current-induced forces acting on offshore structures within acceptable and reliably definable error limits. This, in turn, means a closer look at the accuracy of the kinematic inputs, the method of evaluation of the averaged coefficients, the quality of the data used in their evaluation, sensitivity and the degree of dependence of the predictions of Morison's equation on Reynolds number, Keulegan–Carpenter number, relative roughness, body orientation, proximity effects, parameters specifying the characteristics of the motion of fluid particles, etc. Such an assessment does not allow one to draw conclusions regarding the intrinsic nature of the equation. Both the form of the equation and the uncertainties that go into the characterization of the ocean environment are jointly responsible for the differences between the measured and calculated forces. It is not a meaningful exercise to relegate the errors to a few causes. As far as the ocean environment is concerned, Morison's equation *with calibrated coefficients* is tolerated in light of all other uncertainties and hidden and intentional safety factors that go into the design. *In other words, in spite of the semiempirical nature of Morison's equation, the imperfections of the field data, and the vagaries of nature, it has been possible to create monumental engineering structures, with only occasional failures.* We must also note that the search must go on to understand the nature of a turbulent vortex, occurrence of alternating vortices, the governing laws of turbulence, and the prediction of continuous as well as discontinuous fluid motions.

To a fluid dynamicist, the critical assessment of Morison's equation means the understanding of why, how, and when a linear-quadratic sum represents the force acting on a body immersed in a given time-dependent separated flow. Clearly a wholesale assessment of Morison's equation is not very meaningful. Equally important is the distinction between the desire to improve it by the addition of one or more semi-empirical terms and, as previously noted, the desire to understand the essence of hydrodynamic resistance in terms of the formation, growth, and motion of vortices. The last objective is the most difficult one to fulfill in view of the fact that even a simple, steady, uniform, 2D flow experiences separation, turbulence, alternating vortices, 3D, and, often turbulent, separated flow in encountering a smooth circular cylinder. Existing numerical models require some fine-tuning to match the predictions to the experimental results. It is for all these reasons and more that this discussion is restricted mainly to in-line forces in sinusoidally oscillating planar flows about (mostly) circular cylinders.

Clearly, SOF is the simplest of all oscillating flows, and the determination of its kinematics does not require the use of intermediate theories whose applicability is subject to separate assessment. We do not, however, conclude that the applicability or lack of applicability of Morison's equation to one particular flow situation precludes its use in other flow situations. Morison's equation yields no information about the transverse force and seems to be adapted best to a range of Keulegan–Carpenter numbers smaller than about 8 and larger than about 25, in which the complex problems associated with the motion of a few vortices are not as much pronounced. In fact, Fig. 3.33a (one of many such examples) shows that the residue (the difference between the measured and calculated forces) is a significant fraction of the maximum force.

Numerous attempts have been made either to improve Morison's equation or to devise new equations (see, e.g., Barnouin *et al.* 1979). So far no satisfactory results have been obtained. It appears that it would be rather difficult to abandon the linear-quadratic sum because it works rather well outside a narrow range of the Keulegan–Carpenter numbers. Thus, as far as the engineering design is concerned, it is preferable to attempt to *improve the equation* rather than to devise a new one. In fact, this was the original proposal of Keulegan and Carpenter who (*assuming the diffraction effects to be negligible*) expressed the force as

$$\frac{2F}{\rho D U_m^2} = \frac{\pi^2}{K} C_m \sin\theta - C_d \cos\theta |\cos\theta| + \Delta R \qquad (3.9.16)$$

where ΔR represents the residue given by [see Eq. (3.9.9)]

$$\Delta R = 2[A_3 \sin 3\theta + A_5 \sin 5\theta| + \cdots]$$
$$+ 2[B_3' \cos 3\theta + B_5' \cos 5\theta + \cdots] \qquad (3.9.17)$$

In the range of K and Re values for which Morison's equation fails to represent the measured force with sufficient accuracy, Keulegan and Carpenter approximated AR with two terms involving only A_3 and B_3'. This procedure showed considerable improvement over the two-term Morison equation in the range of $10 < K < 25$. The obvious disadvantage of this expanded form of the equation is that it now requires the evaluation of four coefficients, namely, C_d, C_m, A_3, and B_3'. Even then the calculated and measured forces do not always correspond, partly because of the existence of other harmonics and partly because of the pronounced effects of the randomness of the vortex shedding, coherence along the cylinder, and the motion of a few vortices, vice large number of vortices. This, in turn, requires the addition of two more terms. Obviously this is a tedious exercise of questionable value: The curve-fitting exercise does not add anything new to the understanding of the physics of the force modeling. It is partly for this reason and partly because of the uncertainties of the input parameters that the two-term Morison equation has been tolerated and used over the past 30 years in spite of its known limitations. The inaccuracies resulting from the use of the said equation have been taken care of partly by nature through the mitigating effects of the ocean environment (e.g., reduced spanwise coherence, omnidirectionality of the waves and currents) and partly by the designer through the use of hidden and intentional safety factors.

Ideally, one would like to revise Morison's equation with the following constraints: (a) the revision should be fluid-mechanically satisfying; (b) the revised form of the equation should not contain more than the two coefficients already in use, namely, C_d and C_m; (c) the coefficients of the additional terms should be related to C_d and/or C_m, through a careful spectral and Fourier analysis of ΔR; and (d) the revised equation should reduce to Morison's equation in the drag and in the inertia-dominated regimes of the flow.

Lighthill (1986) rewrote the Morison equation as

$$F = C_m^* \rho \frac{dU}{dt} V_{\text{vol}} + \frac{1}{2} \rho A U^2 C_d \qquad (3.9.18)$$

where C_m^* now is *the ideal value of the inertia coefficient* or better, *added inertia*, for any K and Re and A and V_{vol} are the projected area and the volume of the body, respectively. We have noted previously that the ideal value of *the inertia coefficient is valid only at the start of the motion, prior to the* *onset of separation and/or the viscous effects.* Lighthill's version of the MOJS equation requires only **one experimentally determined coefficient**: C_d, presumably dependent on such parameters as the Reynolds number, Keulegan–Carpenter number, relative roughness, and the direction of the body motion. Sarpkaya (2001b) has shown that it is **impossible** to find a suitable C_d value that enables Eq. (3.9.18) to represent the measured force even for a circular cylinder. The reason for this is that the inertia coefficient in Lighthill's inertial force ($C_m^* \rho \frac{dU}{dt} V_{\text{vol}}$) is not valid and does not remain constant beyond the moment of the impulsive start.

Sarpkaya (2001b) introduced a new parameter Λ, *without introducing another coefficient,* as

$$\Lambda = \frac{\pi^2}{K}(C_m^* - C_m) \tag{3.9.19}$$

based on the reasons described in the paper previously cited and rewrote the Morison equation as

$$F = -C_d \left|\cos \omega t\right| \cos \omega t + \frac{\pi^2}{K} C_m \sin \omega t - \frac{\Delta |\Delta|}{C_d} \sin 3\omega t \tag{3.9.20}$$

Three sample plots shown in Figs. 3.33a, b, c attest to the fact that the Eq. (3.9.20) represents the measured force with very small residue *even in the range of K values where the residue is normally largest.* Furthermore, Eq. (3.9.20) is physics based and no more complicated than the original Morison equation.

In summary, it is suggested that Morison's equation is quite satisfactory in the drag and in the inertia-dominated regimes; it is unlikely that an entirely new equation will ever replace it, and that the addition of one more term (*with no new coefficients*), based on the analysis of the residues, may result in a practical, and more suitable expression, as previously shown, if not fluid-mechanically more satisfactory explanation of the resistance in complex time-dependent flows. Clearly, Lighthill's suggestion is without any merit.

3.9.7 *Oscillatory plus mean flow or the in-line oscillations of a cylinder in steady flow*

The in-line oscillations of a cylinder in uniform flow of velocity V have not been studied extensively. Chen and Ballengee (1971) examined the vortex shedding from circular cylinders in an oscillating free stream of 3 Hz with A/D from 15 to 1000, $D/VT = 0.003$, and the Reynolds numbers up to 40000. They found that the vortex shedding from a circular cylinder

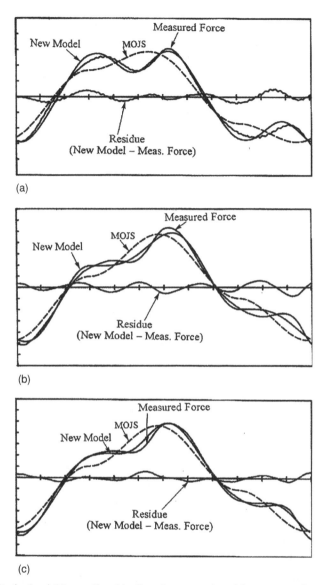

Figure 3.33. (a, b, c) Normalized in-line forces and residue versus time for various values of K, Re, and k/D.

responds instantaneously to the free-stream oscillations and that "the instantaneous Strouhal number stays sensibly constant at 0.2 ± 0.01." This is rather expected because the amplitude of oscillations is many times that of the cylinder diameter and the flow in the absence of significant accelerations exhibits a quasi-steady behavior.

Tanida *et al.* (1973) found that, for a circular cylinder oscillating parallel to the flow ($A/D = 0.14$, D/VT from 0 to 0.5, and $Re = 80$ and 4000), the vortex synchronization can be observed in a range around double the Strouhal frequency, where vortices are shed with a frequency half the imposed one. The results of Tanida and his associates showed that the mean drag reaches its maximum in the middle of the synchronization range, i.e., $D/\overline{V}T$ between 0.2 and 0.4, and that the sign of C_d is such that no energy can be extracted from the fluid to render the oscillations unstable. In other words, in-line oscillations are stable for the two Reynolds numbers at $A/D = 0.14$. Tanida *et al.* (1973) conjectured that instability is likely to occur at much higher Reynolds numbers. Sarpkaya (1993a, 2002a) has shown that the occurrence of the instability depends on the wavelength and the frequency parameter β.

Goddard (1972) carried out a numerical analysis of the drag response of a cylinder to streamwise fluctuations through the use of the Navier–Stokes equations for $Re = 40$ and $D/\overline{V}T = 0.019, 0.12,$ and 3.18, and for $Re = 200$ and $D/\overline{V}T = 0.149$. He found that for very low frequencies the instantaneous values of the drag correspond very nearly to the quasi-steady solution and that for higher frequencies the drag anticipates the free-stream velocity maximum. This work cannot be generalized to higher Reynolds numbers because it is based on the laminar flow assumptions and because the diffusion of vorticity in the concentrated vortices for Re larger than about 200 is primarily turbulent.

From a more practical point of view, in-line oscillations in uniform flow attracted attention partly because of the superposition of waves with currents and partly because of the damaging vortex-induced oscillations encountered in tidal waters. Such oscillations were observed at an oil terminal on the Humber estuary in England during the late 1960s. A brief description of the problems encountered has been given by Sainsbury and King (1971). Subsequently, Wootton *et al.* (1974) conducted full-scale experiments to ascertain the causes of in-line vibrations of the pilings. They found that the Strouhal number of the far wake remains constant at about 0.23. They further found that two distinct flow patterns exist in the immediate wake of the cylinder, depending on the particular value of $f_n D/V$, where f_n is the natural frequency of the piling. The inverse of this frequency parameter is known as the reduced velocity. One would ordinarily expect that the in-line oscillations will occur when the reduced velocity $U_r = V/f_n D$ is equal to about 2.5 because the frequency of the drag oscillations is twice the Strouhal frequency. Wootton *et al.* observed that the in-line oscillations occur over a range of $U_r = 1$ to 3 and in two distinct resonant response regimes separated by $U_r = 2$. At $U_r = 1.7$ a symmetric vortex shedding was observed. At $U_r = 2$, the vortex shedding changed to the commonly observed form of alternate shedding, indicating a radical change in the cause of excitation and dynamic response.

King (1974) and King *et al.* (1973) observed that the in-line oscillations occur for a range of reduced velocity from 1.5 to 4. They too have noted two distinct regimes of vortex shedding separated by $U_r = 2.5$.

Crandall *et al.* (1975) investigated the destructive vibration of trash racks that is due to fluid–structure interaction and found that the excitation mechanism involved synchronization between the fluctuating drag and the in-line motion of the cylinders. The vibrations occurred at the fourth and fifth modes. The reduced velocity ranged from 1.12 to 1.81. They noted a drastic change at $U_r = 1.47$ that corresponded to a change in the mode of vibration from the fourth to the fifth mode. Crandall *et al.* (1975) did not measure the vortex-shedding frequencies but advanced very interesting concepts regarding the cause of synchronization (see Leehey and Hanson 1971).

Griffin and Ramberg (1976) conducted similar experiments in air at a Reynolds number of 190. Their results have essentially substantiated the previous observations of Wootton and King. No information was obtained regarding the forces acting on the cylinder.

Evidently the determination of the forces acting on a cylinder undergoing sinusoidal in-line oscillations is just as important as the understanding of its kinematics. Because the phenomenon is identical to that where the cylinder is subjected to a time-dependent flow characterized by $U = \overline{V} - U_m \cos\theta$, the evaluation of the force might shed some light on the combined effect of waves and currents on the members of offshore structures (Sarpkaya *et al.* 1997). The vortex-induced oscillations are discussed in greater detail in Chapter 6.

It is ordinarily assumed (as recommended by the American Petroleum Institute, API, 1977) that Morison's equation applies equally well to periodic flow with a mean velocity and that C_d and C_m have current-invariant, Fourier, or least-squared averages equal to those applicable to rigid, stationary cylinders in wavy flow. This, in turn, implies that C_d and C_m are independent of the *biased* convection of vortices and its attendant consequences. The fact that this is *not necessarily so* is clearly evidenced by the measurements of Sarpkaya (1977b) and Verley and Moe (1979). Thus the effect of the current–SOF combination on the motion of vortices and on the force-transfer coefficients must be carefully examined in light of available data, and the limits of application of Morison's equation to such flows must be assessed anew. The latter is particularly important in view of the fact that the drag and inertia coefficients in ocean tests (where there are always some currents and body motion) are evaluated through the use of Morison's equation (note that the values of C_d and C_m may vary considerably from one half-wave cycle to another because of the current-induced biasing of the wake and vortex formation and that neither set of coefficients may be identical to those obtained without current). The particular flow previously discussed is one of many realistic and yet insoluble flow problems encountered in the ocean environment.

Obviously there has been an overemphasis between the predictions of the MOJS equation (and its various generalizations to special cases) and the measured values. One needs to be reminded that the MOJS equation is not a solution of the Navier–Stokes equations. It represents the simplest possible linear sum of the drag and inertial forces. In reality, both the laboratory and field experiments represent an enormously complex texture of conditions, and the results are often difficult to interpret. Thus the successes of the offshore engineering world in placing relatively small structures in the vast spaces of ocean and the vagaries of nature might be considered both a source of puzzlement and gratification.

It has been customary to express the in-line force either as

$$F = \frac{1}{2}\rho C_{dc} D(\overline{V} - U_m \cos\theta)|\overline{V} - U_m \cos\theta| + \left(\frac{\pi}{4}\rho C_{mc} D^2\right)\frac{dU}{dt} \quad (3.9.21)$$

or as (Sarpkaya 1977b; Verley and Moe 1979; Sarpkaya and Storm 1985)

$$\frac{2F}{\rho D\overline{V}^2} = \overline{C}_d - C_{dh}\left(\frac{U_m T}{D}\right)^2\left(\frac{D}{\overline{V}T}\right)^2 |\cos\theta|\cos\theta$$

$$+ C_{mh}\pi^2\left(\frac{U_m T}{D}\right)\left(\frac{D}{\overline{V}T}\right)^2 \sin\theta \quad (3.9.22)$$

The coefficients C_{dc}, C_{mc}, C_{dh}, and C_{mh} are given by their Fourier averages and, in general, $C_{dc} \neq C_{dh} \neq C_d$ and $C_{mc} \neq C_{mh} \neq C_m$, where C_d and C_m are the in-line force coefficients for the SOF alone. All of the coefficients appearing in (3.9.21) and (3.9.22), including $\overline{C}_d$, are functions of VT/D, $U_m T/D$ (or A/D), Reynolds number, and the relative roughness (to the best of our knowledge, there are conjectures but no data for the discussion of the cylinder yaw, body proximity, etc., in wave–current flow).

The said coefficients may be determined only experimentally either in a wave channel with a mean current, in which case the additional complexities brought about by the three-dimensionality of the flow tend to obscure the systematic evaluation of the said parameters, or in a uniform flow channel by oscillating the body in the in-line direction, or by moving the body at a constant speed in a SOF. The results obtained by Sarpkaya (1957, 1977b) and Verley and Moe (1979), through the use of one or the other of the preceding methods, have shown that the force coefficients undergo dramatic changes with increasing A/D in the range of D/VT values from 0.1 to 0.5. The question of whether a single drag coefficient C_{dc}, as in Morison's equation for strictly SOF, could suffice to calculate the drag force accurately, for a cylinder oscillating in uniform flow requires additional data at much higher β.

Figures 3.34 and 3.35 show C_{dc} and C_{mc} as functions of $K = U$, T/D for various values of VT/D. The data have been obtained by Verley and

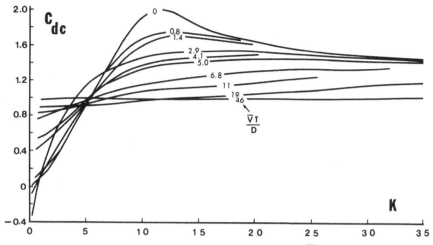

Figure 3.34. Variation of C_{dc} with K for various values of $\overline{V}T/D$, $\beta \cong 300$ (Verley and Moe 1979).

Moe (1979) for $\beta \cong 300$, primarily for the purpose of determining the so-called "fluid damping" for cylinders oscillating in a current. It must be emphasized that the Reynolds number $Re = UD/v$ increases with K along each curve, reaching a maximum value of about 12000 at $K = 40$. Thus the use of these data at large Reynolds numbers is not warranted. Nevertheless, Figs. 3.34 and 3.35 show that the current causes profound changes not only in C_{dc} but also in C_{mc} relative to the no-current case, *at the corresponding K and Re values.* For example, at $K = 25(Re = 7500)$, $C_m = 1.1$ (no current) and $C_{mc} = 1.55$ for $1.4 < VT/D < 30$. A value of $VT/D = 2.5$ at $K = 25$ corresponds to $V/U = 0.1$, a situation that is not unusual for

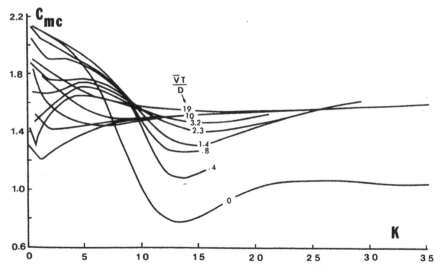

Figure 3.35. Variation of C_{mc} with K for various values of $\overline{V}T/D$, $\beta \cong 300$ (Verley and Moe 1979).

the ocean environment except that the Reynolds numbers are considerably larger. It is known that C_m for a smooth cylinder in SOF increases to about 1.8 at large K and Re. Thus it is possible that the effect of current on C_{mc} at high K and Re numbers may be negligible and C_{mc} may differ little from C_m. Equally important is the fact that as K increases (say K larger than about 25) the inertial component of the in-line force becomes negligible. In the intermediate region of K values, the effect of current is to increase C_{mc} and reduce C_{dc} relative to their no-current values. This may in part account for both the scatter and the relatively smaller value of the drag coefficient obtained from field data through the use of the Morison equation. Clearly the use of one half-wave cycle to calculate C_d and C_m in wave–current flow is not warranted and does not account for the actual behavior of the biased wake (Zdravkovich 1996b).

A simple calculation shows that the semi-peak-to-peak value of the maximum in-line force for $V/U < 0.1$ is about 10% overestimated by using the no-current values of C_d and C_m relative to that in which the actual values of C_{mc} and C_{dc} are used. For only practical purposes, it is tentatively concluded that the use of modified Morison equation (3.9.10) for $VT/D < K/10$, i.e., $V/U < 0.1$, with C_d and C_m values obtained with pure wave or SOF may lead to reasonably conservative in-line forces.

Evidently, profound understanding of the subject is needed to delineate the effect of current on waves (wave–current interaction) and on wave forces (both in-line and transverse) as a function of K, Re, D/VT, relative roughness, yaw and body proximity, particularly at the intermediate and high K and Re values. Additional research is needed on the effect of time-and-depth-dependent current on all aspects of the phenomenon. It appears that much of this information will have to be obtained experimentally, as needed for special cases. In doing so, however, one has to make sure that the test system is indeed fluid-mechanically sound and physically reliable. Otherwise one can arrive at erroneous conclusions and misleading interpretations of the meaning of the data.

3.9.8 *Forced oscillations of a cylinder in a trough*

Garrison *et al.* (1977) conducted experiments in a wooden trough. A photograph of their carriage assembly (Fig. 3.36) and typical force traces (Figs. 3.37a, b) are reproduced herein with the courtesy of John B. Field (1975). Figure 3.38 is a typical *unfiltered* in-line-force trace used by Sarpkaya (1976, 1977, 1986, 1987, 1990).

Figures 3.39 and 3.40 show the large scatter in Garrison's representative data. The waves generated by the shaker and reflected by the end walls gave Garrison and his colleague the *wrong impression that the drag coefficient*

Figure 3.36. Garrison *et al.* (1977) and Garrison (1990) used their carriage to oscillate smooth and rough cylinders. Clearly, the position of their struts and the size of the end plates are far from adequate for the acquisition of reliable data.

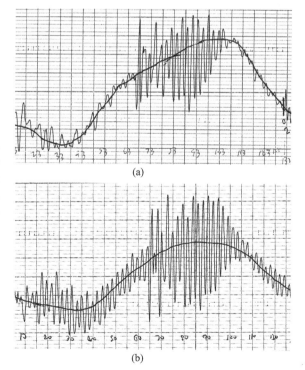

Figure 3.37. Typical in-line force traces used by Garrison *et al.* (1977): (a) $A/D = 5$; (b) $A/D = 4$. (Courtesy of John B. Field).

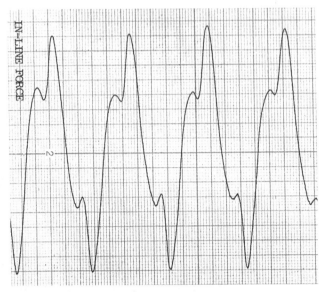

Figure 3.38. Typical unfiltered in-line force traces used by Sarpkaya (1976b, 1977a, 1987, 1990b), $K = 11.5$, $k/D = 0.02$, $\beta = 6836$.

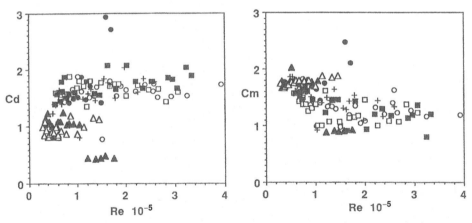

Figure 3.39. (On the left) Garrison *et al.*'s 1977 drag coefficient versus the Reynolds number for a 3-inch rough cylinder, as originally reported by May (1976): $\triangle$, $K = 5.24$; $\blacktriangle$, $K = 6.28$; $+$, $K = 9.42$; $\blacksquare$, $K = 12.57$; $\square$, $K = 15.71$; $\circ$, $K = 18.85$; $\bullet$, $K = 22$.

Figure 3.40. (On the right) Garrison *et al.*'s (1977) inertia coefficient ($C_m = 1 + C_a$) versus the Reynolds number for a 3-inch rough cylinder, as originally reported by May (1976); symbols are the same as in Fig. 3.39.

is decreasing with increasing wave amplitude. The real reason is, of course, the fact that the largest scatter occurred at larger K as seen in Fig. 3.37. As a whole, the large scatter of the data attests to the unsuitability of their carriage assembly (undesirable end conditions of the test cylinder, flexible plates holding the cylinder), *asymmetry* of the oscillations, and the severe vibrations (particularly at higher amplitudes).

3.9.9 *Oscillatory flow in a smaller U-shaped water tunnel*

We will now describe briefly the experiments carried by Skomedal, Vada, and Sortland (1989) in Norway through the use of a smaller U-shaped water tunnel. Figure 3.41 shows a drawing of the subject tunnel and the Fig. 3.42 shows the drag coefficients obtained by Sarpkaya (1976), Bearman *et al.* (1985), and the experimental and numerical results of Skomedal *et al.* for two Beta values (534, and 140). Sarpkaya's measurements (1976) are in excellent agreement with those obtained by Skomedal, Vada, and Sortland (1989) in a smaller U-shaped water tunnel.

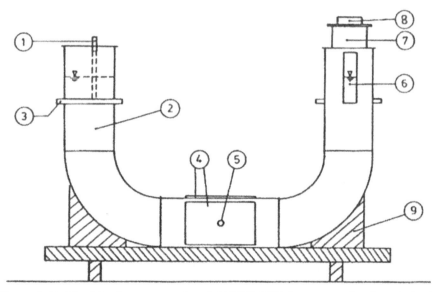

Figure 3.41. The U-tube water tank at MARINTEK. 1, wave probe; 2, main body in aluminum; 3, wood stiffener; 4, Plexiglas window; 5, test cylinder; 6, Plexiglas window; 7, honeycomb section; 8, driving fan; 9, supporting bed.

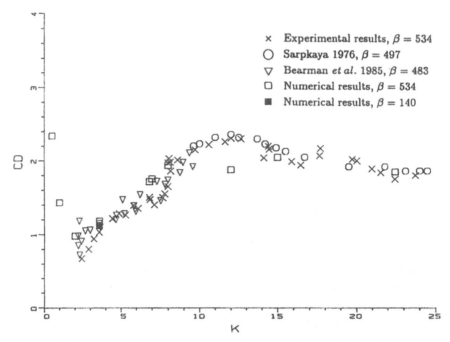

Figure 3.42. Force coefficients from experiments and numerical calculation for one circular cylinder.

4

Waves and Wave–Structure Interactions

The interaction of waves (generated by numerous causes, e.g., wind and earthquakes) with structures makes marine hydrodynamics. Thus, the integrity of the structures depends on the understanding of the interfacial mechanics, which is far more complex than either the waves or flow about bluff bodies for a large number of reasons: turbulence, separation, in-line and transverse forces, empirical equations (e.g., Morison's equation), empirical nature of the lift, drag, and inertia coefficients based on relatively low-Reynolds number experiments (carried out with purely sinusoidal oscillations in a large U-shaped water tunnel or on offshore platforms with large scatter), three-dimensionality and the omnidirectionality of the interaction, flow-induced vibrations, proximity of the bodies, breaking of the waves, refraction, diffraction, reflection, the ever present and ever elusive damping, and myriads of other occurrences limit our ability to design structures with reasonable confidence, as evidenced by occasional failures. With the realization of the fact that the subject phenomena are complex beyond our capacity to fully observe, measure, or to compute, even in the era of nanotechnology and quantum computing, we discuss waves and wave–structure interactions separately but with a keen interest in each simultaneously. This chapter discusses only the most important characteristics of the ocean waves and serves as an integral part of the rest of the topics covered.

4.1 Surface gravity waves

As noted by Billingham and King (2000), "Whenever we see or hear anything, we do so because of the existence of waves." In this chapter our emphasis is on the interaction of *currents* and *design waves* with *offshore structures*. It is rather remarkable that *progressive linear gravity waves* represent the first approximation of more fundamental and advanced versions of waves, aptly described by Wehausen and Laitone (1960); Phillips (1977); LeBlond and Mysak (1978); Geernaert and Plant (1990); Sarpkaya (1992);

Komen *et al.* (1994); Malenica *et al.* (1995); Sawaragi (1995); Young (1999); Goda (2000); Lavrenov (2003); Janssen (2004); Sarpkaya (2006a); Svendsen (2006); Holthuijsen (2007) and many others.

From a practical point of view, one needs to be reminded that the waves in the ocean environment are indeed very complex and are normally described by their *variance density spectrum*. Our objective is not to further the state of the idealized wave theories, no matter how noble the effort may be, but rather to combine the simplified or imperfect wave theories with the elements of equally imperfect force predictions: *drag, lift, and inertia forces, flow-induced vibrations due to currents, waves, wind, gust, and the damping of structures* (Basu 1986; Sarpkaya 2006a).

From a scientific as well as engineering point of view, the understanding of the Morison equation and its limitations are of particular importance toward the understanding of unsteady flow about bluff bodies. For example, it does not account for the fluctuating lift force and fails to represent the in-line force with sufficient accuracy. Clearly, fluid mechanics cannot be tied up to something as primitive and simple as the linear sum of the drag and inertial forces. Preoccupation with the two-term Morison equation and the wave forces on cylinders has deterred the researchers of the past century from making insightful experiments and theoretical investigations to go beyond the Morison formula. An enlightening discussion of this is given by BSRA (1976); Moe and Gudmestad (1998).

4.1.1 *Linear wave theory*

Waves as well as the offshore structures immersed in them differ significantly, and thus the character of the interaction: small structures requiring only some understanding of the separated flows and the so-called Morison's equation, and large and medium structures in need of diffraction analysis.

The waves are the eternal problem. In general, they may be regular or irregular, unidirectional or omnidirectional, linear or nonlinear, and propagate over relatively shallow waters, intermediate depths, and deep waters under the influence of wind, gravity, surface tension, stratification, bottom friction, temperature and density gradients, bottom topography, etc.

We begin with the simplest possible case and assume that the waves are two-dimensional, progressive, permanent gravity waves of period T, amplitude H, and length L (between two successive downward or upward crossings) over a smooth horizontal rigid bed, as shown in Fig. 4.1. They are composed of a continuous, inviscid, incompressible fluid, with no temperature or salinity gradients, no vertical density gradient (even though gravitational force apply), no wind or wind shear, no surface tension, no bottom friction, no growth, and no attenuation. Therefore, the creation of such a wave

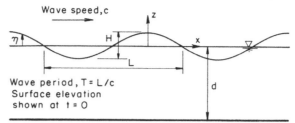

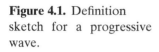

Figure 4.1. Definition sketch for a progressive wave.

requires an inspiration or externally applied impulse in addition to gravitational force. Once created, it sustains itself even if it experiences inviscid separation. Surprisingly enough, *such a wave* serves as *the foundation of all other waves*, from a simple sine wave to Stokes' fifth-order theory.

The wave speed $c(= L/T)$ is usually expressed in terms of $k(= 2\pi/L)$ and $\omega(= 2\pi/T)$ as $c = \omega/k$. The determination of the wave speed c and the particle motion throughout the flow for an irrotational incompressible flow requires a velocity potential that satisfies the Laplace equation

$$\frac{\partial^2 \phi}{\partial x^2} + \frac{\partial^2 \phi}{\partial z^2} = 0 \qquad (4.1.1)$$

subject to the boundary conditions at the seabed:

$$\frac{\partial \phi}{\partial z} = 0 \quad \text{at} \quad z = -d \qquad (4.1.2)$$

The kinematic and dynamic boundary conditions at the free surface are given, respectively, by

$$\frac{\partial \eta}{\partial t} + \frac{\partial \phi}{\partial x}\frac{\partial \eta}{\partial x} - \frac{\partial \phi}{\partial z} = 0 \quad \text{at} \quad z = \eta \qquad (4.1.3)$$

and by

$$\frac{\partial \phi}{\partial t} + \frac{1}{2}\left[\left(\frac{\partial \phi}{\partial x}\right)^2 + \left(\frac{\partial \phi}{\partial z}\right)^2\right] + g\eta = f(t) \quad \text{at} \quad z = \eta \qquad (4.1.4)$$

The exact solution of the above equations is indeed very complex: there are a number of nonlinear terms in (4.1.3) and (4.1.4) and the free-surface conditions have to be applied at the initially unknown free surface. Under these circumstances one's ingenuity turns to the *linearization* of the governing equations. Assuming H is much smaller than L and d and neglecting the nonlinear terms (i.e., adopting a small amplitude wave theory), the boundary conditions are satisfied at z = 0. This simplification represents a relatively sound first approximation and helps to reduce (4.1.3) and (4.1.4) to

$$\frac{\partial \phi}{\partial z} - \frac{\partial \eta}{\partial t} = 0 \quad \text{at} \quad z = 0 \tag{4.1.5}$$

and

$$\frac{\partial \phi}{\partial t} + g\eta = 0 \quad \text{at} \quad z = 0 \tag{4.1.6}$$

These, in turn, lead to

$$\frac{\partial^2 \phi}{\partial t^2} + g\frac{\partial \phi}{\partial z} = 0 \tag{4.1.7}$$

and

$$\eta = -\frac{1}{g}\left(\frac{\partial \phi}{\partial t}\right)_{z=0} \tag{4.1.8}$$

The solution of the above equations becomes relatively trivial through the use of the separation of variables. Then the velocity potential reduces to

$$\phi = \frac{gH}{2kc}\frac{\cosh[k(z+d)]}{\cosh(kd)}\sin[k(x-ct)] \tag{4.1.9}$$

and the *linear dispersion relation* to

$$\omega^2 = gk\tanh(kd) \tag{4.1.10a}$$

or

$$c^2 = \frac{g}{k}\tanh(kd) \tag{4.1.10b}$$

The foregoing is sufficient to evaluate a large number of physical quantities such as horizontal and vertical displacements, horizontal and vertical particle velocities, accelerations, pressure, group velocity, average energy density, energy flux, and radiation stress. These are tabulated in Table 4.1.

The fluid particles move clockwise around small ellipses under the influence of progressive gravity waves. The wave height decreases with depth and the ellipses become progressively thinner and shorter. Normally, the waves are classified approximately as *shallow-water waves* where $d/L < 1/20$; as intermediate-depth waves where $1/20 < d/L < 1/2$; and as deep-water

Table 4.1. *Characteristics of the idealized linear waves*

Velocity potential	$\phi = \dfrac{\pi H}{kT} \dfrac{\cosh(ks)}{\sinh(kd)} \sin\theta$
	$\quad = \dfrac{gH}{2\omega} \dfrac{\cosh(ks)}{\cosh(kd)} \sin\theta$
Dispersion relation	$c^2 = \dfrac{\omega^2}{k^2} = \dfrac{g}{k}\tanh(kd)$
Surface elevation	$\eta = \dfrac{H}{2}\cos\theta$
Horizontal particle displacement	$\xi = -\dfrac{H}{2}\dfrac{\cosh(ks)}{\sinh(kd)}\sin\theta$
Vertical particle displacement	$\zeta = \dfrac{H}{2}\dfrac{\sinh(ks)}{\sinh(kd)}\cos\theta$
Horizontal particle velocity	$u = \dfrac{\pi H}{T}\dfrac{\cosh(ks)}{\sinh(kd)}\cos\theta$
Vertical particle velocity	$w = \dfrac{\pi H}{T}\dfrac{\sinh(ks)}{\sinh(kd)}\sin\theta$
Horizontal particle acceleration	$\dfrac{\partial u}{\partial t} = \dfrac{2\pi^2 H}{T^2}\dfrac{\cosh(ks)}{\sinh(kd)}\sin\theta$
Vertical particle acceleration	$\dfrac{\partial w}{\partial t} = -\dfrac{2\pi^2 H}{T^2}\dfrac{\sinh(ks)}{\sinh(kd)}\cos\theta$
Pressure	$p = -\rho g z + \dfrac{1}{2}\rho g H\dfrac{\cosh(ks)}{\cosh(kd)}\cos\theta$
Group velocity	$c_G = \dfrac{1}{2}\left[1 + \dfrac{2kd}{\sinh(2kd)}\right]c$
Average energy density	$E = \dfrac{1}{8}\rho g H^2$
Energy flux	$P = E c_G$
Radiation stress	$S_{xx} = \left[\dfrac{1}{2} + \dfrac{2kd}{\sinh(2kd)}\right]E$
	$S_{xy} = S_{yx} = 0$
	$S_{yy} = \left[\dfrac{kd}{\sinh(2kd)}\right]E$

waves where $d/L > 1/2$. The wave periods at the corresponding depths are shallow-water waves ($0.0025 > d/gT^2$), intermediate-depth waves ($0.0025 > d/gT^2 < 0.08$) and the deep-water waves ($d/gT^2 > 0.08$). In other words, the amplitude of horizontal velocity (and displacement) decreases with depth according to $\cosh[k(z+d)]$, while the amplitude of

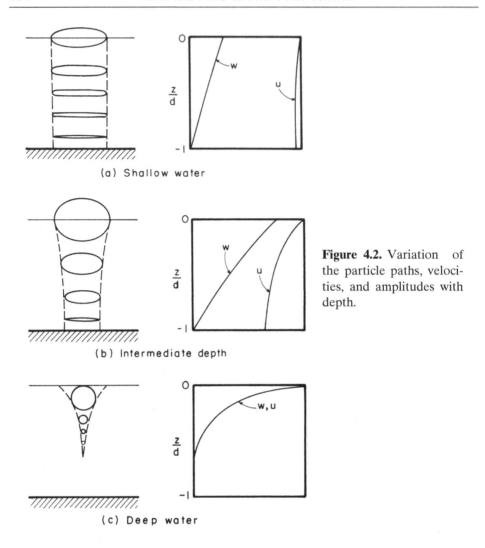

(a) Shallow water

(b) Intermediate depth

Figure 4.2. Variation of the particle paths, velocities, and amplitudes with depth.

(c) Deep water

vertical velocity (and displacement) decreases according to $\sinh\,[k(z+d)]$. Obviously, a submerged body will be most detectable in shallow and intermediate depths. In deep waters, the orbits diminish quickly and the detection of objects (e.g., a submarine) resting on the ocean floor is possible either acoustically (propeller noise) or by electromagnetic means (Higuchi *et al.* 2008).

Typical profiles relating to the shallow, intermediate, and deep-water ranges are sketched in Fig. 4.2. Note that at intermediate depths, the orbits diminish in amplitude with depth and also become flatter until the vertical component vanishes at the seabed in accordance with the seabed boundary condition. In practice, a design wave is usually specified in terms of H, T, and d because they are usually the easiest to measure or estimate from observations. The wave number k may be calculated from (4.1.10a) or (4.1.10b) by a simple iteration.

The shape and the spectrum of the waves cannot be represented by the linear wave theory *when the waves are too steep or the water depths are too shallow.* Obviously, it is neither practicable nor meaningful to deal individually with each and every wave even though one may attempt to extract statistical averages from them for design purposes. Under these circumstances, the intelligent choice is to resort to higher-order approximations.

There are a large number of nonlinear wave theories, each with its own power and limitations. We will mention here only three of them: Stokes (1847) second-, third-, and fifth-order theories, Dean's (1965) streamfunction theory for steep waves, and the cnoidal wave theory of Korteweg and de Vries (1895). However, the important fact is that these waves are *periodic*, not *harmonic*, i.e., unlike the real waves, *they do not naturally evolve as they propagate.* Studies on waves over the years have shown that *no wave theory performs sufficiently well.* In other words, there is neither a 'perfect' wave theory nor a perfect sea. Even the models of Stokes and Dean do not perform well in water depths in the order of wave heights or less. The cnoidal wave model has been devised to fulfill or approximate such a need. Detailed discussions of the wave theories are given by Cokelet (1977), Dean (1974), Fenton (1985), Le Méhauté (1976), Sakai and Battjes (1980), Skjelbreia and Hendrickson (1960), and Holthuijsen (2007), among many others.

4.1.2 *Higher-order wave theories*

As noted above, there are a large number of wave theories and the problem of selecting the most suitable one for a particular application (i.e., for given H, T, and d) is rather difficult. Any comparison must be considered only in relation to the prevailing environmental characteristics and the particular location. Obviously, the degree of sophistication of the theory does not necessarily ensure sounder and more economical engineering.

The results of the Stokes second-order theory show that most time-varying quantities contain second-order components at *twice the wave frequency* superposed on the fundamental components predicted by the linear wave theory. This, as expected, gives rise to wave crests, which are steeper, and troughs, which are flatter than those of a sinusoidal profile as shown in Fig. 4.3. Furthermore, *the particle paths are no longer closed orbits* and there is, in consequence, a gradual *drift* or mass transport, which should not be confused with the so-called *added mass (added inertia)*. If a body were to be accelerated in a higher-order wave it will experience both *drift* and *added inertia* (positive or negative) in addition to all the incalculable consequences of flow separation (McIver 1992).

Stokes (1847 and 1880), Munk (1949), Reid and Bretschneider (1953), De (1955), Skjelbreia and Hendrickson (1960), Dean (1965), Tsuchiya and

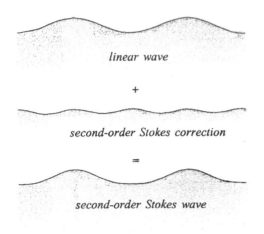

linear wave

+

second-order Stokes correction

=

second-order Stokes wave

Figure 4.3. The surface profile of a second-order Stokes wave (after L. H. Holthuijsen 2007).

time or space ⟶

Yamaguchi (1972), Lambrakos and Brannon (1974), Mei (1989), Massel (1996), Dean and Dalrymple (1998), Mei *et al.* (2005), and others have developed successive approximations for the prediction of the properties of higher-order *waves*. Skjelbreia and Hendrickson (1960) have extended Stokes wave theory to the fifth order and their approach has found widespread usage in engineering practice.

The velocity potential is expressed as the sum of five terms:

$$\frac{k\phi}{c} = \sum_{n=1}^{5} \phi'_n \cosh(nks)\sin(n\theta) \qquad (4.1.11)$$

$$k\eta = \sum_{n=1}^{5} \eta'_n \cos(n\theta) \qquad (4.1.12)$$

The coefficients appearing in (4.1.11) and (4.1.12) are tabulated by Skjelbreia and Hendrickson (1960) and are not shown here for sake of brevity.

Comparisons between various theories may be made on both theoretical as well as experimental grounds and these are now reviewed in turn: Dean (1970) has compared several wave theories on a theoretical basis. The criterion he used was the "closeness of fit" of the predicted motion to the complete problem formulation. Since the Laplace equation and bottom boundary condition are exactly satisfied in all the theories considered, the error of fit to the two nonlinear free-surface boundary conditions was used as the criterion of validity. The wave theories examined included linear wave theory, Stokes third- and fifth-order theories, cnoidal developed by Korteweg and de Vries (1895) where the wave characteristics are expressed in terms of the Jacobian elliptic function **cn** and hence the terminology "cnoidal wave

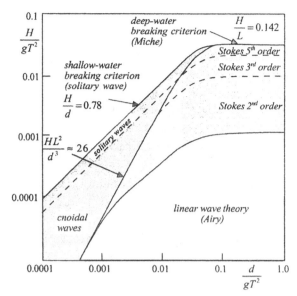

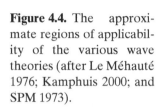

Figure 4.4. The approximate regions of applicability of the various wave theories (after Le Méhauté 1976; Kamphuis 2000; and SPM 1973).

theory," solitary waves (first and second approximations) and the stream-function theories. Dean found that the first-order cnoidal, the linear, the Stokes fifth-order, and the stream-function theories were generally the most suitable ones. He emphasized that the method used to assess the theories does not necessarily imply the selection of the best overall theory. Le Méhauté (1976) has presented a convenient plot showing the approximate limits of validity of the various wave theories. There are strong similarities between Le Méhauté's (Fig. 4.4) and Dean's plots (not shown here).

The emerging fact is that "approximation" is the way of life in marine engineering primarily due to the vagaries of nature; i.e., plots, computers, recommendations, and the predictions are bound to be approximate. Acquisition of inspiration through cumulative experience, development of sound judgment, and, more importantly, passing this wisdom to the next generation are indeed the most important elements of life in marine engineering. It appears that the interaction and interference of human needs with the evolution of our planet will have far-reaching consequences beyond our wisest predictions and wildest expectations.

4.1.3 *Character of the forces predicted*

It has already been intimated that the inherent instability of the oceans gives rise to complex waves and wave, wind, gust, and hurricane forces, which may be substantially different from those assumed to exist for design purposes. In fact, limited data suggest that real waves exert forces, which do not act in uniform directions on members of a structure in space. *It appears that the forces predicted through the use of a design wave will be larger than those produced on the structure by a real wave field of comparable characteristics,* i.e., the real ocean environment has mitigating effects, which can be quantified only approximately (in general, disorder or chaos tends to increase energy loss and minimize the total impact). However, it is very important for a designer to realize that the interaction of ever-changing omnidirectional waves and currents with structures may occasionally give rise to exceptions: *freak waves and larger than expected forces and moments.* Thus, one needs to know the vagaries of the waves and currents and the uncertainties of the fluid–structure interaction (Vengatesan *et al.* 2000). In short, one should not be mesmerized by the sheer beauty of idealized waves or by the smooth plots of the drag and inertia coefficients. The unexpected will happen. The question is, Is there an upper limit? This depends not only on fluid forces but also on the current state (age, roughness, orientation, proximity, etc.) of the structural elements. Where is the most likely region of failure-inception, the possibility of a chain reaction versus dealing with a reparable element, of no catastrophic consequences, and so on? Perhaps no phenomenon requires more fundamental knowledge, practical experience, and the wisdom of ages than marine hydrodynamics.

An approximate quantitative mapping of the various loading regimes, expressed in terms of wave height, structure diameter, and depth below the water surface, may be obtained by assuming some appropriate values for the drag and inertia coefficients, a suitable wave theory, and a wavelength-to-wave-height ratio. Such an exercise was carried out long ago by Hogben (1976) by assuming a linear water wave for deep water. He assumed $C_d = 0.6$ for the post-supercritical regime and $C_d = 1.2$ for the subcritical regime. The inertia coefficient C_m was assumed to be equal to 2 for both regimes. Figures 4.5a and 4.5b show the various loading regimes for a wavelength-to-wave height ratio of 15. Evidently, at depths greater than about half the wavelength, the amplitudes of wave disturbances and corresponding wave forces are very small. It should be emphasized that Fig. 4.5a represents only ballpark values for a specific set of conditions and should not be regarded as a design criterion. The designers should be aware of the various loading regimes and should prepare their own figures, through the use of the parameters and wave conditions most appropriate to their design. We must also emphasize that the knowing of the idealized data or the data obtained under controlled laboratory conditions are extremely important in defining the lower limits of the force and moment coefficients, in enhancing the understanding of the physics of the prevailing phenomena, and in developing reliable ideas. Then, one can assess to what extent the freak waves in a given ocean can exceed the laboratory values.

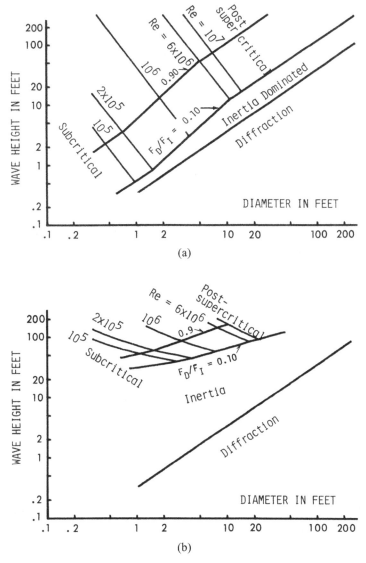

Figure 4.5. Comparative importance of loading regimes at (a) $z = 0$ (after Hogben 1976), (b) $z = 330$ ft (after Hogben 1976).

4.1.4 *Random waves*

As noted earlier, the behavior of the ocean waves is random. Therefore, the understanding of their probabilistic aspects and the prediction of the resulting random loads on structures are of considerable importance (see, e.g., Rice 1944–1945; Longuet-Higgins 1952; Davenport and Root 1958; Kinsman 1965; Bendat and Piersol, 1971; Price *et al.* 2002; Phillips 1977; and Price and Paidoussis 1984).

Sverdrup and Munk (1947) were the first to take into consideration the random characteristics of the waves. They have introduced the concept of the significant wave height, denoted by H_s or $H_{1/3}$. It is defined as the average height of the highest one-third of all the waves for a particular sea state. In 1952 Longuet-Higgins applied the statistical theory of random signals, developed a decade earlier by Rice (1944–1945), to the random elevation of the waves. In the following we describe as concisely as possible the considerable progress made toward the understanding and quantification of the ocean wave statistics.

The knowledge of the *spectral density* and *probability distribution* of wave maxima may not always provide an adequate description of random wave behavior (Bendat and Piersol, 1971; Phillips 1977). A further phenomenon that is considered significant in certain cases concerns the properties of a wave group. This is defined as a succession of consecutive waves with heights larger than a specified value (see, e.g., Goda 1970, 1976; Nagai 1973; Nolte and Hsu 1973; Rye 1974; Ewing 1973; and Siefert 1976).

Numerous approaches have been developed for a one-dimensional frequency spectrum. Any one of these may be extended to two- or three-dimensional spectra. A random sea may be represented as an infinite sum of sinusoidal components traveling with different wave numbers k, frequencies f, and directions θ (referred to a base direction, say that of the predominant wind). The free-surface elevation is thus written as

$$\eta = \sum_{n=1}^{\infty} A_n \exp\{i \left[k_n \cos(\theta_n)x + k_n \sin(\theta_n)y - 2\pi f_n t\right]\} \qquad (4.1.13)$$

where the *complex amplitudes A_n are random quantities*. Obviously, one may adopt a spectrum $S(k, f, \theta)$, such that $S(k, f, \theta)dk\,df\,d\theta$ represents the contribution to the variance σ_η^2 due to component waves with wave numbers between k and $k + dk$, frequencies between f and $f + df$, and directions between θ and $\theta + d\theta$. Then the spectral density or the energy spectrum of x may be written as

$$S(k, f, \theta)dk\,df\,d\theta = \sum_{k_n}^{k_n+dk} \sum_{f_n}^{f_n+df} \sum_{\theta_n}^{\theta_n+d\theta} \frac{1}{2}|A_n|^2 \qquad (4.1.14)$$

and the variance as

$$\sigma_\eta^2 = \int_0^\infty \int_0^\infty \int_{-\pi}^{\pi} S(k, f, \theta)dk\,df\,d\theta \qquad (4.1.15)$$

This three-dimensional spectrum is based on a polar representation of the wave number k. Thus, a random sea may be represented as an infinite sum of sinusoidal components traveling with different wave numbers k, frequencies f, and directions θ. The free-surface elevation, based on a polar representation of the wave number k, is given by

$$\eta = \sum_{n=1}^{\infty} A_n \exp\left\{i[k_n \cos(\theta_n)x + k_n \sin(\theta_n)y - 2\pi f_n t]\right\} \quad (4.1.16)$$

where the complex amplitudes A_n are random quantities. For all intents and purposes, the use of a one-dimensional spectrum is preferable to avoid the unwieldy character of the 2D and 3D spectra.

4.1.5 *Representative frequency spectra*

The following are a few of the most commonly used spectra: *Darbyshire spectrum* (1952), *Neumann spectrum* (1953), *Bretschneider spectrum* (1959), *Pierson–Moskowitz spectrum* (1964), and *Jonswap spectrum* (1973). Only two of them are described here for the sake of brevity.

Pierson–Moskowitz spectrum

This is written as

$$S(f) = \frac{\alpha g^2}{(2\pi)^4 f^5} \exp\left(-\frac{B}{f^4}\right) \quad (4.1.17)$$

where $\alpha = 8.1 \times 10^{-3}$ and is called the *Phillips'* constant and $B = 0.74(g)/2\pi U)^4$. This spectrum depends only on the wind speed U. The Bretschneider and Pierson–Moskowitz spectra differ only in the magnitudes assigned to αg^2 and B.

For *the fully developed sea conditions*, the International Ship and Offshore Structures Congress (ISSC) recommended the use of the modified *Pierson–Moskowitz* spectrum given by

$$\frac{S(\omega)}{H_{1/3}^2 T_1} = \frac{0.11}{2\pi} \left(\frac{\omega T_1}{2\pi}\right)^{-5} \exp\left[-0.44\left(\frac{\omega T_1}{2\pi}\right)^{-4}\right] \quad (4.1.18)$$

where $H_{1/3}$ is the significant wave height defined as the mean of the one-third highest waves and T_1 is a mean wave period defined as $T_1 = 2\pi m_0/m_1$ where $m_k = \int_0^\infty \omega^k S(\omega)d\omega$. $H_{1/3}$ is often redefined as $H_{1/3} = 4\sqrt{m_0}$ and is close to the one defined above.

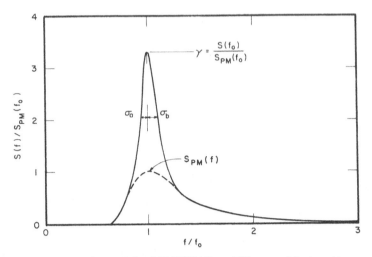

Figure 4.6. Comparison of the JONSWAP and Pierson–Moskowitz spectra.

Jonswap spectrum

It is based on the Joint North Sea Wave Project (Hasslemann *et al.* 1973) and constitutes a modification to the Pierson–Moskowitz spectrum to account for the effect of fetch restrictions and to provide for a more sharply peaked spectrum. It is given as

$$S(f) = \frac{\alpha g^2}{(2\pi)^4 f^5} \exp\left(-\frac{5}{4}\left(\frac{f}{f_0}\right)^{-4}\right)\gamma^a \qquad (4.1.19)$$

where $a = \exp[-(f - f_0)^2/2\sigma^2 f_0^2]$ and $\sigma_a = 0.07$ for $f \leq f_0$ and $\sigma_b = 0.09$ for $f > f_0$. Here f_0 is the peak frequency at which $S(f)$ is a maximum and is found to be related to the fetch parameter $f_0 = 2.84(gF/U^2)^{-0.33}$; σ_a and σ_b relate respectively to the widths of the left and right sides of the spectral peak; α is equivalent to the Phillips' constant but is now taken to depend on the fetch parameter; $\alpha = 0.066(gF/U^2)^{-0.22}$; and γ is the ratio of the maximum spectral density to that of the corresponding Pierson–Moskowitz spectrum. This was found to have a mean value of about 3.3, corresponding to a more sharply peaked spectrum than predicted by the Pierson–Moskowitz spectrum or by the Modified Pierson–Moskowitz spectrum. The profiles of these two spectra are compared in Fig. 4.6. Additional information on wave spectra may be found in the original papers referred to above and in the publications of the International Ship and Offshore Structures Congress (ISSC) and the International Towing Tank Conference (ITTC).

4.2 Wave–structure interaction

The state of the art and the assumptions and uncertainties that go into the prediction of fluid loading on offshore structures have been reviewed numerous times (see, e.g., Hogben *et al.* 1977; Sumer and Fredsøe 1997; Molin 2002 and others). The emerging fact is that the current body of analytical, experimental, and operational knowledge is still inadequate to describe the complex realities of fluid loading and dynamic response of offshore structures as evidenced by often tragic and highly expensive failures.

The primitive state of the description of the wake of bluff bodies is not the total cause of all uncertainties. Winds, gust, waves, currents, ice, earthquakes, soil movement, ship collision, oil spills, and variations in material qualities are random in nature and often difficult to model as inputs. The designer is forced to make idealizations, based on past experience, simplifying assumptions, and extrapolations (with intentional and hidden safety margins) with the hope that the structure will accommodate the uncertainties associated with its design. Clearly, the choice of a couple of hydrodynamic force coefficients is only a small part of the entire design process. It is also clear that what can be written on wave forces on small bodies cannot adequately account for the experience of a seasoned designer. We describe here only the relatively idealized analysis of various components and recent model and field tests, which may be integrated with sufficient experience to describe the actual loading of offshore platforms, composed of small-diameter members. What cannot be described or simulated are the millions of vortices, in all directions with large circulations, trying to push and twist each and every element. Nature has its vagaries and no one knows it better and understands it better than the offshore industry.

4.2.1 *Principal factors of analysis and design*

The development of rules and regulations in specifying the nominal and extreme loading conditions, the translation of these conditions into hydrodynamic loadings, and the execution of formal calculations for the structure constitute the essence of analysis and design. These may be performed through the use of two distinct methods: deterministic (for the extreme loading conditions) and stochastic (for the *nominal condition loadings*). The deterministic method may in turn be pseudo-static or time-dependent. An extensive discussion of the uncertainty of the wave-force peak that is associated with the largest wave height occurring at a particular point and the uncertainty of the largest wave force occurring at a particular sea state, both within a given duration, are discussed extensively by Harland *et al.* 1998.

The extreme conditions are those that give rise to low-cycle fatigue and largest hydrodynamic loadings in general. The nominal conditions are those that give rise to high-cycle fatigue under operational conditions and stem from the waves and currents that constitute the vast majority of the structure's exposure to fluid loading (Sarpkaya and Dalton 1992).

The pseudo-deterministic analysis of the extreme loading conditions is based on a design wave (chosen statistically with specified wave height, period, and direction) and a wave theory (such as Stokes fifth) to calculate the fluid velocities and accelerations along the axis of each structural element. The wave is assumed to be long crested and to propagate without change of form. The instantaneous sectional force is calculated through the use of Morison's equation together with C_d and C_m values appropriate to that section, i.e., C_d $(K, Re, k/D)$ and C_m $(K, Re, k/D)$, and the transverse force through the use of the appropriate lift coefficient, i.e., C_L $(K, Re, k/D)$. The instantaneous force on the element is obtained through the integration of the sectional forces. Then the forces are summed vectorially to get the total force and the overturning moment on the whole structure. The method also allows for the determination of the maximum and intermediate values of the total force and moment on the structure as well as on the individual elements. These in turn are used as input for static strength analysis. A quick and rough estimate of the in-line force and moment acting on a pile may be obtained through the use of the linear wave theory, Morison's equation, and constant drag and inertia coefficients:

$$F = \frac{1}{2}\rho d C_d D \left(\frac{gH^2}{8d}\right)\left(1 + \frac{4\pi d/L}{\sinh(4\pi d/L)}\right)|\cos \omega t|\cos \omega t$$

$$-C_m\left(\rho\frac{\pi D^2}{4}d\right)\left(\frac{gH}{2d}\right)\tanh\frac{2\pi d}{L}\sin \omega t \tag{4.2.1a}$$

and

$$M = \frac{1}{2}\rho C_d D \left(\frac{HL}{4T}\right)^2$$

$$\times \left[\left(\frac{2\pi d/L}{\sinh 2\pi d/L}\right)^2 + \frac{(4\pi d/L)\cosh 2\pi d/L}{\sinh 2\pi d/L} - 1\right]|\cos \omega t|\cos \omega t$$

$$-C_m\left(\rho\frac{\pi D^2}{4}\right)\left(\frac{HL^2}{2T^2}\right)$$

$$\times \left[\frac{(2\pi d/L)\sinh 2\pi d/L - \cosh 2\pi d/L + 1}{\sinh 2\pi d/L}\right]\sin \omega t \tag{4.2.1b}$$

In deeper water, the interaction of time-dependent ocean environment with a dynamically responsive structure leads to complex resonance conditions and gives rise to larger stresses than would be predicted from a pseudo-static analysis. For example, Brannon *et al.* (1974) have shown that the dynamic response can double the static wave load for a 900-ft platform. Ruhl (1976) reported that the dynamic response of a deep-water platform produced by three successive 60-ft, 10-sec waves is larger than that due to a 78-ft, 12-sec design wave. Duggal and Niedzwecki (1995) carried out a large-scale experimental study to investigate the dynamic response of a single flexible cylinder in waves. They have explained the mechanism of transverse wave loading by the depth dependence of K. This resulted in the cylinder being subjected to vortex shedding at several harmonics of the incident wave frequency, decreasing with depth. In general, the force coefficients also vary with depth as a function of their dependence on K, Re, and roughness. Thus, dynamic-response analysis and laboratory tests are not only desirable but also necessary for structures built in deeper waters.

The analysis begins with the synthesis of a representative wave train with a dominant period and significant wave height. The kinematics of the flow field is then determined by using methods such as the linear filtering technique. Then the wave is propagated in space and the time-dependent hydrodynamic force on all elements is calculated with the linearized Morison equation, assuming the structure to remain stationary. Subsequently, the calculated forces are incorporated into the equation of motion of the structure, together with the appropriate mass, stiffness, and damping characteristics, to predict the time-dependent response. The predictions may be improved by successive iterations to a degree afforded by the accuracy of mass, damping, stiffness, sea spectra, and the force coefficients.

The nominal condition loadings often require the use of stochastic spectral analysis for the determination of the fatigue and dynamic response of the structure. For this purpose one needs one or more wave spectra (such as Pierson–Moskowitz spectra; see, e.g., Michel 1967; Borgman 1969a; Mobarek 1965; Borgman and Yfantis 1979; Holthuijsen 2007), and the force-transfer function for each point and wave direction for the entire frequency range. Then the resulting force-response spectra are calculated (Borgman 1965; Pierson and Holmes 1965; Wiegel 1969). The method is valid only when the superposition principle is applicable, i.e., if the nonlinear loads such as drag forces are small in comparison with the linear loads such as inertial forces (St. Dennis 1973). In this approach, all calculations are performed in the frequency domain instead of the time domain as in the case of deterministic time-dependent analysis.

4.2.2 *Design wave and force characterization*

As noted above, a structure must withstand both the forces exerted by an extreme wave and the fatigue due to the accumulated effect of cyclic loading imposed by nominal waves (see, e.g., Marshall 1976; van Koten 1976).

Extreme wave heights are expressed in terms of wave heights having a low probability of occurrence. In the previous years, the probability P that a design wave with a return period of N years will be exceeded in a given duration of n years was given by a number of formulas. In more recent years, the height of the design wave *for a given location* is established on the basis of the data available on wave heights, periods, and directions. These are then used to establish a wave profile and wave loads using packaged codes. **NORSOK Standard N-003 (1999)** has suggested the use of $H^{(100)} = 1.9.H_s$, where $H^{(100)}$ stands for the wave height which is exceeded on the average only once every 100 years. Obviously, this suggestion varies from location to location, designer to designer, and platform to platform (e.g., floating platforms are, as expected, highly period-sensitive.). For the Gulf of Mexico area, experience and the majority of operators have indicated that an appropriate return interval for use is about 100 years. In more recent times, both **API (American Petroleum Institute, 2000)** and DTI have provided numerous rules and recommendations for the design, construction, and operation of offshore structures. One will be well advised to study them as well as many handbooks of offshore engineering.

Design waves are assumed to be uniformly crested and propagate across a hypothetical ocean surface with constant shape and speed. The character of the forces generated by these waves depends on the relative size and shape of the structure. In this chapter it is assumed that in the region near the body the kinematics of the undisturbed flow field do not change in the incident wave direction. Consequently, the force may be "drag dominated" (this is primarily the case with design waves past space frame structures), "inertia dominated" or a combination thereof for intermediate Keulegan–Carpenter numbers. It should be noted in passing that during the period of interaction of a wave with a given element, there is always a time interval during which the inertial forces are predominant and a time interval during which the drag forces are dominant. Evidently, then, the expression "drag dominated" or "inertia dominated" simply refers to the relative magnitudes of the two forces during a given cycle.

As noted earlier, an approximate quantitative mapping of the various loading regimes, expressed in terms of wave height, structure diameter, and depth below the water surface, may be obtained by assuming some appropriate values for the drag and inertia coefficients, a suitable wave theory, and a wavelength-to-wave height ratio (see Figs. 4.la and 4.lb). The designer should be aware of the various loading regimes and should prepare his own figures through the use of the parameters and wave conditions most appropriate to his design. The tension-leg and gravity-based offshore platforms constructed from vertical cylinders may have a resonance period of up to a few seconds. These platforms may in high sea states experience responses of considerable amplitude that are very suddenly generated at the resonance period of the platform. As noted by Huseby and Grue (2000) and in the references cited therein, this is a concern with respect to extreme loading or resonance response (ringing) in the essentially unseparated flow regime. The generation mechanism of the higher-harmonic wave loads to ringing is not well understood; see also Malenica *et al.* (1995), Newman (1996), and Chaplin *et al.* (1997).

The inherent instability or the natural restlessness of the oceans due to a number of forces (moon, wind, earthquakes, global warming, natural currents, temperature gradients, ice, etc.) give rise to complex waves and forces that may be substantially different from those assumed to exist for design purposes. In fact, limited data suggest that real waves exert forces that do not act in uniform directions on members of a structure in space and that change in direction and magnitude as the waves propagate. It appears that the forces predicted through the use of a design wave will be larger than those produced on the structure by a real wave field of comparable characteristics; i.e., the real ocean environment has mitigating effects (with rare exceptions) that can be quantified only approximately.

4.2.3 *Force-transfer coefficients*

As in SOF and uniformly accelerating flows, Morison's equation is used for estimating wave-induced forces on offshore structures. In its simplest form, it may be written as

$$F = \frac{1}{2}\rho C_d A_p |U| U + \rho C_m V_0 \frac{dU}{dt} \tag{4.2.2}$$

in which A_p represents the projected frontal area; V_0 is the displaced volume of the structure; and U is the velocity of the ambient flow. In wave-force calculations U is taken as the horizontal component of the wave particle velocity and the convective acceleration terms are often ignored, i.e., it is assumed that $dU/dt \cong \partial U/\partial t$. Furthermore, it is assumed that in the region near the cylinder the kinematics of the undisturbed flow do not change in the incident wave direction. For cylinder diameters *larger than about* 20% of the wavelength the inertia force is no longer in phase with the acceleration and the *diffraction effects* must be taken into consideration (see Chapter 5).

Considerable uncertainty exists regarding the meaning and application of the inertia force for nonlinear flows in which convective accelerations are not negligible. However, it is easy to show that the inertia forces calculated in the conventional manner will generally overestimate the actual force. In other words, the use of the local acceleration in lieu of total acceleration does not affect the practical application of the Morison equation, which in any case depends on experimental coefficients.

With a little poetic license, Morison's equation for a circular cylinder may be written in a more general form as

$$F = \begin{Bmatrix} F_x \\ F_y \end{Bmatrix} = \frac{1}{2}\rho C_d D \begin{Bmatrix} V_x \\ V_y \end{Bmatrix} \sqrt{V_x^2 + V_y^2} + \frac{1}{4}\rho\pi D^2 C_m \begin{Bmatrix} a_x \\ a_y \end{Bmatrix} \quad (4.2.3)$$

For a horizontal cylinder in a uniform flow field with $V_x = -U_m\cos\theta$ and $V_y = -U_m\sin\theta$, Eq. (4.2.3) reduces to

$$\frac{2F_x}{\rho DU_m^2} = \frac{\pi^2 D}{U_m T}C_m\sin\theta - C_d\left(\cos^2\theta + (V_m/U_m)^2\sin^2\theta\right)^{1/2}\cos\theta \quad (4.2.4)$$

and

$$\frac{2F_y}{\rho DU_m^2} = \frac{\pi^2 D}{U_m T}C_m\sin\theta - C_d\left(\sin^2\theta + (U_m/V_m)^2\cos^2\theta\right)^{1/2}\sin\theta \quad (4.2.5)$$

Equations (4.2.4) and (4.2.5) state that the instantaneous in-line force in the x- and y-directions is equal to the sum of the projections, on the respective axis, of the instantaneous values of the total-velocity-square-dependent drag force and the total-acceleration-dependent inertial force. This implies that the flow over a cycle may be regarded as a juxtaposition of planar flows with instantaneous velocities and accelerations given by

$$q = \left(U_m^2\cos^2\theta + V_m^2\sin^2\theta\right)^{1/2} \quad a = \left(U_m^2\sin^2\theta + V_m^2\cos^2\theta\right)^{1/2} \quad (4.2.6)$$

The vortices do not move with the velocity of the ambient flow and the wake does not rotate about the cylinder at the same rate as the ambient velocity vector. In other words, one must be aware of the fact that the writing of Morison's equation in vectorial form does not necessarily imply that the behavior of the wake can be correctly represented by it. One must also note that the Fourier averages of C_d and C_m are no longer given by (3.9.11a) and (3.9.11b).

Equations (4.2.3) and (4.2.4) cannot be written using only the x- or only the y-component of the velocity in Morison's equation. This will assume that the drag component of the in-line force is proportional to the square of the projected velocity rather than to the square of the instantaneous total velocity (note that U and V are assumed to be normal to the axis of the cylinder). Finally, it is because of the assumptions noted above that the drag components of the forces given by (4.2.3) and (4.2.4) become linear for $U_m = V_m$ (fluid particles undergoing circular orbits).

The integration of (4.2.2) over depth and summation over N vertical piles yield a model equation for total forces as

$$F_T = \sum_0^N \int_{-d}^0 [0.5\rho C_d(K, \text{Re}, k/D)DV_n|V_n|$$
$$+ 0.25\pi \rho D^2 C_m(K, \text{Re}, k/D) a_n] dz \qquad (4.2.7)$$

where a_n and V_n are the accelerations and velocities along the n-th pile.

The MOJS equation gave rise to a great deal of discussion on what values of the two coefficients should be used. Furthermore, the importance of roughness, rotation of the velocity vector, orientation of the cylinder, proximity of other members, spanwise coherence, currents, free surface, etc. has remained in doubt since experimental evidence published over the past 50 years has been quite inconclusive. The problem has further been compounded by the difficulty of accurately measuring the velocity and acceleration to be used in Morison's equation. In general, the nature of the equation rather than the lack of precision of measurements or the difficulty of calculating the kinematics of the flow from the existing wave theories has been criticized (see also Chaplin and Kesavan 1993).

The problem is particularly complicated in oceans, because in general, the sea surface is nonlinear and irregular and the waves may originate from a range of directions. Until about 30 years ago it was possible to adequately measure the kinematics of the waves. This shortcoming required that the kinematics used in the correlation of the measured forces with the Morison equation be established through the use of a suitable theory. An additional complication is that the wave theory simply cannot establish the presence of a steady current. Furthermore, it can be demonstrated that the kinematics of the wave are affected by nonlinear currents at other depths, above or below the point of measurement, even if a current does not exist at the point of measurement. Consequently, it is important to distinguish between the force coefficients obtained with strictly sinusoidal flows, monochromatic waves, irregular two-dimensional waves, and three-dimensional waves. In general, the instantaneous values of wave height and period are insufficient to enable confident prediction of wave forces to be made. A consideration of the previous history is also necessary.

Kinematics of the flow field are often calculated rather than measured. As discussed in some detail above, there are several wave theories,* which might be used in calculating the velocities and accelerations. These theories deal with two-dimensional waves and ignore the effect of the three-dimensionality of the flow.

Stokes fifth-order theory is commonly used to describe the characteristics of steep nonlinear waves in relatively deep water. Airy's theory yields fairly similar results even if it is extended beyond the limits of linear assumptions to predict the velocities above the still water level. Consequently, it has been used on occasion to calculate the wave kinematics and the resulting wave forces by performing the integrations up to the *actual water surface* rather than to the still water level, as it is commonly done in linear analysis. Dean's stream-function method (1965), which essentially is a *Fourier expansion of the stream function*, is used quite often since it provides the best fit to the surface boundary conditions even for irregular waves. Each theory, be it linear or nonlinear, analytical or numerical, has its own limitations and ranges of applicability. Numerous works have been carried out to develop design guidelines for selecting the appropriate wave theory for specific site conditions.

It is very costly to make measurements in the ocean environment. Furthermore, it is not easy to interpret the results particularly in separating the effect of currents from the kinematics of waves (Sarpkaya 1957; Peregrine 1976; Thomas 1979). Thus, it may be generally stated that it is difficult to reconstruct the precise kinematics of the flow at the level of force measurements either through the use of approximate wave theories or through direct measurements even though the latter is clearly more desirable. The designer must be aware of the fact that the interference of the wake with the velocity probe (e.g., a magnetic velocimeter) is a serious problem. If the arm connecting the probe to the pile is too long (so as to avoid interference) then the vibrations of the probe may become important (this is equally true for long wind poles used to measure wind speed and turbulence at the airports). In either case the wake of the probe itself may be important and must be carefully taken into account through proper calibration under similar flow conditions. The combined wave and current flow about a structure (say a vertical cylinder extending from the ocean bottom through the free surface) gives rise to an exceedingly complex, separated, time-dependent flow. The

---

* Lamb (1932); theory, Munk (1949); solitary wave theory, Reid and Bretschneider (1953); Stoker (1957); Stokes fifth-order theory, Skjelbreia and Hendrickson (1960); Wiegel (1964); stream-function theory, Dean (1965); Kinsman (1965); Whitham (1974); extended velocity potential theory, Lambrakos and Brannon (1974); Le Méhauté (1976); Phillips (1977); LeBlond and Mysak (1978); Lighthill (1979); Mei (1989); Herbich (1990); Sorensen (1993); Rahman (1995); Massel (1996); Young (1999); Kamphuis (2000); Goda (2000); CEM (2002); Mei *et al.* (2005); Svendsen (2006).

velocity vector rotates 360 degrees and it is not possible to think in terms of simple separation points, laminar separation bubbles, turbulent reattachment, etc. The flow regime may change from subcritical all the way to supercritical at a given elevation during the passage of a single wave and at a given time along the pile. Detailed discussions of wind loads on structures are ably described by Dyrbye and Hansen (1997), Holmes (2001), Price and Paidoussis (1984), Gurley and Kareem (1993), Zhou and Kareem (2001), Zhou *et al.* (2000), and Zhou *et al.* (2002), and the references cited therein.

The shear layer is not fed equal amounts of vorticity along the cylinder partly because of the variation of the magnitude and direction of velocity and partly because of the variation of the soft and rigid excrescences along the cylinder. Consequently, the vortices do not detach at the same instant along the cylinder. The spanwise coherence is considerably reduced and the mean coherence length (or correlation length) varies with depth, wave, and flow history. This picture is of course further complicated by the fact that the wake is swept back and forth. The periods of flow deceleration precipitate instability and cause profound changes in the vorticity and pressure distribution about the cylinder. The currents are not necessarily uniform or collinear with the waves.

The flow about a *horizontal cylinder* presents equally complex problems. As noted earlier, even the uniform flow about a cylinder does not result in a perfectly two-dimensional flow. The spanwise coherence depends on the length-to-diameter ratio of the cylinder, upstream disturbances, roughness, Reynolds number, etc.

In wavy flows, the velocity vector and hence the wake rotates about the cylinder. Roughness does not consist of regularly shaped and positioned rigid elements glued on a smooth cylinder, as in laboratory experiments. In field tests, one is most likely to encounter large fixed beds, covered with coral roughness (Sarpkaya 1990a; Nelson 1996).

Even if the flow were to be analyzed, say using direct numerical simulation (DNS), to understand the behavior of the prevailing coherent structures, the results could be analyzed only statistically. If the flow were to oscillate sinusoidally, the mission becomes impossible. The size, shape, and distribution of the structures vary with time, element shape, and position on the cylinder (even for a steady flow). This was just a little glimpse into the complexities of the flow about a simple circular cylinder with well defined, regularly shaped, and uniformly distributed roughness elements. In reality, marine hydrodynamics deals with flow about circular cylinders and other rigid bodies immersed in omnidirectional waves and/or currents varying with depth. These bodies may be covered with various kinds excrescences (rigid or soft; see Figs. 4.7a and 4.7b). This is an exceedingly complex phenomenon and can be dealt with using empirical equations (e.g., Morison equation),

(a)

(b)

Figure 4.7. A marine-roughened pipe with (a) rigid excrescences, and (b) soft excrescences.

past accomplishments, and failures. In short, a purely theoretical approach is beyond our capacity.

We will now discuss briefly the methods of evaluation of a given set of data. It will become clear that there is not a unique method of evaluation of a given set of kinematic and dynamic data and that there is always some bias in data interpretation. Some of the most frequently used methods are:

A. Consider only one wave period, e.g., the wave between two crests or two troughs. Then calculate the kinematics of the flow using an appropriate wave theory, if it has not already been measured. The evaluation of the force coefficients and their meaning then depend on the methods of evaluation and on the force measured:

1. Calculate C_d and C_m using either the Fourier-averaging technique or the method of least squares. Evidently, neither of these methods ensure that the maximum calculated force be equal to the maximum measured force. Furthermore, the quality of representation of the measured force by the linear-quadratic sum with constant coefficients depends on K, Re, relative roughness, highly variable vortex

effects, and the irregularities and nonlinearities of the incident wave train.

2. Calculate C_d and C_m using the force at the points corresponding to maximum velocity and maximum acceleration [note that for nonlinear waves $U = 0$ does not necessarily correspond to $\acute{U}$ (maximum) and vice versa]. This method may yield reliable results for C_d in the drag-dominated region and small differences between the measured and calculated forces (because the phase difference between the maximum velocity and the maximum forces becomes very small in the drag-dominated region). This method also yields accurate results in the inertia-dominated region for the C_m values. It is not, however, recommended in a wide range of K and Re values where both drag and inertia may be of equal importance and highly dependent on the history effects. For sake of reference it should be noted that the ratio of the maximum drag force to the maximum inertial force, at a particular depth for a given wave and cylinder or for a given K and Re, is given by

$$F_d(\text{max})/F_i(\text{max}) = (C_d/C_m)(K/\pi^2) \qquad (4.2.8a)$$

When the above ratio is larger than 0.5, F_d/F_i at the instant of maximum force is given by

$$F_d/F_i = 2\left[(F_d(\text{max})/F_i(\text{max})\right]^2 - 0.5 \qquad (4.2.8b)$$

3. Calculate C_d and C_m from Morison's equation by writing it once for the maximum force and once for the zero force together with the corresponding velocities and accelerations. This method may not always yield stable results as far as the correlation of the drag and inertia coefficients with K and Re is concerned.

The methods cited above give a measure of the drag and inertia coefficients for each wave or half-wave cycle. They may also be regarded as time-invariant averages for the particular wave and section of the cylinder on which the forces were measured.

From time to time the force acting on the entire cylinder is measured or the force is deduced from the bending moment acting at the bottom of the cylinder through the use of a suitable wave theory. In this case there are more alternatives and less certainty in the calculation of the drag and inertia coefficients. One can use the surface values of the velocity and acceleration or the velocity and acceleration at a suitable point along the cylinder. In any case, *the coefficients calculated represent time- and space-averaged quantities.*

B. Consider a series of waves over a suitable time interval. In this case also one can adopt several methods to deduce the drag and inertia coefficients:

1. Use a spectral analysis to achieve the best correspondence between the spectra of the measured and calculated forces. Borgman (1967, 1969b), Brown and Borgman (1966), and Wilson (1965) used Airy's linear wave theory and the linearized version of the drag force. The resulting drag and inertia coefficients showed wide scatter possibly as a result of the linearization of the drag force term.

2. Use of the probabilistic distribution in time of wave forces. Pierson and Holmes (1965) pursued this particular method and found that Morison's equation is appropriate for small diameter cylinders and the drag and inertia coefficients are not constant along the length of a pile. The method involves some uncertainty due to the non-Gaussian nature of wave-force time records.

3. Represent the wave train with the stream function and calculate the kinematics of the flow field. Then, using any one of the methods cited in part A (preferably the Fourier averaging or the method of least squares) calculate the drag and inertia coefficients for one particular wave in the middle of a few successive waves. Subsequently, move the window of the wave train to repeat the procedure. Such calculations can be performed quickly with a simple code.

The true purpose of relatively idealized experiments (e.g., sinusoidally oscillating flow about a circular cylinder) is to emphasize the physics of the phenomena and to determine whether the linear combination of a linear inertial force with a nonlinear drag force (*a two-term equation for one of the most complex unsteady flows in fluid dynamics*) can predict, with sufficient accuracy, the measured time-dependent forces under the most idealized circumstances (i.e., SOF about a simple smooth circular cylinder!). Should this prove to be the case, one can then determine the role played by each controllable parameter in the variation of the coefficients quantifying the drag and inertial forces. This by no means ensures that the said two-term linear superposition will continue to hold true for more complex flow kinematics and body shapes to the same degree of accuracy as in idealized experiments in the flow regimes defined by K and Re.

At sufficiently high Reynolds numbers, alternate vortex shedding gives rise to pseudo-periodic transverse forces. The combination of the entire phenomenon gives rise to one of the most fascinating and intriguing un-solved problems of fluid mechanics.

It is rather surprising that Morison equation works as well as it does in spite of all of its limitations and the complexities of separated flows. It works sufficiently well (for engineering purposes) only when the inertia alone or the drag alone is the dominant force. This happens in the regions where K is smaller than about 8 and larger than about 25. In the intermediate region (i.e., *the drag/inertia dominated region*) the linear sum of the drag and inertial forces *is not a true representation of the interaction of the drag and inertial forces in nature*.

Clearly, both the form of the Morison equation and the uncertainties that go into the characterization of the ocean environment are jointly responsi-ble for the differences between the measured and calculated forces. Mori-son's equation could continue to provide an approximate answer to an approximately defined problem, in spite of the uncertainties in its form and input, only when it is carefully calibrated and fine-tuned with respect to a wave theory and local conditions and tolerated in light of all other uncertainties and hidden and intentional safety factors that go into the final design of a structure. As to the future, it is only through the com-bination of well-documented and thought-out laboratory and field exper-iments, sensitivity analyses, and experiments that the rules for the design of safe marine structures can be continuously refined (Sarpkaya and Cakal 1983).

4.2.4 *A brief summary of the literature giving explicit C_d and C_m values*

We will describe here only the results of the most important wave projects. The differences in the test conditions, methods of measurement, and data evaluation do not permit a critical and comparative assessment of the drag and inertia coefficients obtained in each investigation.

Some of the data came from the measurements carried out in the actual sea conditions: Reid (1958), Wiegel *et al.* (1957) (see Figs. 3.2a and 3.2b in Chapter 3), Wilson (1965) (30-inch pile in confused-sea conditions in the Gulf of Mexico), Thrasher and Aagaard (1969) (whose results have demon-strated clearly that the force-transfer coefficients strongly depend on the particular wave theory used in the evaluation of the wave kinematics), and Kim and Hibbard (1975) (used test piles 38 ft long and 12.75 inches in diam-eter). The overall mean value of the drag coefficient under crests (0.72) was found to be significantly larger than that under the troughs (0.54). With cor-rections to the measured velocities, the overall average of C_d was found to be 0.61 and $C_m = 1.20$. The agreement between the measured and calculated forces was good in the drag-dominated part of the wave cycle, and fair in the inertia-dominated region.

A large-scale experiment was undertaken by Exxon Production Research Company using a highly instrumented ($20 \times 40 \times 120$ ft) platform in 66-ft water in the Gulf of Mexico. Data obtained included local wave forces on clean and barnacle-covered sensors, local wave kinematics, total base shear and overturning moment on the structure, forces on a simulated group of well conductors, and impact forces on a member above mean water level (Lin and Shieh 1997 and Cao *et al.* 2007).

Heideman *et al.* (1979) used two methods to evaluate the drag and inertia coefficients. The first was the least-squared-error procedure for each half-wave cycle. The instantaneous in-line velocity in Morison's equation included both the wave velocity and the projection of the current velocity. The second method consisted of the evaluation of C_d over short segments of waves in which drag force was dominant and of C_m over short segments in which inertia force was dominant.

Heideman *et al.* (1979) concluded that (1) Morison's equation with constant coefficients can be made to fit measured local forces and kinematics satisfactorily over individual half-wave cycles; (2) most of the scatter in the C_d results can be explained by the random wake encounter concept; (3) local deviations in apparent C_d are not spatially correlated in any given wave; (4) C_d results from Sarpkaya's experiments (1976b) represent an upper band to C_d values that may be expected in random three-dimensional oscillatory flow; (5) for $Re > 2 \times 10^5$, the apparent C_d depends on surface roughness and, for members that are nearly in the orbit plane, on K; (6) asymptotic C_d results from the test data in random three-dimensional oscillatory flow are consistent with steady flow data for the same relative roughness; and (7) C_m is greater for smooth cylinders than for rough cylinders, while the reverse is true for C_d.

Ohmart and Gratz (1979) reported the results obtained from wave forces and horizontal particle velocities measured during Hurricane Edith. The data base included a Reynolds number range of 3×10^5 to 3×10^6. Overall, the use of an inertia coefficient of 1.5 and a constant drag coefficient of 0.7 provided the best fit of measured and predicted forces. *For the peak forces, the best fit was obtained using $C_d = 0.7$ and $C_m = 1.7$. These values are in good agreement with those obtained by Sarpkaya* (1976a) *at $Re = 1.5 \times 10^6$.*

Gaston and Ohmart (1979) measured the total wave force and overturning moment on a smooth and roughened 15-ft long, 1-ft diameter, vertical cylinder under conditions of periodic and random waves in a wave tank. The *average values of C_d and C_m were found to be $C_d = 0.77$ and $C_m = 1.81$ for the* smooth cylinder; $C_d = 1.34$ and $C_m = 1.87$ for the cylinder with $k/D = 1/96$;

$C_d = 1.41$ and $C_m = 1.99$ for the cylinder with $k/D = 1/32$; and $C_d = 1.42$ and $C_m = 2.01$ for the cylinder with $k/D = 1/24$.

Sarpkaya's data (1976b) with sand-roughened cylinders have clearly demonstrated that *the effect of roughness on resistance is not the same in steady and sinusoidally oscillating flows, even at relatively high K and Re values*. In response to the discerning few, Sarpkaya enlarged the test section of his U-shaped water tunnel from 3 ft by 3 ft to 3 ft by 4.7 ft (height) in 1980 and repeated the experiments with sand-roughened cylinders, which he originally reported earlier (1976b). The most recent data did not deviate more than ± 2% (see Sarpkaya 1980) from those obtained previously, showing conclusively that the blockage effects were indeed negligible (see also Chaplin 1988).

Heaf (1979) presented a detailed discussion of the effect of marine growth on the performance of fixed offshore platforms in the North Sea. He has pointed out that the marine growth influences the loading of an offshore platform in at least five ways: (1) increased tube diameters, leading to increased projected area and displaced volume and hence to increased hydrodynamic loading; (2) increased drag coefficient, leading to increased hydrodynamic loading; (3) increased mass and hydrodynamic added mass, leading to a reduced natural frequency and hence to an increased dynamic amplification factor; (4) increased structural weight, both in the water and above the water level in air; and (5) effect upon hydrodynamic instabilities, such as vortex shedding and vortex induced vibrations (VIVs). One might also add that *roughness decreases the separation angle and increases the correlation length, vortex strength, and the lift coefficient*. Equally important is the fact that roughness accumulates most in the uppermost region of the structure where the wave- and current-induced velocities and oxygen are largest (Bradshaw 2000; Boukinda *et al.* 2007).

Heaf (1979) reported experiments with cylinders held vertically in a towing tank where the wave loading was due to a highly irregular wave train. He concluded that *"the values of C_d and C_m predicted by Sarpkaya's work (1976b) seem to be the most appropriate data for predicting the effect of different heights of surface roughness on C_d and C_m, and thereby in the wave loading."*

The tests by Burnett (1979) have shown that a cylinder covered in seaweed experiences significantly larger forces (as much as 220% that of the smooth cylinder) than a similar cylinder covered in hard roughness (in *Burnett's experiments K was smaller than about 15*). The reason for this increase is partly due to the inertial effect of the *sweeping back and forth of the seaweed fronds*, partly due to the increased form drag and skin-friction drag, and partly due to the increase of turbulence introduced into the incident flow. Clearly, additional research is needed for all practically significant values of K and Re in order to quantify the effect of soft excrescences on wave loading on structures (Heaf 1979, Houghton 1968). Wolfram and Naghipour (1999) analyzed the data obtained with "heavily roughened" circular cylinders (D = 0.513 and 0.216 m) and concluded that the weighted least square method generally gave the best predictive accuracy. The drag and inertia coefficients were found to be about 1.7 and 2. However, lack of better quantification of "*heavily marine roughened*" cylinder left much to be desired. Verley (1982) devised a simple quasi-steady empirical model of vortex-induced forces in waves and oscillating currents. It appears that the energy due to vortex shedding forces spreads over a range of several multiples of the wave frequency to either side of a modified Strouhal frequency defined by $U_m St/D$. Oscillating flow about two- and three-dimensional Bilge Keels has been investigated by Sarpkaya and O'Keefe (1996). The results have confirmed the fact that the bilge keels do, in fact, provide a high degree of damping, particularly at small amplitudes of oscillation.

4.2.5 *Suggested values for force-transfer coefficients*

Considerable attention has been devoted to the differences between the uniform SOF, monochromatic waves, and the waves in an ocean environment. Not only the kinematics of the flow but also the force-transfer coefficients become increasingly uncertain as one comes closer to the conditions of the ocean environment (see Figs. 4.7a and 4.7b). Consequently, there is substantial latitude in design practices as evidenced by the recommendations of various authoritative sources and guideline agencies ($C_d = 0.6$ to 1.0 and $C_m = 0.5$ to 2.0). It is recommended that either the data given in Figs. 3.20 through 3.24 in Chapter 3 or the plots of the data provided by API-2000 be used. The former will lead to more conservative forces. Common sense and some limited data suggest that real waves exert forces that do not act in uniform directions on members of a structure in space and that change in direction and magnitude as the wave propagates. This in turn would suggest that the drag coefficients (*not the inertia coefficients*) obtained from ocean tests might be somewhat smaller than those that might have resulted from monochromatic waves of similar height and period. It is on the basis of the foregoing that we will make a distinction between the values suggested by the analysis of the measurements made at sea and those for the design of space-frame structures subjected to idealized waves. Hopefully, the analysis will lead to the refinement of the methods of evaluation of the coefficients and to some additional understanding of the effects of three dimensionality of the flow. This will depend in part on the number of structures built and measurements made at a given site over a number of years.

Wave-tank tests as well as the field measurements yield somewhat larger C_m values, presumably due to *larger rates of change of kinetic energy* (see Chapter 2) than those obtained with planar oscillatory flows in the range of K values from about 8 to 20 (interactive or critical or the drag inertia dominated regime). This is also the region where almost all experiments, particularly those conducted in the ocean environment, show the largest scatter. This is partly because of the inherently random nature of the shedding and subsequent mutual interaction of the vortices, partly the fractional shedding of vortices, and partly the reduced spanwise coherence of vortices. In fact, *the smaller the spanwise coherence the larger the C_m in the said region.* It is primarily because of these reasons that the development of a fluid-mechanically satisfying unified wave-force equation that would account for the randomness of shedding, memory effects, etc., is rather difficult. It appears that such an equation will have to have both a probabilistic and a deterministic character.

The values for the force-transfer coefficients suggested herein are based partly on the foregoing considerations, partly on a careful perusal of the pertinent literature, and partly on the suggestions of the authoritative sources (API RP 2A Section 3) and reflect the author's best current judgment and experience.

4.2.6 *Effects of orbital motion, coexisting current, pile orientation, interference, and wall proximity*

General comments

It must first be noted that the effects noted above have been studied to varying degrees of success by a number of investigators (Davis and Ciani 1976; Sarpkaya 1977d; Bearman and Zdravkovich 1978; Jensen and Mogensen 1982; Sarpkaya *et al.* 1982; Grass *et al.* 1985; Ali and Narayanan 1986; Fredsøe and Hansen 1987; Allender and Petrauskas 1987; Bearman and Obasaju 1989; Wolfram *et al.* 1989; Wolfram *et al.* 1993; Hansen 1990; Jonsson 1990; Bryndum *et al.* 1992; Stansby and Star 1992; Sumer *et al.* 1992; Chaplin 1993; Kozakiewicz *et al.* 1995; Zdravkovich 1996b; and the numerous references cited in Sumer and Fredsøe, 1997).

It is not a simple matter to isolate the effect of orbital motion on cylinder resistance in wavy or sinusoidally oscillating flows, particularly in the range of K and Re values of practical significance. In wavy flows, the cylinder has to be placed horizontally at various depths below the free surface and the particle velocity together with the instantaneous forces be measured. In doing so, one has to vary the wave height systematically so as to obtain nearly identical K and Re values. This is a nearly impossible task. Consequently, the measurements made with waves of nearly identical characteristics reflect not only the effect of the orbital motion but also the effects of the variation of K and Re with depth. In other words, one can speak only of isolated examples. Finally, one must also bear in mind the three-dimensional nature of the flow about the cylinder due to spanwise and chordwise instabilities.

In SOFs, the cylinders will have to be oscillated sinusoidally in the transverse direction to simulate the orbital motion. In this case also there are limitations as to the ranges of the K and Re values and the ratio of the amplitudes of the vertical and horizontal velocities.

Koterayama (1979) measured the wave forces on small horizontal cylinders (0.4 to 3.2 inch diameter) and found that C_d and C_m, are smaller than those obtained with planar sinusoidal motion and with vertical cylinders in waves at relatively small Reynolds numbers. His V/U ratios were not specified. It must be emphasized that it is not wise to use data obtained with small cylinders at Reynolds numbers smaller than about 20 000 (*unsteady shear layers are not fully developed*) *unless the design is in the transition region.*

The vortex shedding from a circular cylinder near a *moving* wall (as done by the automobile industry) has been numerically simulated by Huang and Sung (2007) and the drag and lift forces and the pressure coefficient are calculated and the role of a *critical-gap ratio* (at which vortex shedding disappears) is described.

A more detailed discussion of the effects of coexisting current, angle of attack, orbital motion (about vertical and horizontal cylinders), and the wall-proximity effects are carefully compiled and presented by Sumer and Fredsøe (1997). It is rather unfortunate that the information gleaned from the type of motions noted above are, out of necessity, confined to limited ranges of the geometric as well as fluid-mechanical parameters such as Re, K, and k/D. The wisdom acquired from small-scale experiments must be complemented and verified with large-scale experiments in suitable facilities.

Effect of currents

There are numerous physical circumstances in which interactions between waves, currents, and structures occur. The analysis of the interaction of waves with preexisting and/or wind- or wave-generated currents and the interaction of the modified wave–current combination with rigid or elastic structures and their components require different mathematical approaches, relevant observations, and experiments that are applicable to all or some of these physical circumstances.

Basically, the wave–current combination may be treated either as a complex fluid-mechanical phenomenon where the interaction of waves and current is taken into consideration or as a relatively simple phenomenon where *the interaction is ignored and the current is simply superimposed on waves*. If the current is in the direction of wave propagation, the wave amplitude decreases and its length increases. If the current opposes the wave, the wave becomes steeper and shorter. One must also bear in mind that if there is a current varying with depth it may affect the waves on the water surface since the wave motion also varies with depth.

Elegant mathematical reviews of the interaction of water waves and currents have been presented by Peregrine (1976) and Rahman (1995), where an extensive list of references may be found (see also Jonsson 1976). Measurements of wave–current interaction phenomena are scarce. Among the first to perform substantial controlled experiments of this nature was Sarpkaya (1955). He made measurements of wave-phase velocity, amplitude, wavelength, and shape of ascending and descending waves in a long wave–current flume and determined the neutral stability conditions for waves propagating with constant amplitude. Additional experiments were conducted by Inman and Bowen (1963) in connection with sand transport by waves and currents.

Dalrymple and Dean (1975), following the simple superposition principle, related waves of maximum height on currents to equivalent waves in still water. Their procedure does not address the problem of wave–current interaction. Their objective can be stated as follows: Given a wave propagating in still water, determine its height and length on a constant current. It is easy to show that the wave celerity C' and the wavelength L' (both with the presence of a current V) are related by

$$(C' - \overline{V})^2 = \frac{gL'}{2\pi} \tanh \frac{2\pi d}{L'} \qquad (4.2.9)$$

where d represents the water depth. Dalrymple and Dean gave a number of examples to illustrate the use of the superposition method to account for the presence of uniform and steady currents. In general no wave propagation is possible when the celerity of energy transmission is equal to the mean velocity of flow. For small-amplitude waves, this yields (Sarpkaya 1955)

$$\frac{\overline{V}}{\sqrt{gd}} = \frac{1}{2} \left[1 + \frac{4\pi d/L'}{\sinh 4\pi d/L'} \right] [(L'/2\pi d) \tanh(2\pi d/L')]^{1/2} \qquad (4.2.10)$$

which shows that the waves could be stopped against a current of $\overline{V}/C' = 1/3$ for large values of d/L'. This conclusion is an approximate one, however, since the waves near breaking become more peaked at the crest and slow down as shown by Longuet-Higgins and Stewart (1961) and Longuet-Higgins 1976.

Little information exists on the effect of current plus wave interaction on hydrodynamic loading of offshore structures. The complexity of the problem stems from several facts. First, an analytical solution of the separated, time-dependent flow is not yet possible even for relatively idealized situations. Secondly, waves and currents are omnidirectional and the directional distribution of energy is anisotropic. Finally, an experimental investigation of the problem in the practically significant range of Reynolds numbers, Keulegan–Carpenter numbers, relative current velocities, and suitably defined current gradients is practically impossible. Some tests may be carried out in the ocean by towing cylindrical models in head, following, beam and quartering seas with various forward speeds for a range of wave heights and periods. In doing so one should measure in-line and transverse forces, in addition to the kinematics of the waves, and make sure that the model does not undergo transverse oscillations.

The results obtained with superposed mean and two-dimensional SOF about a cylinder have been discussed in Section 3.9. It has been shown, for at least relatively small Reynolds numbers, that for sufficiently small values of V/U, the drag coefficient for a current–SOF combination may be considerably smaller than that for a SOF alone. Qualitatively, this corresponds to a region where the effect of the current on the convection of vortices is most important, (i.e., small K). Evidently, these conclusions are based on the implicit assumption that C_d and C_m remain constant whether there are interactions or not and whether the current is positive or negative. As noted earlier, there is not enough experimental data for an intelligent assessment of the proper values of the drag and inertia coefficients.

The conventional method of including the current effects on wave loading is by vectorially superimposing current profiles over the wave velocity field generated in the absence of a current. In doing so, the acceleration to be used in the calculation of the inertial force should be written as

$$a_x = \frac{Du}{Dt} + \overline{V}\frac{\partial u}{\partial x} = \frac{\partial u}{\partial t} + (u + \overline{V})\frac{\partial u}{\partial x} \qquad (4.2.11)$$

Furthermore, it is suggested that the Reynolds number and the Keulegan–Carpenter number be defined as

$$Re = (U_m + \overline{V})D/\nu \quad \text{and} \quad K = (U_m + \overline{V})T/D \qquad (4.2.12)$$

The drag and inertia coefficients to be determined from a systematic experimental investigation may not necessarily correlate with $\overline{V}/U_m$ and Re and K (as defined above). For most practical applications, however, the variations of C_d and C_m, with large K and Re (for no current) are rather small.

In summary, many laboratory and field investigations remain to be carried out for a better quantification of the wave–current and wave–current–structure interactions. Small-scale laboratory experiments hint that Morison's equation may not be sufficient to describe the fluid loading through a vectorial superposition of wave and current velocities. For V/U smaller than about 0.3 and K larger than about 20, the use of Morison's equation, as stated above, together with the drag and inertia coefficients suggested for use in the absence of current will, in general, lead to conservative in-line forces. It is assumed that the presence of the current or wave–current combination does not lead to in-line and/or transverse hydroelastic oscillations. For a model study of wave and current forces on a fixed platform, the reader is referred to Sekita (1975).

Wave–current interaction and the scatter component of an incident wave are of some concern in the diffraction analysis of large bodies. It appears that such effects are in general small (Hogben, 1976) and are somewhat obscured by flow separation (ordinarily ignored) on diffraction analysis. The presence of currents may be more important for large square-shaped bodies than those for bodies with no sharp corners.

Effect of pile orientation

The effect of body orientation (in the plane of flow) on resistance particularly for bodies of finite length (e.g., a missile at an angle of attack) has been the subject of extensive investigation in steady flows. It has not been possible to correlate the in-plane normal force and the out-of-plane transverse force with a single Reynolds number. Evidently, for a zero angle of attack (flow parallel to the axis of the body), the appropriate Reynolds number is based on the length measured along the body. For a 90° angle of attack (flow normal to the axis of the body), the appropriate Reynolds number is based on the diameter of the body. Between the two flow situations, it is not possible to define a simple characteristic length and hence a universal Reynolds number that will correlate the force-transfer coefficients for all flow regimes.

Hoerner (1965) proposed the independence or crossflow principle or the "cosine law," which states that the normal pressure forces are independent of the tangential velocity for subcritical values of *Re*. Such a principle has been used even for examining the galloping instability of yawed transmission lines or cables of guyed towers. Skarecky (1975) commented that the independence principle may have at least limited validity for smooth and roughened bodies. Van Atta (1968) found a marked deviation of the vortex-shedding frequency from the $\cos\phi$ relation. Furthermore, by varying the tension in the yawed wire, he showed that the effects observed by Hanson (1966) could be attributed to synchronization or lock-in between the natural vortex-shedding frequency and the fourth, fifth, and sixth harmonics of the wire. In doing so, van Atta has shown that yawed cylinders undergo synchronized oscillations in the range $5.9 < U \cos \phi / f_n D < 6.4$ (in this connection see also King 1977b).

The studies cited above have been conducted with steady flows and are sufficient to illustrate the complexity of the problem. The time-dependent flows, in general, and the wave motion, in particular, about oblique cylinders present even more complex problems. The use of the independence principle or the assumption that shedding frequency is proportional to $\cos \phi$ and that the normal component of drag is proportional to $\cos^2 \phi$ may be a *gross simplification* of the behavior of flow in the near wake. Under these circumstances only experiments can lead to some understanding of the problem and to the evolution of approximate calculation methods.

In what follows, we will deal with waves about oblique cylinders. It will become evident that neither the method of decomposition of velocities and/ or forces nor the drag and inertia coefficients appropriate to each method of force decomposition are clear. Furthermore, there are not enough systematic experiments either with waves or with SOFs to guide the analysis. Some experimental data are presented in Sumer and Fredsøe (1997).

Assuming that Morison's equation may be applied to a cylindrical member oriented in a random manner with respect to the mud line, Eq. (5.9) may be written as

$$\begin{Bmatrix} F_x \\ F_y \\ F_z \end{Bmatrix} = \frac{1}{2} \rho C_d D |W_n| \begin{Bmatrix} u_{nx} \\ u_{ny} \\ u_{nz} \end{Bmatrix} + \frac{1}{4} \pi \rho C_m D^2 \begin{Bmatrix} \dot{u}_{nx} \\ \dot{u}_{ny} \\ \dot{u}_{nz} \end{Bmatrix} \qquad (4.2.13)$$

in which W_n is the velocity vector normal to the pipe and C_d and C_m are assumed to be known. The Reynolds number and the Keulegan–Carpenter numbers are defined using the normal velocity, pipe diameter, and the wave period. Thus, for consistency, the values of C_d and C_m recommended in Section 4.2.5 should be used in Eq. (4.2.13).

Wade and Dwyer (1976) examined four methods for calculating wave forces (base shear and overturning moment) acting on tubular members. Each method is within procedures generally accepted by the industry. Horizontal and vertical wave-induced water-particle kinematic vectors were used in each of the wave force methods on two deep-water platforms to compare the horizontal base shear and overturning moments. Sarpkaya *et al.* (1982) presented data for forces acting on yawed smooth and rough circular cylinders in SOF. Such comparisons, however valuable, are not sufficient to assess the validity of one method over the others since the base shear and overturning moment represent the sum of forces and moments over many members at various angles of inclination. More recently, Kozakiewicz *et al.* (1995) tested three angles of inclination and three values of clearances between the cylinder and the wall. Their results indicated *that the data are practically independent of the angle of inclination.* In summary, considerable additional work is required in order to acquire some understanding of the wave forces on oblique members and, hopefully, to establish uniformly accurate and acceptable design criteria. Until then the use of the method outlined in the *"Commentary on Wave Forces," described in the Section 2.3.1 of the 20th edition of the American Petroleum Institute design guide API-RP2A* (2000) should be used.

Obviously, one cannot carry out laboratory experiments for every possible situation that might possibly arise at a particular platform site. Thus the simplifications and gross assumptions have been and will continue to be the integral part of the offshore engineering. It is hoped that the assumptions are either on the safe side or mercifully incorrect without leading to loss of life. The designer should keep in mind the fact that oblique members seldom occur in isolation. The effect of the wake of one member on the others and the alterations in the velocity field may give rise to significant variations in the resultant forces from those predicted for idealized situations through the use of semi-empirical methods.

Interference effects

A body's resistance to flow is strongly affected by what surrounds it. When two bodies are in close proximity, not only the flow about the downstream body but also that about the upstream body may be influenced. For example, a simple splitter plate can cause dramatic changes in flow about the cylinder upstream of the plate. Such a situation may arise in any multimember structure. Examples include condenser and boiler tubes in heat transfer, a variety of columns in pressure suppression pools of nuclear reactors, risers and other tubular structures in offshore engineering, turbine and compressor blades in mechanical or aerospace engineering, and high-rise buildings and transmission lines in civil engineering. The quantification of the interference effects in terms of the pressure distribution, lift and drag forces, vortex shedding frequency, and the dynamic response of the members of the array in terms of the governing flow and structural parameters constitute the essence of the problem.

Contrary to common belief, considerable work has been done on flow interference particularly between circular cylinders in various arrangements and between cylinders and a rigid or free surface. Most of the investigations have been prompted by the need to solve problems of immediate practical interest. Obviously, there are infinite number of possible arrangements of two or more bodies or cylinders positioned at right or oblique angles to the approaching flow direction. Consequently, there is a danger, however undesirable, of ad hoc testing leading to a proliferation of undigested and uncorrelated data. Such data and observations provide phenomenological explanations, but the intrinsic nature of the flow patterns remains a mystery.

The subject of flow interference may be classified in many categories: separated and unseparated flows, steady or time-dependent flows, partial or mutual interference in all types of flows, etc. In the following, we will first discuss interference effects in separated steady flows. Then we will take up the case of unseparated potential flows. Finally, relatively few studies on interference in wavy flows will be presented.

A careful review of flow interference between two circular cylinders in various arrangements has been presented by Zdravkovich (1977) and Heideman and Sarpkaya (1985) where an extensive list of references may be found. Numerous studies for the tandem arrangement (one cylinder behind the other) have shown that the changes in drag, lift, and vortex shedding are not necessarily continuous. In fact, the occurrence of a fairly abrupt change in one or all flow characteristics at a critical spacing is one of the fundamental observations of flow interference in cylinder arrays.

It has been shown experimentally that there is strong interference between two cylinders in tandem for spacing ratios L/D smaller than about 3.5, where L is the distance between the centers of the two cylinders. At a spacing ratio of about 3.5 there is a sudden change of the flow pattern in the gap. The critical spacing appears to depend somewhat on the Reynolds number. It is below 3.5 at $Re = 5.8 \times 10^4$, equal to 3.5 for $Re = 8.3 \times 10^4$, and slightly larger than 3.5 at $Re = 1.1 \times 10^5$. There seems to be no data for Re larger than about 2×10^5. At critical spacing the discontinuous change of the flow patterns causes the following: a jump in drag coefficient of the upstream cylinder, the commencement of the vortex shedding, and a drop in the base pressure. For the downstream cylinder, the base and side pressure coefficients drop, vortex shedding frequency jumps, and the gap pressure and the drag coefficient increase suddenly. Drag coefficient data show (Zdravkovich, 1977) that the upstream cylinder takes the brunt of the burden (see Fig. 4.8). The main feature for all Reynolds numbers (smaller than about 2×10^5) is that, beyond the critical spacing, the downstream cylinder has no effect on the upstream one. In other words, there is mutual interference for spacings *less* than critical and partial interference for spacings *larger* than critical. The drag coefficient of the downstream cylinder shows a strong dependence on the Reynolds number. At high subcritical Reynolds numbers, the wake turbulence from the upstream cylinder induces a supercritical flow around the downstream cylinder and hence the drag remains small even at large spacing.

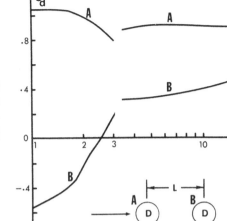

Figure 4.8. Drag coefficient for tandem cylinders at subcritical Reynolds numbers (a representative plot) (after Zdravkovich 1977).

Wardlaw and Cooper (1973) systematically measured drag forces on stranded cables in tandem. However, not much is known about the effect of roughness on interference and additional work is needed. The side-by-side as well as the staggered arrangement has again been investigated by a number of people (see Zdravkovich 1977; Sagatun *et al.* 2002; Coceal *et al.* 2007 for extensive references).

The most striking qualitative feature of such arrangements is the bistable nature of the flow. For spacing ratios from 1.1 to 2.2, the wakes of the two cylinders interfere and are alternately entrained by each other. This gives rise to changes in the base pressures of both cylinders from one steady value to another. The additional feature of the biased flow is that the gap flow biased to one side produces a resultant force on the cylinder, which is deflected relative to the free-stream direction. Consequently, there is a component of the force acting perpendicular to the free-stream direction, which may be called a lift force.

The results for the staggered arrangement show that the upstream and downstream cylinders may be subjected to significantly different lift and drag forces. Depending on their relative positions, the cylinders may experience negligible or strong lift and reduced or enhanced drag force.

There is very little data on wake interaction (Fuji and Gomi 1976) and vortex shedding frequencies (Fitz-Hugh 1973; Norberg 2003) and on lift and drag forces in the range of supercritical Reynolds numbers. Quadflieg (1977) reported some data for the side-by-side arrangement at $Re = 1.51 \times 10^5$. He also used the potential flow theory and the vortex street model to predict the base pressure. A somewhat similar potential-flow analysis, based on

Rosenhead's (1929) analysis of the wake of a cylinder in a channel, has been presented by Weihs (1977) for the side-by-side arrangement. Weihs found that there is a range of flow parameters where the drag force on each cylinder in the row is less than that of a single cylinder in unbounded flow. He noted that this might serve as a drag reduction criterion for barriers, etc. Experiments of Biermann and Herrnstein (1933) have shown that the interference drag coefficients (difference between the drag coefficient measured on one of the cylinders and the drag coefficient of the single cylinder at the same Reynolds number) may be negative for spacing ratios from 1.2 to 2. However, unlike the underlying assumptions of Weihs' analysis, the flow in this region is bistable and may give rise to severe vibrations. Thus, it turns out that such a critical arrangement must be avoided.

An experimental investigation of the effects of spacing, orientation, and Reynolds number on the drag of each cylinder in a group of two and three cylinders was reported by Dalton and Szabo (1976). They have found that the middle and downstream cylinder drag coefficients are smaller and noticeably more dependent on orientation than is the upstream cylinder. The drag coefficient for the upstream cylinder remained nearly constant. In other words, there is a strong mutual interference between the middle and downstream cylinder but only a partial interaction between the upstream and downstream cylinders.

The unsteady wake behind a group of three parallel cylinders has also been investigated by Mujumdar and Douglas (1970). The steady flow about multitube arrangements has been extensively studied in connection with heat exchangers (for a detailed discussion and some photographs see Knudsen and Katz 1958). The purpose of these earlier investigations was to increase the heat transfer and reduce the pressure drop or the energy loss. In general no information has been obtained on the effect of flow interference or shielding on the forces acting on the tubes and on the undesirable hydroelastic oscillations of the tubes. For extensive references and a detailed discussion of the effect of tube spacing in heat exchangers, the reader is referred to Chen (1977).

Hoerner (1965) suggested that the interaction effects in steady subcritical flow are negligible for the side-by-side arrangement of the cylinders if they are more than three diameters apart. For the tandem arrangement, the same effect appeared to be negligible if the cylinders were more than 4 diameters apart. Chen (1972) reported average lift coefficients for a number of arrangements and Reynolds numbers. He found that if the cylinders are 3 diameters apart, there is practically no lift force since the wake does not develop sufficiently. For cylinders at relatively smaller spacing, Chen found that the lift on a cylinder in the wake of another can be very large and does not tend to its single-cylinder value until the space between the two cylinders is increased to about 20 diameters. The above is an example of partial interference since the lift on the first or the front cylinder remains practically constant.

Arita *et al.* (1973) considered a six-pile arrangement and carried out extensive measurements in a calm sea. The outstanding features of their data may be summarized as follows: For the said arrangement, the interference effects on the drag coefficient strongly depend on the Reynolds number. At subcritical Reynolds numbers, the middle cylinders experience the least resistance. At supercritical Reynolds numbers, the drag coefficient of the cylinders in the front row is reduced somewhat. All other cylinders experience the same low drag coefficient, which is considerably lower than that of the front row. It should be emphasized that the foregoing is only for a particular combination of six cylinders. Generalizations to other combinations should be done with extreme care. Furthermore, all of the works cited above dealt with steady flows, mostly in the subcritical regime.

For unseparated two-dimensional flow about a group of cylinders the potential flow theory may be used to determine the inertia coefficient for each cylinder through the use of the method of images and complex variables (see, e.g., Robertson 1965). For additional details the reader is referred to Dalton and Helfinstine (1971) and Yamamoto (1976).

Dalton and Helfinstine (1971) gave an expression for the C_m values on circular cylinders, arbitrarily spaced in groups in a uniform flow. The C_m values were found to vary significantly with cylinder spacing and configuration. Two general rules emerge from their calculations. For a tandem arrangement (the direction of flow along the line joining the axes of the cylinders), the inertia coefficient decreases with decreasing spacing and attains its minimum value when the cylinders touch (the combined body becomes more or less streamlined). For a side-by-side arrangement (the flow normal to the line joining the axes of the cylinders), the inertia coefficient increases with decreasing spacing and attains its maximum value when the cylinders touch (the combined body becomes more or less like a vertical streamlined block). Similar conclusions hold true for two spheres (Sarpkaya 1960).

Spring and Monkmeyer (1974) developed a means of calculating the pressures and forces on a cluster of vertical circular cylinders through the use of the method first employed by MacCamy and Fuchs (1954). Their results are more general than those of Dalton and Helfinstine. Spring and Monkmeyer have shown that the force on a given cylinder is significantly affected by the presence of neighboring cylinders. The inertia coefficient may range from 1.19 to 3.38, departing significantly from the often-assumed value of 2.0. They have also shown that the force perpendicular to the direction of wave advance may be quite significant when the cylinders are close together, rising in one case to 69% of the force component in the direction of wave advance.

The analyses and the works cited above do not deal with the effects of separation and vortex shedding. Consequently, the results are more appropriate to the determination of earthquake forces on large bodies rather than to the evaluation of the inertial component of the force in the drag inertia dominated regime. In fact, there is every reason to believe that the inertial force acting on a cylinder in an array is as much affected by vortex shedding as the drag component of the force. Thus, it is recommended that special arrays, requiring a careful evaluation of the drag and inertial forces, be subjected to tests simulating the prototype conditions.

Gibson and Wang (1977) carried out two different experiments to determine the added mass of a series of tube bundles. The bundles consisted of tubes of uniform diameter d, arranged either in a square configuration or a circular configuration. In the first series of experiments, they towed the model of pile cluster under linear acceleration. In the second series, they vibrated the model at its own natural frequency. For both cases, they calculated the added mass through the use of the measured force and acceleration and plotted them as a function of the *"solidification ratio"* defined by $\sum d/\pi D$ where D is the pitch diameter of the bundle. In an impulsively started separated flow environment, the foregoing may yield *an ideal added-mass coefficient only at the initial instants of motion*. Otherwise, it is an exercise in inviscid flows with no theoretical or practical significance to the understanding of a complex viscous fluid motion.

Relatively few studies have been carried out with oscillating tube bundles (Tanida *et al.* 1973; Bushnell 1977; Sarpkaya 1979a; Sarpkaya *et al.* 1980) where the separation of flow is important. Tanida *et al.* (1973) reported an experimental study with two cylinders oscillating in still water. The results, however important for in-line oscillations, are of little relevance to the subject under consideration. Bushnell (1977) investigated the interference effects on the drag and transverse forces acting on a single member of two cylinder configurations through the use of a pulsating water tunnel. He did not evaluate the drag and inertia coefficients through the use of a suitable method, e.g., Fourier averaging. Instead, Bushnell picked out the maximum force values, which occurred in each half-cycle and averaged them over 10 consecutive values so as to obtain a mean maximum force for each flow direction. The drag and transverse force coefficients were obtained by normalizing the force by $\rho L D U_m^2$. The results have shown that the presence of neighboring cylinders significantly affects the forces on an individual cylinder of an array and the interference effect increases with increasing relative flow displacement. The maximum drag force on shielded cylinders was reduced relative to an exposed cylinder by up to 50%. Bushnell has suggested, on the basis of the foregoing, that a design using a high Reynolds number single-cylinder drag coefficient applied throughout the array would have an extra margin of safety against maximum drag loading due to interference effects. He found that the transverse force could be 3 to 4 times larger for interior array positions than that of a single cylinder. Consequently, the total force on each member of the array and the frequency and amplitude of the oscillation of this force become extremely important. In fact, such a cylinder array supported at regular intervals may exhibit very complex dynamic behavior. Some of the segments between supports may undergo in-line oscillations whereas the others may undergo violent transverse oscillations. Furthermore, the behavior of each tube in the bundle segment may be significantly different.

Sarpkaya (1979a) determined the drag and inertia coefficients for various multiple-tube riser configurations. Each configuration consisted of a number of outer pipes of diameter D_o (uniformly spaced on a circle of diameter D_p) and one central pipe of diameter D_c. The arrays have been subjected to SOF in a U-tunnel. The analysis of the in-line force was based on Morison's equation, written as

$$2F/(\rho D_e L U_m^2) = -C_d(D_a/D_e)|\cos \omega t| \cos \omega t$$
$$+ [\pi^2/(U_m T/D_e)]C_m \sin \omega t \qquad (4.2.14)$$

The Fourier averages of C_d and C_m are given by

$$C_d = -(3/4)(D_e/D_a) \int_0^{2\pi} \left[(F_m \cos \omega t/\rho L D_e U_m^2) \right] d\omega t \qquad (4.2.15)$$

and

$$C_m = (2U_m T/\pi^3 D_e) \int_0^{2\pi} \left[(F_m \sin \omega t/\rho L D_e U_m^2) \right] d\omega t \qquad (4.2.16)$$

in which F_m represents the measured in-line force; T, the period of flow oscillation; and U_m the maximum velocity in a cycle ($U = -U_m \cos \omega t$).

The experimental results for one particular array are shown in Fig. 4.9 as a function of K. The drag coefficient decreases gradually with increasing K and reaches an almost constant value for K larger than about 90. The inertia coefficient increases with increasing K and reaches a terminal value of about 6. The average inertia coefficient defined by

$$C_m^* = \left(\sum C_{mi} D_i^2 \right) / \sum D_i^2$$

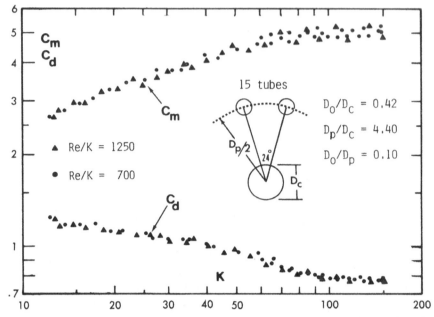

Figure 4.9. Drag and inertia coefficients for a particular tube bundle as a function of K (Sarpkaya 1979a).

was found to be 2.17, for the particular array shown in Fig. 4.9, through the use of the potential-flow theory. The comparison of this value with that obtained experimentally shows that as K approaches zero the experimental value of C_m approaches C_m^*. As K increases, $(C_m - C_m^*)$ increases, showing that some fluid mass is entrapped within the array and *that neither the potential-flow theory nor the diffraction analysis* can adequately describe the behavior of the complex separated flow through the bundle. The data for two different values of *Re/K* show that the force coefficients are independent of the Reynolds number within the range of *Re* and *K* values shown in Fig. 4.9.

The reason for the dependence of C_d and C_m on K is thought to be the dependence of the interaction of the wakes of the outer and inner pipes. The vortices in the wake of a given cylinder lose about 70% of their strength within 10 cylinder diameters. Thus, for small values of K, the vortices generated by a small tube at the center front of the bundle arrive at the central tube as weak vortices. Consequently, each tube behaves more or less as if it were independently subjected to a turbulent SOF. As K increases, not only the turbulence level but also the interaction between the wakes of the various cylinders increase. There is certain amplitude of oscillation beyond which neither the interaction of the wakes nor the increase of the

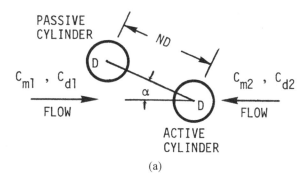

(a)

Figure 4.10. Definition sketch for the (a) drag and inertia coefficients for two cylinders in SOF (Sarpkaya *et al.* 1980), (b) lift coefficients of a shielded cylinder in SOF, and (c) lift coefficients for the front cylinder in SOF.

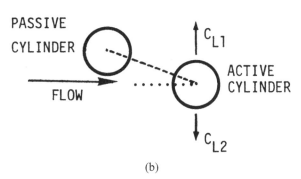

(b)

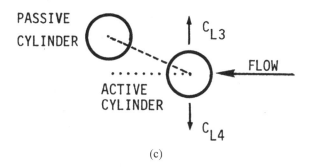

(c)

turbulence level affects the overall force acting on the bundle. A comparison of the total drag force acting on the bundle with the sum of the drag forces acting on each cylinder in isolation in SOF (at the corresponding K and Re values appropriate to each tube) shows that the former is about 10% smaller. For the hydrodynamic loading of risers see also Loken *et al.* (1979).

Sarpkaya *et al.* (1980) determined the lift, drag, and inertia coefficients for a pair of cylinders subjected to SOF. The line joining the centers of the cylinders was rotated at suitable steps relative to the flow direction. The spacing between the cylinder centers was varied from 1.5 diameters to 3.5 diameters (see Fig. 4.10).

The results have been presented in terms of two drag, two inertia, and four lift coefficients because of the fact that the force acting on a given cylinder is not independent of the direction of the flow. The results have shown that the drag and inertia coefficients for the tandem arrangement depend on both K and the relative spacing of the two cylinders. As the amplitude of flow oscillation becomes comparable to or smaller than the gap between the two cylinders, the drag and inertia coefficients gradually approach those corresponding to an isolated cylinder. The drag and inertia coefficients for the side-by-side arrangement exhibit a similar behavior. For a relative spacing larger than about 2.5 diameters, the cylinders behave as if they were independent.

The quantification of the lift force for an arbitrary arrangement requires four different lift coefficients, depending on the direction of the flow and the direction of the force. The results have shown that the lift force acting on the active cylinder may triple at angles of inclination of about 30 degrees. This result is of great practical significance since the cylinders are subjected to rapidly varying unsteady hydrodynamic loads, both in magnitude and direction, especially for smaller spacings.

It follows from the foregoing that the question of flow interference in closely spaced tube bundles is a complex one and only reliable experiments can guide the designer. For cylinder arrays simulating platform legs, etc., three facts must be taken into consideration. Firstly, the length of the ideal, finite, vortex street on either side of the cylinder cannot exceed about half the wave height or the amplitude of flow oscillations. Thus, one may assume that the interference may be negligible if the spacing between the piles is greater than about half the wave height. Secondly, in a wake comprising turbulent vortices, the strength of the vortices decays rapidly and in a distance of about 10 diameters the vortex strength decreases to about 10% of that generated in the boundary layer over a shedding period. Thirdly, the velocity at a point is not constant and varies from zero to U, during a half-cycle. Thus, the use of half-wave height as a criterion in determining the interference-free spacing may be too conservative. For example, for a 100-ft wave passing by a 2-ft pile, the $H/2$ criterion would give a distance of 50 ft, whereas the vortices must have almost completely dissipated within a distance of 20 to 30 ft. Thus, it is suggested that one should compare $H/2$ and $10–15D$ distances and use the smaller of the two as the appropriate distance in neglecting the interference effects. In rare cases, however, the flow about the upstream cylinder may be in the range of high subcritical Reynolds numbers. Then the wake turbulence from the upstream cylinder induces a super-critical or post-supercritical flow around the downstream one and hence the drag remains small even at spacings as large as $H/2$. Thus, the consequences of this type of flow interference are identical to those of an ambient turbulent flow about a single body at sufficiently high subcritical Reynolds numbers.

4.2.7 *Pipe lines and wall-proximity effects*

Pipelines are used for several purposes: to convey oil and gas for power generation, seawater for desalination, and sewage for disposal at sea, and for the protection of communication cables (see, e.g., American Science and Engineering Co. 1975; Olunloyo *et al.* 2007).

There are numerous problems associated with the design and installation of pipelines in deep-water offshore. The subject is too broad to cover entirely, so the purpose of this section is to deal with the prediction of hydrodynamic forces. A comprehensive review of wave forces on submerged pipelines with nearly 100 references has been presented by Davis and Ciani (1976), Fredsøe and Hansen (1987), Jensen and Mogensen (1982), Kozakiewicz *et al.* (1995), and Stansby and Star (1992).

The proximity of a cylinder near a plane boundary gives rise to various hydrodynamic and environmental problems. A boundary layer is established near a bottom plane within which the horizontal water particle velocity varies from zero at the wall to the free-stream velocity at some elevation above the wall. The boundary layer has an important effect on cylinder lift and drag. However, in tests made near a bottom boundary, no one has measured the boundary-layer thickness and related it directly to the measured cylinder forces.

Secondly, the flow asymmetry created when a cylinder is placed near a plane generates a lift force normal to the flow that is velocity dependent and that can be of considerable magnitude for small gaps. This lift force that acts in a downward direction (in an unseparated flow) is, in general, different from that due to separation and vortex shedding. However, the two types of lift forces cannot be separated experimentally. Furthermore, there seems to be no need to do so. When the cylinder touches the boundary, a net force exists away from the wall. Thus, a cylinder that is not restrained on a bottom or freed by scour may become unstable, i.e., it can be alternately raised and lowered by the lift force due to flow asymmetry and vortex shedding about the cylinder.

In unsteady oscillatory flow, such as that induced by waves, considerable complexity is added to the flow phenomena around horizontal cylinders. The force coefficients depend not only on the Reynolds number and the relative gap between the cylinder and the plane boundary but also on the Keulegan–Carpenter number and the time-dependent laminar or turbulent boundary-layer characteristics. Additional complexities arise from the variation in the amplitude and frequency of the waves, orientation of the pipe and of the waves and/or currents, temperature and marine life at the pipe location that determine the kind and type of soft and rigid excrescences on the pipe, the scour and deposition of sediment around the pipe, the hydroelastic oscillations of the pipeline, and the effect of wall porosity. Experimental studies, which examine these effects in detail, are very limited. In fact, the existing data in steady and oscillatory flows about such pipes consist of horizontal and vertical force measurements, which, with an applicable wave theory, are used to calculate the force coefficients. The added mass of submerged perforated pipes are reported in the literature by Muto *et al.* (1979), Sinha and Moorthy (1999), and Molin (2001). However, there is considerable disagreement as to what are the appropriate force coefficients for use in design primarily because of the multitude of variables affecting the flow.

In summary of the foregoing, no analysis exists for accurately describing the separated wave flow about cylinders with or without bottom effects. Furthermore, measurements of water particle kinematics; vortex shedding; drag, inertia, and lift forces; and the bottom boundary layer are rare in the range of parameters of practical importance.

The special case of unseparated flow about a single or N-number of cylinders situated near a plane boundary has been studied through the use of the potential theory by several investigators (see, e.g., Yamamoto *et al.* 1974, and Yamamoto and Nath 1976). The results show that C_m reaches its maximum theoretical value of $\pi^2/3 = 3.29$ for $e/D = 0$, where e is the gap between the cylinder and the wall (see also Kennered 1967). Secondly, both the lift and inertia coefficients nearly reach their ideal values in an infinite fluid ($C_m = 2$ and $C_L = 0$) for $e/D \cong 1$. In other words, for $e/D \approx 1$, the wall-proximity effect on unseparated flow about a cylinder is negligible. Thirdly, there is a significant difference between the case of a cylinder touching the boundary and that of a cylinder slightly away from the boundary. When the cylinder touches the boundary a net force exists away from the wall. This is because the maximum velocity and the minimum pressure occur on the cylinder at a point farthest from the plane. However, if even a very small gap exists between the cylinder and the wall then a large net force exists toward the wall. This is merely a consequence of the inability of the potential theory to deal with the consequences of separation. In an ideal case, the smaller the gap, the larger the velocity through it. Thus, the net force is directed toward the wall. It is worth noting that the case of $e = 0$ cannot be obtained from the case of $e \neq 0$ by letting $e \to 0$. For $e = 0$, $C_L = \pi(\pi^2 + 3)/9 = 4.49$ (von Muller 1929). For $e \to 0$, $C_L = -\infty$. In both cases, $C_m = -\pi^2/3$.

The foregoing results are based on the potential-flow theory and as such they are more appropriate to the determination of fluid loading due to vibrations generated by earthquakes. In real flows, the wall-proximity and separation effects can play significant roles in both the in-line and transverse forces.

Few studies have been carried out with steady flows over cylinders in the vicinity of a wall. Beattie *et al.* (1971) measured the pressure distribution over smooth and rough cylinders in the range $8 \times 10^4 < Re < 2 \times 10^6$. They used the fluid velocity at the top of the cylinder as the reference velocity in calculating the force coefficients. Their results showed as much as 35% scatter.

Wilson and Caldwell (1971) conducted experiments with two parallel cylinders at Reynolds numbers below 8×10^4. They obtained $C_d = 1.1$ at $Re = 5.7 \times 10^4$ and $C_d = 1.6$ at $Re = 3.3 \times 10^4$.

A series of careful experiments were conducted by Goktun (1975) in a wind and also water tunnel in the range of $9 \times 10^4 < Re < 25 \times 10^4$ for relative gaps of $e/D = 0.1, 0.125, 0.25, 0.5, 1.0, 1.5, 2$, and 2.66. Goktun measured the surface pressures and calculated the lift and drag coefficients. The variation of the drag coefficient with the gap size exhibited an interesting and unexpected trend. *The drag was a minimum when the cylinder was resting on the boundary and was a maximum for $e/D = 0.5$. Furthermore, the drag coefficient showed rapid variations as e/D was increased from 0.125 to 0.25.*

Bearman and Zdravkovich (1978) investigated experimentally the flow round a circular cylinder placed at various heights above a plane boundary. Distributions of mean pressure around the cylinder and along the plate were measured at a Reynolds number, based on cylinder diameter, of 4.5×10^4. They have concluded that the only influence of the plate on vortex shedding is to make it a more highly tuned process as the gap is reduced down to about $e/D = 0.3$.

The foregoing investigations are not directly applicable to the determination of wave forces on pipelines. Numerous studies have been conducted with waves or oscillatory flows on the determination of force coefficients for cylinders near a plane boundary (Johansson and Reinius 1963; Johnson 1970; Wilson and Caldwell 1971; Beattie *et al.* 1971; Priest 1971; Al-Kazily 1972; Grace 1973; Littlejohns 1974; A.S.E.C. 1975; Davis and Ciani 1976; Sarpkaya 1976a; Bowie 1977; Sarpkaya 1977; Grace *et al.* 1979; Graham 1979; Layton and Scott 1979; Wright and Yamamoto 1979; Graham and Machemehl 1980; Sarpkaya and Rajabi 1980).

Grace and Nicinski (1976) used a 17.5-ft long, 16-inch-diameter steel pipe, supported 3 inches above a rigid block placed on the ocean floor at a depth 37 ft below the mean level. They have determined the lift, drag, and inertia coefficients through the use of the force and wave records. The data exhibited considerable scatter primarily due to the difficulty of quantifying the complex wave conditions. *This is and will always be one of the predicaments of marine hydrodynamics.*

Sarpkaya (1977d) measured the in-line and transverse forces on cylinders placed at various distances from the bottom of the U-shaped water tunnel. The drag and inertia coefficients for the in-line force have been calculated through the use of the Fourier analysis and the method of least squares. The lift coefficient has been expressed in the usual manner by normalizing the amplitude of the first harmonic of the lift force by $0.5\rho LDU_m^2$.

Sarpkaya has shown that the drag, inertia, ad lift coefficients are functions of K, e, Re, e/D, and the depth of penetration of the viscous wave or the boundary layer thickness, i.e.,

$$[C_d, C_m, C_L] = f_i (K, \ Re, \ e/D, \ \delta/D) \tag{4.2.17}$$

In these experiments δ/D was approximately 0.073 and the boundary-layer effects have been ignored for $e/D > 0.1$. Flow visualization experiments with dye have shown that for very small values of e/D, ($e/D \rightarrow 0.1$), a jet-like flow exists between the cylinder and the plate. The flow separating from the top of the cylinder contains high-frequency oscillations but does not curl up into vortices immediately behind the cylinder. In fact, the immediate wake is essentially free from large vortices. These observations have shown that a plate does not have to be placed near the rear stagnation point of a cylinder, in the form of a splitter plate, in order to interfere with the vortex-shedding process. It is apparent that the gap blocks the flow and gives rise to earlier separation over the top of the cylinder. This in turn increases the transverse force as well as the in-line force. No attempt is made here to simplify an extremely complex separated flow situation. The emerging experimental facts are that the wall proximity ceases to affect the flow and the force-transfer coefficients for e/D larger than about 0.5 and that for smaller values of e/D the frequency of oscillations in the two shear layers are decoupled. The interruption of the regular vortex shedding, earlier separation, and the effect of the vortices shed in the previous cycles lead, as a whole, to larger force-transfer coefficients for decreasing values of e/D. The increase of the inertia coefficient in separated flow under consideration follows the same trend as in inviscid, steady, unseparated flow as far as the effect of wall proximity is concerned. For large values of K, the drag component of the in-line force dominates and the increase of the inertia coefficient is not of practical importance since the total inertial force is relatively insignificant.

Figures 4.11 through 4.14 show the drag and inertia coefficients for two representative values of K, namely, $K = 40$ and $K = 100$, for $e/D = 0.1, 0.2,$

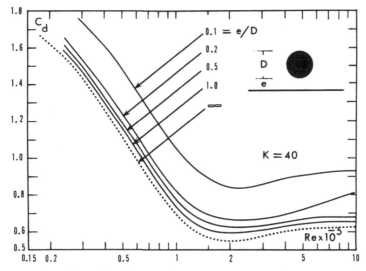

Figure 4.11. The effect of wall proximity on the drag coefficient, $K = 40$ (Sarpkaya 1977d).

0.5, 1, and the free cylinder. Evidently, the effect of the wall proximity is to increase both the drag and inertia coefficients for e/D values smaller than about 0.5. For larger values of e/D, the effect of wall proximity is practically negligible as evidenced by the comparison of the force coefficients with those obtained for cylinders at larger wall distances.

The transverse force in a given cycle is composed of two parts. One part is toward the wall (expressed in terms of the lift coefficient C_{LT}) and the other part is away from the wall (expressed in terms of a lift coefficient C_{LA})

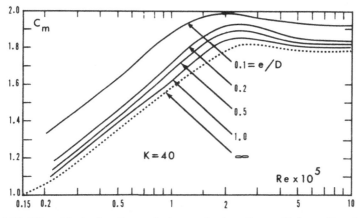

Figure 4.12. The effect of wall proximity on the inertia coefficient, $K = 40$ (Sarpkaya 1977d).

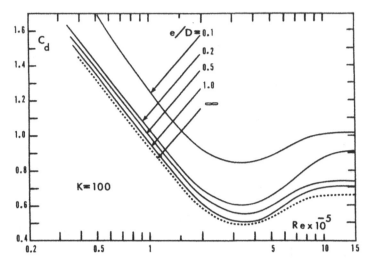

Figure 4.13. The effect of wall proximity on the drag coefficient, $K = 100$ (Sarp-kaya 1977d).

(see Figs. 4.15 and 4.16). The former occurs during the periods of flow where the velocity and hence the separation effects are relatively small. Evidently, had there been no separation, the lift force would have been always toward the wall. Nevertheless, the separation effects and the effect of the vortices shed in the remainder of the cycle are not entirely eliminated even during the periods of low velocity. The transverse force toward the wall is relatively small and fairly independent of e/D.

The transverse force away from the wall reaches its maximum during the periods of large velocities and separation. The position of the separation point, high-frequency oscillations in the upper shear layer, and the subsequent formation of the vortices are strongly influenced by the relative spacing of the cylinder for e/D values smaller than about 0.5. Hence, C_{LA} varies

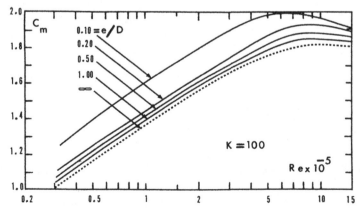

Figure 4.14. The effect of wall proximity on the inertia coefficient of a circular cylinder, $K = 100$ (Sarpkaya 1977d).

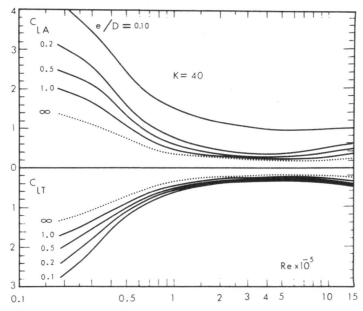

Figure 4.15. The effect of wall proximity on the lift coefficient of a circular cylinder, $K = 40$ (Sarpkaya 1977d).

considerably with e/D in the range of $e/D < 0.5$ and is significantly larger than the free-cylinder lift coefficients. It is obvious that the shear layer emanating from the lower side of the cylinder is not as free as that emanating from the topside of the cylinder because of the boundedness of the wall jet

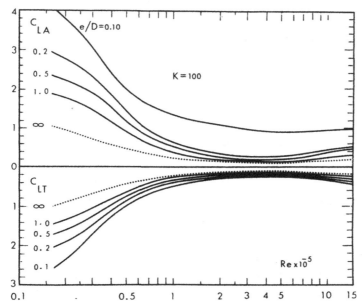

Figure 4.16. The effect of wall proximity on the lift coefficients of a circular cylinder, $K = 100$ (Sarpkaya 1977d).

between the cylinder and the plane wall. Thus, the lift force toward the wall is not as much affected by the variations in e/D. It is also evident that the transverse force toward the wall is about 90 degrees out of phase with that away from the wall.

For cylinders near a plane wall ($e \neq 0$) it may be concluded that

1. The drag and inertia coefficients for the in-line force acting on the cylinder are increased by the presence of the wall. This increase is most evident in the range of e/D values smaller than about 0.5. Both coefficients depend on the Reynolds number, Keulegan–Carpenter number, and e/D. The effect of the boundary layer or the penetration depth of the viscous wave is small provided that the boundary layer remains laminar. For turbulent oscillatory boundary layers the characteristics of the wall jet and separation over the cylinder may be significantly affected.

2. The proximity of the wall helps to decouple the frequency of oscillations in the top and bottom shear layers. This decoupling effect prevents the occurrence of regular vortex shedding for small values of e/D.

3. The transverse force toward the wall is relatively small and fairly independent of e/D. It occurs during the periods of low velocity or high acceleration. The transverse force away from the wall is quite large and dependent on e/D, particularly for e/D smaller than about 0.5. The two forces are about 90 degrees out of phase.

4. The use of Morison's equation to decompose the in-line force into two components is quite sound. The lumping of the entire in-line force into a single coefficient is not justified and obscures the mechanics of the flow.

As noted earlier, the case of $e = 0$ (no gap) differs significantly from that with a gap as small as $e/D = 0.1$. Sarpkaya and Rajabi (1980) conducted a series of experiments with smooth and rough cylinders and determined the drag, inertia, and the lift coefficients. The small gap between the cylinder and the plane boundary (bottom of the U-shaped tunnel) was sealed with a very thin plastic wrapping sheet, attached both to the plane boundary and the bottom of the cylinder. Furthermore, the gaps between the tunnel walls and the cylinder ends were sealed with foamy material. The drag and inertia coefficients for smooth cylinders are shown in Figs. 4.17 and 4.18 for two values of the frequency parameter.

Clearly, C_d can reach very high values relative to the case of e/D and is a function of the Reynolds number for a given K. The inertia coefficient does not appear to depend on Re and increases with increasing K. For very small values of K where the separation effects are negligible, C_m approaches

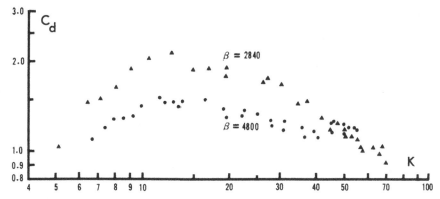

Figure 4.17. Drag coefficient for a bottom-mounted smooth cylinder (Sarpkaya and Rajabi 1980).

its theoretical potential flow value of 3.29. No generalizations can be made regarding the relative theoretical and experimental values of C_m. For a cylinder sufficiently away from a boundary, C_m is always smaller than its theoretical value of 2. In the present case ($e/D = 0$), the theoretical value of C_m is smaller than the experimental value, at least within the range of Reynolds numbers encountered.

The drag and inertia coefficients for the rough cylinders ($k/D = 1/100$) are shown in Figs. 4.19 and 4.20. The effect of roughness on C_d is quite significant. Once again, the inertia coefficient is very little affected by the Reynolds number or by roughness.

The maximum and minimum values of the lift coefficient (the lift force is always away from the wall) are shown in Figs. 4.21 and 4.22 for the smooth cylinders. The lift coefficient reaches very high values at relatively small K values. The potential flow value of C_L for $e/D = 0$ is given by von Muller (1929) as $C_L = \pi(\pi^2 + 3)/9 = 4.493$. Clearly, the experimental values for the smooth cylinders are relatively larger than the theoretical value, at about $K = 7$. The effect of separation at this value of K is such as to increase the lift. It is expected that the lift coefficient will reduce to about 4.5 as K

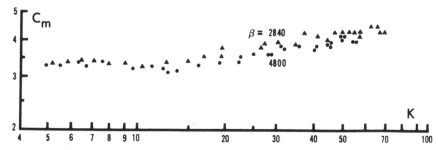

Figure 4.18. Inertia coefficient for a bottom-mounted smooth cylinder (Sarpkaya and Rajabi 1980).

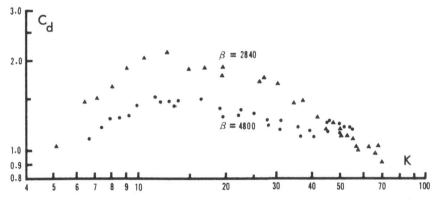

Figure 4.19. Drag coefficient for a bottom-mounted rough cylinder in SOF (Sarpkaya and Rajabi 1980).

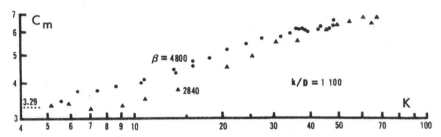

Figure 4.20. Inertia coefficient for a bottom-mounted rough cylinder in SOF (Sarpkaya and Rajabi 1980).

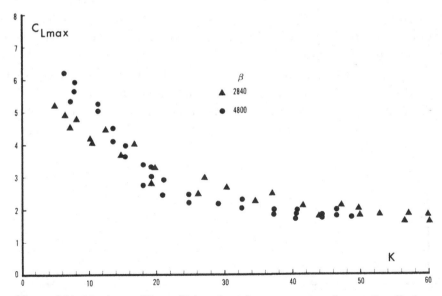

Figure 4.21. Maximum lift coefficient for a bottom-mounted smooth cylinder.

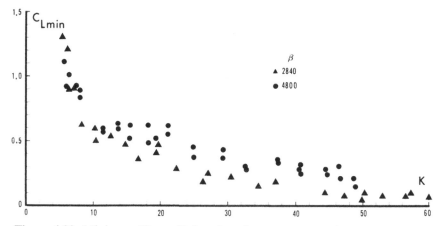

Figure 4.22. Minimum lift coefficient for a bottom-mounted smooth cylinder.

approaches zero. The maximum and minimum values of C_L for rough cylinders are shown in Figs. 4.23 and 4.24. The maximum value of C_L at about $K = 7$ is slightly smaller than that corresponding to the smooth cylinder. Otherwise, there is very little difference between the lift coefficients for the smooth and rough cylinders. The ratio of the lift-force frequency to the flow-oscillation frequency was found to remain constant at a value of 2 for both the smooth and rough cylinders. This shows that it is the separation of flow over the cylinder at each half cycle of flow and not the subsequent shedding of vortices that determine the fluctuations of the lift force for the type of flow–cylinder combination considered herein.

It appears that separation over a bottom-mounted cylinder occurs at smaller K values than that for a cylinder away from the wall. In general the effects of separation even at small K values are quite profound and the potential-flow values of the inertia coefficient and the lift coefficient tend to

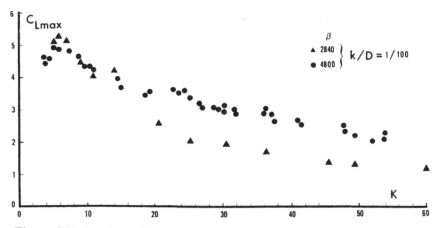

Figure 4.23. Maximum lift coefficient for a bottom-mounted rough cylinder.

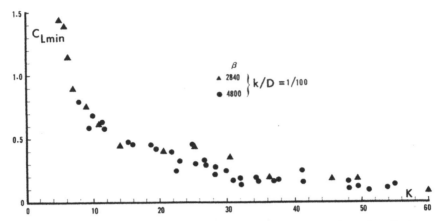

Figure 4.24. Minimum lift coefficient for a bottom-mounted rough cylinder.

underestimate the forces acting on a bottom-mounted cylinder. Evidently, the potential-flow theory gives no clues about the drag force and the calculations must be based on the results obtained experimentally.

4.2.8 *Wave impact loads*

Theory

Information about the forces acting on bluff bodies subjected to wave slamming is of significant importance in ocean engineering and naval architecture. The design of structures that must survive in a wave environment depends on knowledge of the forces that occur at impact, as well as on the dynamic response of the system. Two typical examples include the structural members of offshore drilling platforms at the splash zone and the often encountered slamming of ships.

The general problem of hydrodynamic impact has been studied extensively, motivated in part by its importance in ordnance and missile technology. Extensive mathematical models have been developed for cases of simple geometry, such as spheres and wedges (Szebehely 1959). These models have been well supported by experiment. Unfortunately, the special case of wave impact has not been studied extensively.

Kaplan and Silbert (1976) developed a solution for the forces acting on a cylinder from the instant of impact to full immersion. Miller (1977) presented the results of a series of wave-tank experiments to establish the magnitude of the wave-force slamming coefficient for a horizontal circular cylinder. He found an average slamming coefficient of $C_s = 3.6$ where $C_s = 2F/(\rho D L U_m^2)$.

Faltinsen *et al.* (1977) investigated the load acting on rigid horizontal circular cylinders (with end plates and length-to-diameter ratios of about 1) that were forced with constant velocity through an initially calm free surface. They found that the slamming coefficient ranged from 4.1 to 6.4. They also conducted experiments with flexible horizontal cylinders and found that the analytically predicted values were always lower (50 to 90%) than those found experimentally. Zhou *et al.* (1991) made measurements of the pressure distributions on surface-piercing vertical cylinders due to breaking waves and found that the highest impact pressures are subject to considerable variability. The high-impact region is found to be localized in space and time, and the variability is attributed to the random dynamics of the breaking wave front and the entrapped air. In other words, the prediction of the largest pressures is essentially a stochastic problem. Sun and Faltinsen (2006) used the boundary-element method (BEM) to simulate the water impact on horizontal cylinders and cylindrical shells. The calculated structural responses are compared with the experimental results and reasonable agreement is achieved by avoiding some numerical difficulties (see also Zhu *et al.* 2007 on water entry and exit of a horizontal circular cylinder).

Experiments

Sarpkaya (1978c) conducted slamming experiments with SOF impacting a horizontal cylinder and found that (a) the dynamic response of the system is as important as the impact force (i.e., one cannot be determined without accounting for the other); (b) the initial value of the slamming coefficient is essentially equal to its theoretical value of π; (c) the system response may be amplified or attenuated, depending on its dynamic characteristics; (d) the buoyancy-corrected normalized force in the drag-dominated region reaches a maximum at a relative fluid displacement of about 1.75; and (e) roughness increases the rise time of the force and tends to decrease the amplification factor.

The general case of hydrodynamic impact usually is described by using incompressible potential-flow theory. The compressibility of water and air and the cushioning effect of air (air boundary layer, depression of the water surface just before impact, etc.) are ignored. For a moving body with mass M and velocity v_0 impacting a quiescent surface, the system momentum is Mv_0. Neglecting nonconservative forces, the momentum of the system is unchanged during penetration. However, the mass of the system increases because of the fluid set in motion near the body. Also known as added mass, m results in reducing the velocity. Thus, the system momentum after penetration is $(M + m)v = Mv_0$. The impact force at any instant is a function of

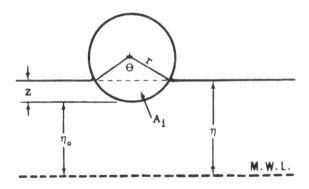

Figure 4.25. Definition sketch for wave slamming.

m and $m\ \dot{m}$. Therefore, the solution requires knowledge of the added mass and its time derivative. The determination of the added mass is not a simple matter and the results depend on the assumptions made (Moran 1965). The primary source of difficulty is the mathematical singularities encountered at the spray root (Chou 1946; Fabula 1957; Karman and Wattendorf 1929; Schnitzer and Hathaway 1953). Spray formation, air cushioning, and the flexibility of the impacting body (dynamic response) are the major sources of error in experiments.

The following analysis is based on the added mass calculated by Taylor (1930), which ignores the spray root problem. Kaplan and Silbert (1976) have shown that the force acting on a horizontal cylinder by a wave system that propagates normal to it is equal to the sum of the buoyant force and the time rate of change of momentum. Thus, one has

$$F/L = \rho g A_i + (m + \rho A_i)\ddot{\eta} + \frac{\partial m}{\partial z}\dot{\eta}^2 \qquad (4.2.18)$$

in which F represents the force acting on the cylinder; L, the length of the cylinder; ρ, the density of fluid; g, gravitational acceleration; A_i, the immersed area; m, the added mass per unit length; η, the instantaneous height of the wave surface above the mean water level; and z, the instantaneous depth of immersion (see Fig. 4.25). The first and second derivatives of η with respect to time are denoted by $\dot{\eta}+$ and $\ddot{\eta}$. The added mass is given by Taylor (1930) as

$$m = 0.5\rho r^2 \left[\frac{2\pi^3}{3} \frac{(1 - \cos\theta)}{(2\pi - \theta)^2} + \frac{\pi}{3}(1 - \cos\theta) + (\sin\theta - \theta) \right]$$

in which r represents the radius of the cylinder, and θ is defined as shown in Fig. 4.25.

The motion of the free surface is related to the maximum amplitude by

$$\eta = A \sin 2\pi t / T \qquad (4.2.19)$$

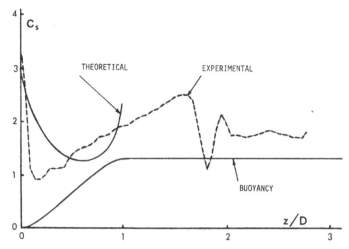

Figure 4.26. Comparison of theoretical and experimental slamming coefficients (Sarpkaya 1978c).

where A and T represent the amplitude and period of the free surface in a SOF. It is easy to show that at the instant of impact the slamming coefficient is given by

$$C_s = 2F / \left(\rho D L U_m^2 \right) = \pi \qquad (4.2.20)$$

Thus, (4.2.20) indicates that C_s, and consequently the impact force, is of an impulsive nature beginning with a finite value at the instant of impact. Since viscous forces are neglected, one would expect the solution to deviate from the actual situation as the cylinder becomes more fully immersed. Where this becomes the case can only be determined experimentally. A comparison of the experimental and theoretical results is shown in Fig. 4.26. Evidently, the predictions of the theoretical model are not valid beyond very small values of z/D. However, this is not of major concern since the largest impact force occurs at the instant of impact. It is not realistic to assume that the impact force rises from zero to π instantaneously. Several factors, specifically the *compressibility of air* between the cylinder and water surface, entrapped gases in the water, surface irregularities, and water droplets on the surface of the cylinder would account for some finite rise time. For inclined cylinders the rise time may be quite significant. Nonetheless, the rise time can be expected to be short, i.e., in the order of milliseconds. The exact nature of the rise is an interesting question for further study.

The realization that the impact force is of an impulsive nature requires consideration of the fact that this force does not act on a perfectly rigid body, but rather on a cylinder, which is supported elastically. The response of such a system approaches that of a rigid body only if its natural frequency approaches infinity. Additionally, the response of the system to an impulsive force is heavily dependent on the exact nature of the force itself as well as on the system's natural frequency. Sarpkaya (1978c) has shown that the slamming coefficient may lie between 0.571π and 1.7π, depending on the rise time and the natural frequency of the elastically mounted cylinder. The significance of the foregoing is that the values of C_s, determined experimentally from the measured reaction forces at the supports of a cylinder, may show wide scatter depending on the dynamic response of the cylinder and the rise time.

If the surface is not perfectly plane (which it never is), rise time may vary from experiment to experiment, resulting in an apparent nonrepeatability. This is particularly true for the conditions in the ocean environment. Evidently, controlled laboratory experiments help to establish the ideal value of the impact coefficient and to explain the reasons for the observed scatter in the data. The determination of the *impact-force magnification factor* for the cylinder in the ocean environment must necessarily consider the random nature of the disturbances at the wave surface, the orientation of the structural member relative to a given wave, currents, and 3D nature of the waves, surface roughness, spray, etc. There are no deterministic means to predict the reaction forces acting on the supports of a member due to wave impact even when the structural characteristics of the member (damping, natural frequency, etc.) and the ideal value of the impact force are known and understood. This is so because the rise time depends on all the nondeterministic conditions just cited. It is on the basis of the foregoing that the following design recommendations are made: (a) use $C_s^0 = 5.5$ *if a dynamic response analysis is not to be carried out*, and (b) use $C_s^0 = 3.2$ if the impact-force magnification is to be determined through a dynamic-response analysis.

Kaplan (1979) described some of the problems associated with the measurement of high-frequency impact forces on offshore structures and reported C_s values that ranged from 1.88 to 5.11, with just as many values above the theoretical value of π as below, and a mean value of $C_s = 2.98$. While this type of comparison is not quite precise, the general range of magnitudes may be considered to lie within the range of values reported by others, as noted by Kaplan. Finally, it should be noted that the velocity-square-dependent part of the drag may be added to the force expression given by (4.2.18), which in general is expected to vary with the degree of immersion. Clearly, the determination of the drag coefficient for such a complex, time-dependent free surface flow is not possible. In general, the contribution of the velocity-square-dependent drag is quite small and may be ignored in view of the rather uncertain nature of the determination of the rise time, structural and dynamic characteristics of the member, and the slamming coefficient. Miller (1980) suggested that for most structures the current practice of applying a constant slamming coefficient of 3.5 is conservative when estimating extreme stresses. For fatigue estimation this appears to be the case irrespective of the member geometry.

Breaking wave loads

Numerous experiments have been carried out on the prediction of the impact pressure induced by breaking waves on vertical cylinders in regular and random waves. Ochi and Tsai (1984) considered two relatively distinct conditions: waves approaching the structure after they broke and the waves breaking close to the structure. Our analysis of the Ochi and Tsai data has shown that the magnitude of the pressure for the two cases is nearly identical and is given by

$$p = 1.65\rho(gT/2\pi)^2 \tag{4.2.21}$$

A comprehensive study of the water impact on a cylinder is carried out by Lin and Shieh (1997) using a digital imaging system and a high-speed data-acquisition system to measure simultaneously the impact pressure and the flow-field velocity distribution.

5

Wave Forces on Large Bodies

5.1 Introduction

The interaction of a wave with a solid boundary leads to *reflection, refraction*, and *diffraction*. The water waves, like the light waves, may be reflected, refracted (turned or bent when they pass from one medium into another of different density) and diffracted (scattered in all directions when they impinge upon a barrier). The magnification of these phenomena depends on the shape and dimensions of the body (or bodies) impacted by the incident wave as well as on the motions of the barrier. As succinctly stated by Billingham and King (2000), the reflected/scattered/diffracted field is the difference between the unaffected incident wave and the actual solution.

The wave–body interaction is a complicated problem and cannot be solved without some simplifying assumptions. It is necessary to assume the fluid to be inviscid and incompressible, the flow to be unseparated, and the effects of surface tension, dissolved gases, cavitation, and vertical density and temperature gradients to be negligible. Then the flow field is represented by a scalar velocity potential, satisfying the Laplace equation within the fluid domain.

For a small body in ocean waves, D/L (*the diffraction parameter*) and the diffraction are negligibly small. For large bodies (with a typical volume ten times that of the great pyramid of Khufu), the relative fluid displacement UT/D and hence the viscous energy dissipation is small. Thus, the only source of energy dissipation is the propagation of the gravity waves, which undergo significant scattering or diffraction when the structure (*devoid of sharp corners*) spans more than about a fifth of the incident wavelength. This calls for a non-Morison type of analysis. In other words, the flow separation, coherence length, vortex shedding, wall roughness, wall turbulence, hydroelastic oscillations, and the Reynolds number become irrelevant or of negligible importance and the motion falls in the inertia-dominated regime. Then it can be shown through the use of the linear wave theory that K for a vertical circular cylinder of diameter D reduces to $K = \pi H/D \tanh(kd)$ at the still-water level and does not exceed about 2.2 (often much smaller). With very little or no separation (see Chapter 3), the drag force due to viscous effects becomes negligible and *the assumption of an inviscid, unseparated, unsteady flow becomes fully justified.*

Review of the numerical inviscid-flow methods in water-wave diffraction and radiation is presented by several investigators (Whitham 1974; Raman *et al.* 1975; Mei 1978; Garrison 1978; Isaacson 1978, 1979a; Meyer 1979; Lozano and Liu 1980; and Molin 2002), just to mention a few. The analysis of any one case requires lengthy manipulations of the transcendental functions, now taken care of by "off-the-shelf" computer codes.

The impetus for numerous contributions to diffraction of water waves on bodies came partly from the construction of large concrete structures (in the mid-70s) at medium depths and partly from the simplicity of the analyses ensured by the absence of separation and Reynolds number: *nonseparated inviscid flow*. There are, to be sure, a number of relatively more challenging large-body problems that deal with the combination of diffraction with refraction (Liu *et al.* 1979), waves with floating porous cylinders (Williams *et al.* 2000), nonlinear interactions between waves and floating bodies (Wu and Hu 2004), and diffraction with 2D and 3D *irregular waves* (Skourup 1994).

The force acting on a body (e.g., a vertical circular cylinder) defined by depth parameter D/L, wave steepness H/L, and the *diffraction* parameter D/L may be written as

$$F/(\rho g H D^2) = f(d/L, H/L, D/L) \tag{5.1}$$

which *for a given structure* located at a given depth (d/L), subjected to *infinite number* of identical waves (*no history effects, no dispersion, no randomness*) of relatively small steepness (H/L), as in waves described by the linear wave theory, the normalized force may be approximated by $F/(\rho g H D^2) = f(D/L)$. The simplifications are fully justified for a number of reasons: for large structures (increasing D), the diffraction effects increase and the flow becomes increasingly inertia dominated (smaller Keulegan–Carpenter numbers and smaller chances of separation, except at localized separation pockets at sharp corners). There are no sharp demarcation lines between the various degrees of dominance of diffraction/inertia/drag/wave-breaking: *Inertia-dominated regime* $(0.01 < H/D < 0.25, \pi D/L < 0.5$, negligible diffraction); *small-drag-, large-inertia-dominated regime* $(0.25 < H/D < 1.5, \pi D/L < 0.5)$, *Morison region* (or comparable drag and inertia region); ($H/D > 20$, large drag region), and the *diffraction region* ($\pi D/L <$ 0.5). Understandably, all of these approximately defined domains are bounded by regional facts, experience, breaking of deep-water waves, *memory–pressure effects* on the body (resulting from the evolution of propagating waves), and the reasons flowing from *needs, ages, and experience*. Clearly, diffraction simplifies the analysis because *it does not have to deal with the consequences of viscosity and separation.*

A typical gravity platform may have a base section with a diameter of the order of 300 ft and columns with diameters of the order of 60 ft. A general description of the challenges and innovations in offshore technology, including gravity platforms, is given by Clauss (2007) and an in-depth discussion of large structures in the context of North Sea operations is given by Bruun (1976).

As the extraction of offshore oil moved from near-shore to off-shore, and then to very large depths, the gravity structures fell out of favor for a number of reasons, the two most important ones being the *even* or *noneven subsidence* of the gravity-based platforms (installed in mid-70s in the North Sea) and the more frequent recurrence of *very high waves* toward the end of the 20th century (possibly beyond). Both of these have prompted massive programs to *elevating, removal,* or *replacement* of some of the platforms.

As noted by Terzaghi (1927) "Foundation problems are of such character that a strictly theoretical mathematical treatment will always be impossible. The only way to handle them efficiently consists of finding out, first, what has happened on preceding jobs of a similar character; next, the kind of soil on which the operations were performed; and finally, why the operations have led to certain results. By systematically accumulating such knowledge, the empirical data being well defined by the results of adequate soil investigations, foundation engineering could be developed into a semi-empirical science." This is true not only for the foundations or large gravity structures, but also for all other aspects of the offshore engineering. As noted by Lacasse (1999), "Geotechnical engineers need to communicate better with the related fields of geology, geophysics, structural engineering, and hydrodynamics."

The long-term foundation and structural effects led to the evolution of thinner and slimmer platforms and to the gradual abandonment of gravity structures. The designs are now based on newer concepts, slimmer structures, *virtual modeling,* and on an integrated understanding of the *structural design and fluid dynamics*. In view of the foregoing, we will present only the fundamentals of the diffraction analysis, keeping on mind the fact that it is unencumbered by the more challenging consequences of viscosity and separation, needing only a suitable computer program.

5.2 The case of linear diffraction

The analysis of the linear diffraction load on a surface-piercing vertical cylinder, with the Sommerfeld (1896) radiation condition, was pioneered by Havelock (1940), Omer and Hall (1949), and MacCamy and Fuchs (1954). This led to a large number of papers, evolution of novel numerical schemes (e.g. finite element, boundary integral, mixed Eulerian–Lagrangian, just to mention a few), and to numerical wave tanks and off-the-shelf computer codes: AQWA (WS Atkins plc., Surrey, UK), HOBEM (FCA International, Inc., Houston TX), NEPTUNE (Zentech, Inc., Houston TX), SESAM, and DNV (Hovik, Norway), among many others.

It is evident from what we have seen so far that the analysis requires the determination of a velocity potential, which satisfies the equation of Laplace, i.e., $\nabla^2 \phi = 0$ within the fluid domain, subject to appropriate boundary conditions representing the linearized kinematic and dynamic free-surface conditions, given by

$$\frac{\partial^2 \phi}{\partial t^2} + g \frac{\partial \phi}{\partial z} = 0 \quad \text{at} \quad z = 0 \tag{5.2}$$

$$\text{and} \quad \eta = -\frac{1}{g} \left(\frac{\partial \phi}{\partial t} \right)_{z=0} \tag{5.3}$$

The kinematic boundary conditions (no penetration) at the seabed and at the body surface are given, respectively, by

$$\frac{\partial \phi}{\partial n} = 0 \quad \text{at} \quad z = -d \tag{5.4}$$

$$\frac{\partial \phi}{\partial n} = 0 \quad \text{at the body surface.} \tag{5.5}$$

It must be emphasized that the foregoing is possible only because of the assumptions that the *fluid is incompressible and inviscid, the flow is unseparated, the wave amplitude (wave curvature) is small, and the structure is non-deformable.*

The analysis necessarily begins with the introduction of an "incident-wave" potential ϕ_w and a "scattered-wave" potential ϕ_s. The sum of the two potentials ($\phi = \phi_w + \phi_s$) must satisfy Laplace's equation within the fluid domain. At large distances, ϕ_s must necessarily represent the outgoing wave

(Sommerfeld 1949; Stoker 1957), i.e.,

$$\text{Limit } r^{1/2} \left[\frac{\partial \phi_s}{\partial r} - i k \phi_s \right] = 0 \quad \text{as} \quad r \to \infty \tag{5.6}$$

which is known as the Sommerfeld radiation condition.

The incident-wave potential (see 4.1.9) is specified in complex form as

$$\phi_w = -\frac{igH}{2\omega} \frac{\cosh[k(z+d)]}{\cosh(kd)} e^{i(kx-\omega t)} \tag{5.7}$$

where both ϕ_s and ϕ_w satisfy the boundary conditions as well as the radiation condition (5.6). The body-surface boundary condition, with n denoting distance in a direction normal to the body surface, may be expressed as

$$\frac{\partial \phi_s}{\partial n} = -\frac{\partial \phi_w}{\partial n} \tag{5.8}$$

Equations (5.2) through (5.8) establish the dependence of the incident- and scattered-wave potentials whose sum, as noted above, is the desired velocity potential ϕ. Then the pressure in the fluid domain is obtained using the *linearized* Bernoulli equation

$$p = -\rho g z - \rho \frac{\partial \phi}{\partial t} \tag{5.9}$$

Relatively simple computer programs enable one to find the pressure distribution and the components of the force, and moments. As noted earlier, there are a large number of commercially available codes to carry out the desired calculations. Often an attempt is made to compare the forces and moments calculated with the Morison equation. There is no justification for it for a number of reasons: (i) the diffraction forces depend on the order of the diffraction theory (it is not unusual to encounter large differences between the predictions of the first- and second-order theories); and (ii) the drag and inertia coefficients in Morison equation are unknown, particularly where a free surface involved. Thus, it is not a justifiable assumption that the diffraction forces cannot exceed those predicted through the use of the full Morison equation.

5.3 Froude–Krylov force

The force that would act on the body if it were *transparent* to the wave motion (i.e., as if the structure were not there to distort the wave field) is called the Froude–Krylov force. In certain cases it yields the leading-order

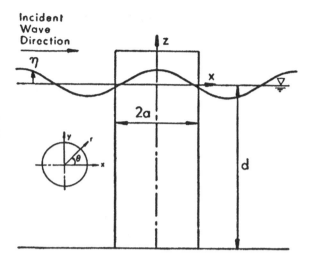

Figure 5.1. Wave diffraction around a circular cylinder.

contributions (Peters and Stoker 1957) in the vertical plane (surge, heave, and pitch) and for slender bodies provided that the wavelength is large compared to the beam. It was expedient in the *pre-computer era* as a gross approximation of the effects of diffraction about a rigid body. It requires additional ad hoc assumptions and approximations for the force coefficients and it can hardly be justified in the present era in view of the existing computer power.

5.4 The case of a circular cylinder

The classic example of diffraction due to a vertical circular cylinder extending from the seabed and piercing the free surface has been an obligatory part of practically every book on marine hydrodynamics. We will make no exception here but present the formulation and the results as concisely as possible. Its application to arbitrary shapes, multiple bodies, and numerous other body combinations are scattered throughout the literature of the last half of the past century.

The Laplace equation for the wave diffraction and the boundary conditions around a circular cylinder in cylindrical coordinates (see Fig. 5.1) reduce to

$$\frac{\partial^2 \phi}{\partial r^2} + \frac{1}{r}\frac{\partial \phi}{\partial \theta} + \frac{1}{r^2}\frac{\partial^2 \phi}{\partial \theta^2} + \frac{\partial^2 \phi}{\partial z^2} = 0 \qquad (5.10)$$

and

$$\frac{\partial \phi}{\partial r} = \frac{\partial \phi_w}{\partial r} + \frac{\partial \phi_s}{\partial r} = 0 \quad \text{at} \quad r = a \qquad (5.11)$$

where a is the cylinder radius. The incident-wave potential, introduced in (5.7), may be reduced to (e.g., Abramowitz and Stegun 1965)

$$\phi_w = -\frac{igH}{2\omega}\frac{\cosh[k(z+d)]}{\cosh(kd)}\left[\sum_{m=0}^{\infty} \beta_m J_m(kr)\cos(m\theta)\right]e^{-i\omega t} \tag{5.12}$$

where $\beta_m = 1$ for $m = 0$, $\beta_m = 2i^m$ for $m \geq 1$, and $J_m(kr)$ is the Bessel function of the first kind of order m and argument kr.

The scattered-wave potential, based on the Hankel function of the first kind with initially unknown complex coefficients, is given by

$$\phi_s = -\frac{igH}{2\omega}\frac{\cosh[k(z+d)]}{\cosh(kd)}\left[\sum_{m=0}^{\infty} \beta_m B_m H_m^{(1)}(kr)\cos(m\theta)\right]e^{-i\omega t} \tag{5.13}$$

which satisfies Laplace's equation and the bottom and free-surface boundary conditions. The asymptotic form of $H_m^{(1)}(kr)$, which readily satisfies the far-field condition, is given by

$$H_m^{(1)}(kr) \rightarrow \sqrt{\frac{2}{\pi kr}}\,\exp\left[i\left(kr - (2m+1)\frac{\pi}{4}\right)\right] \tag{5.14}$$

Inserting (5.13) and (5.14) into (5.11), one obtains

$$B_m = -J_m'(ka)/H_m^{(1)'}(ka) \tag{5.15}$$

where the primes denote differentiations with respect to the argument. Then the desired velocity potential (i.e., the complete solution) is given by

$$\phi = -\frac{igH}{2\omega}\frac{\cosh[k(z+d)]}{\cosh(kd)}$$
$$\times \left[\sum_{m=0}^{\infty} \beta_m \left(J_m(kr) - \frac{J_m'(ka)}{H_m^{(1)'}(ka)}H_m^{(1)}(kr)\right)\cos(m\theta)\right]e^{-i\omega t} \tag{5.16}$$

For $r = a$, part of the quantity in the above parentheses may be simplified to (Abramowitz and Stegun 1965)

$$J_m(ka) - \frac{J_m'(ka)}{H_m^{(1)'}(ka)}H_m^{(1)}(ka) = \frac{2i}{\pi ka\, H_m^{(1)'}(ka)} \tag{5.17}$$

The foregoing analysis shows the rather cumbersome but conceptually simple nature of the transcendental functions. The results of numerous solutions

Table 5.1. *Results of the diffraction theory for a vertical circular cylinder*

$$\frac{\phi}{gH/\omega} = -\frac{1}{2}\frac{\cosh(ks)}{\cosh(kd)}$$

$$\times \left\{ \sum_{m=0}^{\infty} i\beta_m \left[J_m(kr) - \frac{J'm(ka)}{H_m^{(1)'}(ka)} H_m^{(1)}(kr) \right] \cos(m\theta) \right\} e^{-i\omega t}$$

$$\left(\frac{\eta}{H}\right)_{r=a} = \left\{ \sum_{m=0}^{\infty} \frac{i\beta_m \cos(m\theta)}{\pi ka\, H_m^{(1)'}(ka)} \right\} e^{-i\omega t}$$

$$\left(\frac{p}{\rho gH}\right)_{r=a} = -\frac{z}{H} + \frac{\cosh(ks)}{\cosh(kd)} \sum_{m=0}^{\infty} \frac{i\beta_m \cos(m\theta)}{\pi ka\, H_m^{(1)'}(ka)} e^{-i\omega t}$$

$$\frac{\partial F/\partial s}{\rho gHa} = 2\frac{A(ka)}{ka}\frac{\cosh(ks)}{\cosh(kd)} \cos(\omega t - \delta)$$

$$\frac{F}{\rho gHad} = 2\frac{A(ka)}{ka}\frac{\tanh(kd)}{kd} \cos(\omega t - \delta)$$

$$\frac{M}{\rho gHad^2} = 2\frac{A(ka)}{ka}\left[\frac{kd\sinh(kd) + 1 - \cosh(kd)}{(kd)^2 \cosh(kd)} \right] \cos(\omega t - \delta)$$

$$\frac{F_k}{\rho gHad} = -\pi J_1(ka)\frac{\tanh(kd)}{kd} \sin(\omega t)$$

$$C_h = \frac{2A(ka)}{ka J_1(ka)}, \quad C_m = \frac{4A(ka)}{\pi (ka)^2}$$

where the real parts of complex expressions are understood, and

$$s = z + d,$$
$$\beta_0 = 1, \beta_m = 2i^m \text{ for } m \geq 1,$$
$$A(ka) = [J_1'^2(ka) + Y_1'^2(ka)]^{-1/2},$$
$$\delta = -tan^{-1}[Y_1'(ka)/J_1'(ka)].$$

have been tabulated in the literature. Here we will present the results only for a vertical circular cylinder (see Table 5.1).

The total force and the moment exerted on the cylinder are given by

$$\frac{F}{\rho gHad} = 2\frac{A(ka)}{ka}\frac{\tanh(kd)}{kd} \cos(\omega t - \delta) \tag{5.18}$$

$$\frac{M}{\rho gHad^2} = 2\frac{A(ka)}{(ka)}\left[\frac{kd\sinh(kd) + 1 - \cosh(kd)}{(kd)^2 \cosh(kd)} \right] \cos(\omega t - \delta) \tag{5.19}$$

in which $A = -igH/2\omega$ and δ is the phase angle by which the force lags behind the incident-wave crest passing through $x = 0$.

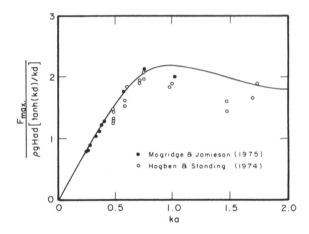

Figure 5.2. Comparison of the measured and calculated forces on a vertical cylinder. (Hogben and Standing 1974; Mogridge and Jamieson 1975).

Mogridge and Jamieson (1976) suggested the use of plots of force and moment maxima and the corresponding phase angle δ for computational convenience. Here we show only the maximum force and moment,

$$F_{\max} = \frac{\pi^2 \rho HLD^2}{4T^2} C_m(ka) \tag{5.20}$$

$$M_{\max} = \rho g HLD^2 C_m(ka) f(kd) \tag{5.21}$$

where $f(kd) = [kd \tanh(kd) + \operatorname{sech}(kd) - 1]/16$.

Numerous comparisons have been made between the measured (e.g., Hogben and Standing 1974; Mogridge and Jamieson 1975) and the predicted normalized force (MacCamy and Fuchs 1954) acting on a circular cylinder as shown in Fig. 5.2.

Comparisons of the experimental and numerical results for truncated, surface-piercing, circular, and square columns are shown in Fig. 5.3: (a) circular, $h/d = 0.7$, (b) square, $h/d = 0.7$, (c) circular, surface piercing, and (d) square, surface piercing. For all cases, the differences between the measured and predicted forces are relatively small (Hogben and Standing 1975). More complicated cases involving wave diffraction by a cylinder array have been presented by Malenica *et al.* (1999), Maniar and Newman (1997), and Ohl *et al.* (2001), and others.

The case of bodies with arbitrary geometry has been of considerable interest. Detailed descriptions of numerous works carried out in the 1970s are given by Hogben *et al.* (1977). A comparison of computer predictions with the experimental results of Mogridge and Jamieson (1976) and Isaacson (1979b) for the wave force on a square-section cylinder at orientations of $\alpha = 0°$ and $45°$ is shown in Fig. 5.4.

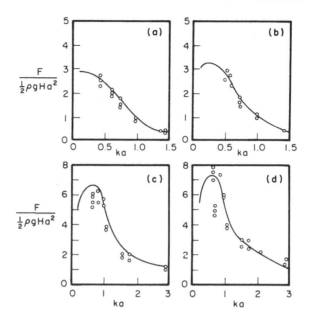

Figure 5.3. Comparison of experiments and computer predictions of the horizontal wave forces on circular and square columns: (a) circular ($h/d = 07$), (b) square ($h/d = 0.7$), (c) circular, surface piercing, and (d) square, surface piercing. Hogben and Standing (1975).

5.5 Higher-order wave diffraction and the force acting on a vertical cylinder

The inhomogeneous free-surface boundary condition is the primary obstacle in developing higher-order diffraction theories in the frequency domain. However, there are circumstances that require the use of the higher-order theories. In fact, Kriebel (1990) developed a second-order solution based on velocity potential decomposition of the diffraction of a monochromatic wave in the presence of a bottom-mounted circular cylinder. His calculations in cases of steep waves have shown that the maximum second-order run-up on the cylinder can exceed that from linear theory by more than 50%. More recently, this fact has been shown numerous times for single as well as multiple circular and noncircular cylinders. It is a means to predict the surface elevations, velocity potentials, forces, and motions with greater accuracy by

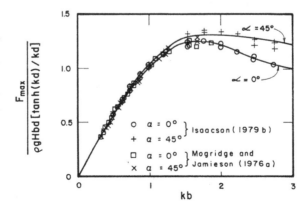

Figure 5.4. Comparison of numerical predictions with the experimental results of Mogridge and Jamieson (1975, 1976) and Isaacson (1979b) for the wave force on a square-section cylinder at orientations of $\alpha = 0°$ and $45°$.

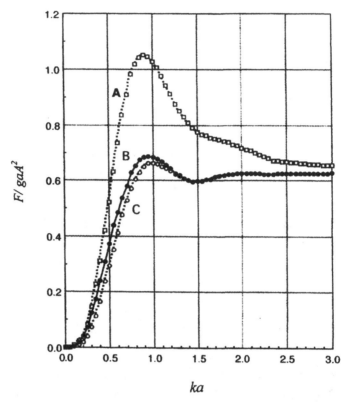

Figure 5.5. Force on a vertical circular cylinder at three values of h/a : A, 1; B, 3; C, infinite.

splitting the Stokes expansions into a series of linear states:

$$F = F^{(0)} + \varepsilon F^{(1)} + \varepsilon^2 F^{(2)} \tag{5.22}$$

where ε is a small parameter and the second-order approximation is represented by the third term, i.e., the amplitude and frequency become, respectively, the square of the first order and twice the frequency of the incident wave. One can then calculate, for example, the force acting on a vertical cylinder (Molin 1979; Rahman 1984; Eatock Taylor *et al.* 1989; Kim and Yu 1989; Li *et al.* 1991; Malenica *et al.* 1995, 1999; Huseby and Grue 2000; Bai and Eatock Taylor 2006, 2007):

$$\frac{F_d}{\rho g a \, A^2} = \frac{4}{\pi^2 (ka)^3} \left(1 + \frac{2kh}{sh\, 2kh} \right) \sum_{m=0}^{\infty} \frac{[1 - m(m+1)/(k^2 a^2)]^2}{\left(J_m'^2 + Y_m'^2 \right) \left(J_{m+1}'^2 + Y_{m+1}'^2 \right)} \tag{5.23}$$

where the Bessel functions are evaluated at ka. Figure 5.5 shows the mean value of the normalized force for three different depths. At large values of ka, $F_d/\rho g a \, A^2$ approaches 2/3. Clearly, the diffraction force is *minimum* at infinite depth and *increases* with decreasing h/a. Furthermore, the force at any h/a approaches 2/3 as ka increases.

In more recent times, more sophisticated diffraction analyses have been carried out. For example, Bai and Eatock Taylor (2007) analyzed the fully nonlinear and focused wave propagation and diffraction around a vertical circular cylinder in a numerical tank. They have used a mixed Eulerian–Lagrangian approach to update the moving boundary surfaces in a Lagrangian scheme, in which a higher-order BEM is applied to solve the wave field based on a Eulerian description at each time step. The results have shown that their domain decomposition technique is very efficient, and can provide accurate results when compared with the experimental data. Soylemez and Goren (2004) presented a complete analytical solution for the linear diffraction of oblique waves by horizontal rectangular cylinders either fixed at the free surface or mounted on the seabed in a finite depth of water.

Kagemoto and Yu (1986) expressed the interactions among multiple three-dimensional bodies in waves in terms of an exact algebraic method. Interaction of waves with arrays of vertical cylinders are analyzed by Linton and Evans (1990), Linton and McIver (1996), Maniar and Newman (1997), and in the numerous references cited therein through the use of various methods. For example, the case of trapped modes about multiple cylinders in a channel is analyzed using the multipole method, in which singular solutions of the Helmholtz equation satisfying an antisymmetry condition on the channel center plane are modified to include the boundary conditions on the channel walls.

The time-domain analysis of second-order wave diffraction by an array of vertical cylinders was employed by Wang and Wu (2007). The radiation condition was imposed through a combination of the damping-zone method and the Sommerfeld–Orlanski equation. Results of various configurations, including two cylinders, four cylinders, an array of ten cylinders and two arrays of eight cylinders are provided to show the effect of the interaction and their behavior near the trapped mode. The results presented are consistent with those from the frequency-domain analysis. All these are of course based on the assumption that the perturbation method is valid.

The recent progress in the computation and understanding of wave interaction with arrays of offshore structures has been reviewed by McIver (2002), focusing primarily on new developments in computational methods, resonant effects in infinite arrays and the consequences for finite arrays, and nonlinear effects. The complexity of the geometry of relatively large structures requires numerical rather than closed-form solutions. Such an approach also helps to carry out rapid simulations to cover a relatively large parameter space. This is where the power of methods such as the boundary-element method (BEM) manifests itself. The geometry of the structure is expressed in terms of a large number of flat or curved panels, as in finite-element analysis (e.g. Houston 1981). Often two or three discretizations (with smaller and smaller panels) are desirable to determine or to optimize the panel size, and hence the accuracy of the forces predicted. The design and construction of an offshore structure are such monumental and expensive undertakings that neither the computer size nor the time to repeat the simulations should be of logical concern. What is important is the determination of the forces and moments (i.e., the exciting force) acting on the body. This, of course, is identical to what it would be if the body were fixed. The exciting force could be determined in the same manner as was adopted in the fixed-body case, but the calculation does not now require ϕ_s to be determined explicitly.

By applying the Green's theorem, it is possible to express the exciting force directly in terms of the incident and forced potentials. Such expressions are the so-called *Haskind relations*, which Newman (1962) has popularized. They play an important role in checking the accuracy of the numerical codes. The *Haskind relations* may be written as

$$F_i^{(e)} = \rho \int_S \left[\phi_w \frac{\partial \varphi_i^{(f)}}{\partial n} - \phi_i^{(f)} \frac{\partial \phi_w}{\partial n} \right] dS \tag{5.24}$$

Thus once the force potential components $\phi_j^{(f)}$ have been calculated to obtain the added-mass and the damping coefficients, it may be simpler to use (5.24) directly rather than first to calculate ϕ_s'.

Using Green's theorem, the above integral can be expressed in terms of S_∞, a control surface in the fluid at a large distance from the body, as

$$F_i^{(e)} = -\rho \int_S \left[\phi_w \frac{\partial \varphi_i^{(f)}}{\partial n} - \phi_i^{(f)} \frac{\partial \phi_w}{\partial n} \right] dS \tag{5.25}$$

It follows that an asymptotic form of $\phi_j^{(f)}$ may be employed, and considerations based on the principle of energy conservation may then be used to relate the damping coefficients of the body to the asymptotic behavior of ϕ_f at S_∞, and subsequently to the exciting force. This approach is particularly useful in the case of an axisymmetric body where the amplitudes (but not the phases) of the exciting force may be expressed explicitly in terms of the damping coefficients. For additional details see Newman (1962).

5.6 Closing remarks

We have not discussed a number of related topics, partly for the sake of brevity, partly to remain true to the real purpose of the book (*presentation of the fundamental concepts of marine hydrodynamics*), and partly because the analysis of bodies of arbitrary shape is described in great detail by Van Oort-merssen (1972), Hogben and Standing (1974), Faltinsen and Michaelsen (1974), among many others, and handled effectively through the use of a large number of codes, e.g. (AQUADYN, DIODORE, HYDROSTAR, etc.). We have also not discussed the role of irregular waves and currents (often treated using the linear theory), and situations that involve floating bodies, vibration, and damping. The former is considered outside the scope of this book. The latter is covered in detail in Chapter 7.

6

Vortex-Induced Vibrations

6.1 Key concepts

Vortex-induced vibrations (VIVs) are part of the family of flow–structure interactions and thus of great practical importance in people's interactions with nature. They occur in many engineering situations, such as bridges, stacks, transmission lines, aircraft, offshore structures, thermo-wells, engines, heat exchangers, marine cables, flexible structures in petroleum production, and other hydrodynamic and hydroacoustic applications.

Numerous contributions to flow-induced vibrations (FIVs) in general and to VIVs in particular toward the understanding of the fundamental flow mechanisms and the acquisition of design data through insights into marine hydrodynamics, physical and numerical experiments, and theoretical analyses have collectively defined the objectives of the FIV and VIV research.

During the past century, a great deal of work has been done on flow-induced vibrations and fluid-elastic instability. The number of contributions has increased exponentially. Thus, the amount of time required for any one researcher to comprehend the literature and to plow through the empirical morass became an increasingly larger fraction of his research time. As one of the pioneers of our subject (G. V. Parkinson) stated (1974), "An elastic structure exposed to a fluid flow may vibrate under the action of flow for a variety of causes. If the incident flow is oscillatory, either in an organized form, as with the vortex street from an upstream body, or in the form of random turbulence, the structure will develop an oscillatory response. If the structure is streamlined, like an airfoil, and the incident flow is steady, the elastic characteristics in two or more degrees of freedom may permit the structure to *extract energy* from the flow so as to develop a catastrophic flutter."

Here, we will consider only the vortex-induced vibrations. It is part of a number of disciplines, incorporating fluid mechanics, structural mechanics, vibrations, computational fluid dynamics (CFD), acoustics, wavelet transforms, complex demodulation analysis, statistics, and smart materials. They occur in many engineering situations, such as bridges, stacks, transmission lines, aircraft control surfaces, offshore structures, thermo-wells, engines, heat exchangers, marine cables, towed cables, drilling and production risers in petroleum production, mooring cables, moored structures, tethered structures, buoyancy and spar hulls, pipelines, cable laying, members of jacketed structures, and other hydrodynamic and hydroacoustic applications. The most recent interest in long cylindrical members in water ensues from the development of hydrocarbon resources in depths of 1000 m or more. In fact, the depths reached during the past 60 years or so increased as $h \approx (1/540)N^{3.5}$, where h is the depth and N is the number of years, starting with $N = 0$ in 1949.

It cannot be emphasized strongly enough that the current state of the laboratory work concerns the interaction of a rigid body (mostly and most importantly for a circular cylinder) whose degrees of freedom have been *reduced from six to often one* (i.e., transverse motion) with a 3D separated flow, dominated by large-scale vortical structures. A few exceptions in which a second degree of freedom is allowed (in the in-line direction) will be discussed later. The restrictions imposed on the physical and numerical experiments are a measure of the complexity of the self-regulated motion. It is not yet clear as to how the additional degrees of freedom will change some of the observations, measurements, numerical simulations, and the contemplated applications in all ranges of Reynolds numbers (subcritical, critical, transcritical, and supercritical).

Numerous contributions to flow-induced oscillations in general and to vortex-induced vibrations in particular have collectively defined the objectives of the current VIV research and have guided the acquisition of design data through physical and numerical experiments, theoretical analyses, and physical insight. As stated by William Froude in 1886, "I contend that unless the reliability of small-scale experiments is emphatically disproved, it is useless to spend vast sums of money on full-scale trials which after all may be misdirected unless the ground is thoroughly cleared beforehand by an exhaustive investigation on a small scale." The ultimate objective is, of course, the understanding, prediction, and prevention of vortex-induced vibrations (preferably, without drag penalty) partly through the use of one of the numerous numerical models or commercial codes such as FLUENT and AcuSolve (discussed in Section 6.13).

As most aptly noted by Fischer and Patera (1994), "Fluid dynamics is, of course, not simply the solution of the Navier-Stokes equations for a particular configuration. Most broader problem statements, applied or fundamental, involve a vector of physical and *technology-related parameters,* which must be averaged over, eliminated by optimization, or varied. In all these examples, the typically rather large parameter space precludes a purely numerical solution; analytical, heuristic, and experimental data, as well as intuition, must be brought to bear if the final goals are to be achieved." In other words, the numerical simulations are to be guided and inspired by ground-breaking measurements and flow visualization, mostly with nonintrusive techniques: digital particle image velocimeter (DPIV), laser-Doppler velocimeter (LDV), time-resolved PIV, pressure-sensitive paints, and other means which will surely emerge in the years to come. These must be augmented with large-scale benchmark experiments to guide the numerical simulations at very large Reynolds numbers.

Much progress has been made during the past decade, both numerically and experimentally, toward the understanding of the kinematics (vice dynamics) of VIV, albeit in the *low*-Reynolds *number regime*. The fundamental reason for this is that VIV is not a small perturbation superimposed on a mean steady motion. It is an inherently nonlinear, self-governed or self-regulated, multi-degree-of-freedom phenomenon. It presents unsteady flow characteristics manifested by the existence of two unsteady shear layers and large-scale structures. There is much that is known and understood and much that remains in the empirical/descriptive realm of knowledge: What is the dominant response frequency, the range of normalized velocity, the variation of the phase angle (by which the force leads the displacement), and the response amplitude in the synchronization range as a function of the controlling and influencing parameters?

Industrial applications highlight our inability to predict the dynamic response of fluid–structure interactions. They continue to require the input of the in-phase and out-of-phase components of the lift coefficients (or the transverse force), in-line drag coefficients, correlation lengths, damping coefficients (to be discussed in the next section), relative roughness, shear, waves, and currents, among other governing and influencing parameters, and thus also require the input of relatively large safety factors. Fundamental studies as well as large-scale experiments (*if disseminated in the open literature*) will provide the necessary understanding for the quantification of the relationships between the response of a structure and the governing and influencing parameters.

The difficulties experienced in describing the nature, identifying the occurrence, and predicting the characteristics of vortex-induced oscillations of bluff bodies and galloping (to a lesser extent) have been reviewed by Parkinson (1974), Sarpkaya (1979b), Bearman (1984), Chen (1987), Parkinson (1989), Pantazopoulos (1994), Sarpkaya (1995, 2004), and in a number of books (Blevins 1990; Sumer and Fredsoe 1997; Au-Yang 2001; and Zdravkovich 1997, 2003).

The flow about bluff bodies (*fixed* or in *motion*) requires and/or gives rise to circumstances (influencing parameters) mostly beyond the capacity of the experimenter to control. Some of these are the finiteness of the body or the aspect ratio (as in the case of aircraft wings and ensuing end conditions), mobile separation points on curved surfaces, 3D behavior over 2D bodies, unpredictable (or difficult to predict) spanwise correlation, nonquantifiable growth of the disturbances in the wake, in the shear layers, and in the boundary layers (before and after separation). Furthermore, one may need to consider the distribution of the ambient velocity and turbulence (intensity and integral length scales), blockage ratio, surface roughness, yaw, body deformation, temperature gradients, stratification, bottom and free-surface effects, and the impossibility to perform DNSs at industrially significant Reynolds numbers.

In spite of most of these complex circumstances, the Strouhal number ($St = f_{st} D / U$) emerges as the most robust parameter. It is followed by the mean base pressure whose instantaneous values can be as high as –0.2 and as low as –3.5 (Zdravkovich 1997, p. 133). Even the crudest numerical simulations or experiments predict the Strouhal number with sufficient accuracy. However, as noted by many researchers over the years, this is as much an advantage as it is an obvious shortcoming because the robust parameters do not serve as unique identification cards for the integrated effects of the individually nonquantifiable parameters. In fact, one or two parameters into which we can lump our inability to account for all the *influencing* parameters do not seem to exist. This is partly because some of the influencing parameters can become governing when they exceed certain critical values. For example, ambient turbulence may change the transition in the wake, in the separated shear layers, in the *boundary layers in the vicinity of the separation zone* (leading to possible reattachment of the flow), and in the boundary layers upstream of the mobile separation points. Thus, ambient turbulence (possibly quantifiable in terms of its four integral length scales and eddy-dissipation rate) may significantly affect the flow around the VIV suppression devices, and hence, may affect their performance. Then the question naturally arises as to what are a handful of parameters that could possibly serve (albeit imperfectly) as unique identification parameters for the intended purpose. The most obvious candidates are those that exhibit large scatter in every experiment (confining ourselves to steady flow about smooth circular cylinders).

Measurements during the past century identified the fluctuating lift in steady flow about a cylinder at rest as the most likely quantifier of the combined effect of the influencing parameters (see, e.g., Norberg 2003 and references cited therein). Thus, having a representative time record of the lift force and its complex demodulation analysis, in addition to the spectra and the rms (root-mean-square) value of lift, would be most desirable. As noted above, it would be equally desirable to have a measure of the turbulence distribution, eddy-dissipation rate, and the integral length scales (x, y, z, and t) of the ambient flow. As to the flow around the cylinder, the extension of the current understanding of the transition waves and their disappearance in the range of Reynolds numbers from about $Re \cong 2 \times 10^4$ to 5×10^4 (Bloor and Gerrard 1966; Gerrard 1966, 1978; Zdravkovich 1997; Norberg 2003) to unsteady flows would be quite valuable.

Measurements similar to those noted above are equally desirable for all bodies capable of giving rise to VIV undergoing forced or self-excited oscillations in steady uniform or sheared flows. It appears that for such flows the fluctuating lift and the phase angle (between the total force and the displacement) or the "in-phase" and "out-of-phase" components of the transverse force could serve as suitable identification cards for the integrated effects of some of the important parameters, e.g., to assess the effects of Reynolds number on the correlation length along the cylinder, on the wake modes, and on the state of transition to turbulence in the free shear layers emanating from a cylinder undergoing VIV at a given A/D for a range of Re or at a given Re for a range of A/D. However desirable, the sensitivity of the transition waves to VIV has not yet been investigated. Apparently, direct measurements will be a formidable task. Evidence, albeit indirectly, must be gathered to identify, for example, wake and response phenomena, which occur only at relatively small Reynolds numbers.

The size and shape of the afterbody, the frequency spectrum of the fluctuations of the separation angle, surface condition, the upstream turbulence, and the motion of the cylinder require serious investigation, particularly for comparison with direct numerical simulations of VIV *using simulated ambient turbulent flow*. In summary, it is interesting to note that a phenomenon as robust as vortex shedding gives rise to forces as unpredictable as the lift force whose power can be fully appreciated only when one tries to eliminate VIV without excising the after body. In fact, it may be more advantageous to predict and thereby to avoid the vortex-induced vibrations than to attempt to eliminate them. After all, the fluctuating lift will always be there, with or without VIV, and the pure circular cylinder will always be the preferred shape for basic research and engineering applications.

6.1.1 *Nomenclature*

The advent of powerful computers has increasingly forced the formulation of the VIV problems in mathematical and CFD terms. With these thoughts in mind, we purposely choose two- or three-letter subscripts to enhance the instant recognition of the most important symbols:

f_{vac}: the frequency measured in vacuum, as the only natural frequency $[f_{vac} = (1/2\pi)(k/m)^{1/2}]$, where k denotes the linear spring constant and "m" the mass of the oscillating body. Its relation to the frequency measured in air and water will be discussed later.

f_{com}: the common frequency at which *synchronization* or lock-in occurs at a given velocity, i.e., $f_{ex} = f_{com}$.

f_{ex}: the frequency of oscillation of a (forced or self-excited) body, meaning *the excitation frequency*, regardless of whether there is lock-in or not. At lock-in $f_{ex} = f_{com} = [(1/2\pi)[k/(m + \Delta m)^{1/2}]$, where Δm is the added mass.

f_{st}: the vortex shedding frequency (or the Strouhal frequency) *of a body at rest*. It is uniquely related to the velocity of the flow and the characteristic size of the body through the Strouhal number $St = f_{st}D/U$ where U is the steady ambient velocity of the uniform flow.

f_{vs}: the vortex shedding frequency of *a body in motion* (forced or self-excited). In the lock-in range, f_{vs} becomes increasingly smaller than f_{st} until the lockout.

A cylinder may be *forced* to oscillate at any frequency and amplitude within reason. Furthermore, one is at liberty to change the frequency and/or the amplitude content of the oscillations. Outside the synchronization region(s) the force experienced by the body will have both the Strouhal and body oscillation frequencies. In other words, it is understood that in the periodic and quasi-periodic non-lock-in regions, the two frequencies (f_{st} and f_{ex}) will appear for a while, and then, at a slightly higher frequency, the synchronization will occur, leaving only one frequency, i.e., $f_{com} = f_{ex}$. However, one must hasten to note that in self-excited oscillations one should not expect perfect synchronization because of the continuous interaction between the body and the fluctuations of added mass, separation line, amplitude, correlation length, and the phase angle. This is distinct from the *beating motion* of a cable undergoing amplitude-modulated oscillations by excitation at two or more frequencies wherein the vortex-shedding frequency alternates between the imposed frequency and the Strouhal frequency.

The following are identified as the most important dimensionless parameters: f_{com}/f_{st}, f_{ex}/f_{st}, f_{ex}/f_{vac}, $V_r = U/f_{ex}D$, and $Re = UD/\nu$ where ν is the kinematic viscosity. Finally, f_{wtr} and f_{air} will be introduced for those in dire need of a frequency in still water or in still air. Additional parameters, controlling and influencing VIV, will be discussed later.

6.2 Introduction

An all-inclusive definition of a self-exciting or, better, self-regulating "bluff body" does not exist. It can be described only in general terms by relying on the readers' imagination. It is an elastic or elastically mounted fore-and-aft body of proper mass, material damping, and shape whose cross-section facing the ambient flow at high-enough Reynolds numbers gives rise to separated flow and hence to two shear layers, which interact with each other and bound an unsteady wake. A body with no aft section (e.g., a D section with a flow from *right to left*) behaves like a bluff body but does not present an aft profile on which the alternating lift force can act. In this text, a body capable of giving rise to VIV will be called a "VIV-body."

Numerous experiments have shown that a VIV-body (e.g., a circular- or square-section cylinder) with material damping determined in vacuum ζ_{mv} (especially for sharp-edged bodies), and proper reduced mass $m^* = (m/L)/(\rho_f \pi D^2/4)$, or $m^* = \rho_m/\rho_f$ with $\rho_m = (m/L)/(\pi D^2/4)$ as the "mean mass density," may be excited by the vortices it sheds if it is mounted on springs and exposed to a steady uniform flow. When the prevailing vortex-shedding frequency f_{vs} (not always close to f_{st}) and the excitation frequency f_{ex} of the body approach a common frequency f_{com}, the body begins to experience relatively small vortex-induced oscillations. These are controlled by the spring constant, body mass, structural damping, the density and the motion of the fluid surrounding the body, plus a number of difficult-to-quantify influencing parameters (discussed in detail by Schewe 1983a). This is followed by a substantial increase of the coherence length (Koopmann 1967a; Ramberg and Griffin 1976, for oscillating cylinders; Mansy *et al.* 1994 for stationary cylinders; and Novak and Tanaka 1975, in smooth flow).

The vortex-excited oscillations increase the vortex strength (Atsavapranee *et al.* 1998) when the amplitude in the transverse direction exceeds a threshold value of about $0.1D$ (it is about $0.02D$ for the in-line oscillations; see, e.g., Okajima *et al.* 2002; Sugimoto *et al.* 2002, and the references cited therein). The ratio of the circulation of the nascent vortex to that of the shed vortex is about $\Gamma/\Gamma_0 = 0.51 \pm 0.08$ for the stationary cylinder (Atsavapranee *et al.* 1998), in agreement with those reported in the literature. For the cylinder oscillating in the lock-in range ($V_r = U/f_{ex}D \approx 4-8$), the ratio of the vortex strength to the total supply of circulation in one shedding period Γ/Γ_0 is found to be 0.66 ± 0.09. This conforms Sarpkaya's findings (1963) in accelerating flow about cylinders.

If the velocity U and hence the amplitude are increased (gradually) to new values, the diameter "seen" by the flow (or the spacing between the shear layers defining the virtual body) increases initially, *at least for small amplitudes.* The apparent "increase" in D is compensated by the real increase in velocity U, thus keeping the vortex-shedding frequency nearly constant. This represents a departure from the vortex-shedding frequency of the fixed cylinder at the new (increased) ambient flow velocity. In other words, the self-excitation begins with a *real decrease in the frequency of vortex shedding* to a value close to f_{com}, which may be close to but not equal to that obtained from pluck tests in still water. The vortex-shedding mode and frequency change most to match f_{com}. Apparently, the body motion is dominant (but not invariant) in the sense that it accommodates somewhat the changes in vortex shedding by letting the flow change its virtual mass and hence its frequency and acceleration as both the flow and the body arrive at a common frequency to which the body responds with exuberance: the matching of the frequency of the prevailing dynamics of the vortical wake with the frequency of the body oscillation.

Figures 6.1 and 6.2 show two interesting examples. The first (Fig. 6.1, after Feng 1968) was conducted in air with a single-degree-of-freedom flexible cylinder with a relatively large mass ($m^* = \rho_m/\rho_{air} = 248, \zeta = 0.00103, m^*\zeta = 0.255$) and relatively large Re, *varying with* $V_r = U/f_{air}D$, from 10^4 to 5×10^4.

In self-excited cases, the Reynolds number increases along the V_r axis because f_{air}, f_{wtr}, and D are kept constant and only U or Re is allowed to vary as prescribed. This is unlike the forced oscillation experiments in which *Re is kept constant* and the frequency of excitation is varied. Khalak and Williamson (hereafter KW) (1999) stated that "our studies here and in Govardhan and Williamson (2000), for $Re = 3500$–10000, indicate that it is principally the parameter $(m^*\zeta)$, which influences whether the Upper branch [see Fig. 6.2] will appear or not." The existing facts do not support their assumption. For *steady flow about a stationary* cylinder, Basu (1986) noted that "the Re number range 1350 to 8000 overlaps the Re number range in which the principal feature of the flow field is the upstream movement of transition in the free shear layer with increasing Re number." The "eddy formation length," L_f, shrinks from $L_f \approx 1.9D$ at $Re = 5 \times 10^3$ to $L_f \approx 1.1D$ at $Re = 14 \times 10^3$ and the distance to the center of the transition region (transition from a laminar to a turbulent free shear layer) decreases from about $L_t \approx D$ to $L_t \approx 0.4D$ (Zdravkovich 1997). Clearly, the physics of the shear layers, in general, and that of the unsteady shear layers, in particular, is very complex. For this reason the sensitivity of the transition waves due to VIV (even for a single degree of freedom) has not yet been directly investigated. However, the experiments of Carberry (2002) with oscillating cylinders (with Reynolds numbers in the same range as that of KW) provide direct evidence to show that the phase angle φ and, in particular, the lift coefficient C_L, significantly increase (as much as 100%) with increasing

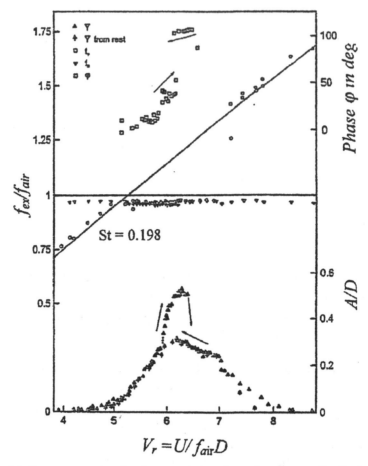

Figure 6.1. Response and wake characteristics of a spring-mounted cylinder freely oscillating in air: $m^* = 248$, $\zeta = 0.00103$, $m^*\zeta = 0.255$, and Re (varying with $V_r = U/f_{\mathrm{air}}D$) from 10^4 to 5×10^4 (Feng 1968).

Reynolds number ($Re = 2300$, 4400, and 9100) for a given A/D, and with A/D for a given Reynolds number (up to A/D values of about 0.5).

Feng's data (A/D and phase φ versus $U/f_{\mathrm{air}}D$), at higher Reynolds numbers (10^4 to 5×10^4), have only two branches (initial and lower). The KW (1999) data (A/D versus $U/f_{\mathrm{wtr}}D$), at lower Reynolds numbers (3500–10000) have three branches (initial, upper, and lower), much larger peak amplitude, and a broader synchronization range. It must be emphasized that the horizontal axis in both figures denotes $U/f_{\mathrm{air}}D$ and $U/f_{\mathrm{wtr}}D$, as well as the increasing Reynolds number, $Re = UD/\nu$. In all such free VIV experiments the frequency f_{air} (or f_{wtr}) and D are kept constant (along with m^* and ζ) and the variation of A/D with U or with the Reynolds number is plotted from a minimum to a maximum Re. Evidently, Re may be kept constant by maintaining U, D, L/D, k, ζ, ρ_f, and ν constant and varying only m in $\rho_m = 4m/(\pi L D^2)$ and hence f_{vac} in $V_r = U/f_{\mathrm{vac}}D$ or in $f_{\mathrm{ex}}/f_{\mathrm{vac}}$. This

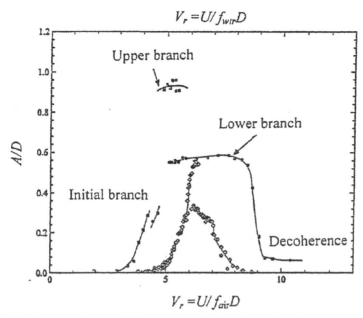

Figure 6.2. A comparison of Feng's amplitude data (maximum A/D versus f_{ex}/f_{air}, the lower axis) with those obtained by Khalak and Williamson (1999) in water (the upper axis) with a single-degree-of-freedom flexible cylinder ($m^* = 10.1$, $\zeta = 0.0013$, $m^*\zeta = 0.013$, and Re varying from 3×10^3 to 10^4). Feng's data at higher Re has only two branches (initial and lower, as in Fig. 6.1). The Khalak and Williamson data (A/D versus $U/f_{wtr}D$) have three branches (initial, upper, and lower), larger peak amplitude, and a broader synchronization range.

will define a 3D space showing A/D versus f_{ex}/f_{vac} in each plane of constant Re. This will enhance the understanding of the various regimes and their dependence on Reynolds number, shear layer instability, mass ratio or ρ_m/ρ_{wtr}, material damping, aspect ratio, modes of vibration, modes of vortex shedding, abrupt as well as gradual changes in the phase, and the degrees of freedom, along with several other parameters, to be discussed later. However, before we proceed further, we will emphasize that there are significant differences between the low- and high-Reynolds-number flows in the laboratory and ocean environment. For example, none of the high-Reynolds-number experiments ($Re > 20000$) show such phenomena as "initial branch" (seen only at Reynolds numbers smaller than about 5000). The vortex modes of 2P, 2S, and others have been mapped only at Reynolds numbers below 1000. Blackburn and Henderson (1999) did not find the 2P mode at $Re = 500$. Evangelinos and Karniadakis (1999) at $Re = 1000$ found multiple vorticity concentrations and transient mixtures of (P + S) and 2P modes in the near wake and general wake instability further downstream. Brika and Laneville (1993) found 2P and 2S modes in the range of Reynolds numbers from 3.4×10^3 to 11.8×10^3. As we have noted earlier, the mean position of the line of transition to turbulence does not reach upstream enough for *Re* less than about 15×10^3 to 20×10^3. Lastly, it should be noted that at *Re* larger than about 20 000, it might not be possible to photograph coherent vortex structures in the wake.

In free oscillations at sufficiently large *Re*, the oscillations are not sinusoidal, as evidenced by many experiments and numerical simulations. Consequently the flow does not become fully established, say "periodic," because the amplitude, added mass, frequency, phase angle, vortex structures, and shear layers never become fully established. Each cycle is affected by the character of the previous cycle. Therefore, the ever-changing topology of the flow prevents it from exhibiting sharp changes (as branches) in the A/D versus V_r plots at high *Re*.

As noted by Huerre (2002), "vortex shedding is represented by a *global mode*, i.e., a self-sustained time-periodic state characterized by a specific spatial structure and a frequency, which is the same throughout the flow domain." The freedom provided to a body through elastic mounting is capable of modifying the character of the global mode, i.e., modifying both the frequency and the spatial structure of the *near wake*. In any case, the body vibrates neither at f_{wtr} (for VIVs in water), nor at f_{air} (for VIVs in air), nor at f_{st} of the fixed body. Surprisingly, synchronization (sometimes referred to as the lock-in, lock-on, vortex capture, or frequency capture) occurs not only near f_{st}, but also over a wide range of flow velocities. It appears that "lock-in" or "lock-on," "vortex capture," or "frequency capture" are misnomers. In fact, one needs to redefine VIV to make a distinction between the two distinct roles, ***excitation and driving***, played by the vortices.

Excitation means that the vortices can and do excite the body, even when the out-of-phase component of their lift force is relatively small (in comparison to mechanical restoring force), provided that their shedding frequency f_{vs} is close to the prevailing frequency of the body. This is like a single-degree-of-freedom system with small (but non-zero) viscous damping. *Thus, even weak vortices in some regions of f_{ex}/f_{st} can excite a body to large amplitudes when the body and the vortices arrive at a common frequency f_{com} (in the region of f_{ex}/f_{st} from about 0.5 to 0.9).*

The driving ability of the vortices ensues from their particular modal dynamics to give rise to a sufficiently large out-of-phase lift component in certain regions of f_{ex}/f_{st} (from about 0.9 to 1.0). Thus, the excitation function of the vortices is a mandatory requirement for the inception and maintenance of VIV. However, the effectiveness of their driving function depends on a number of parameters (the range of f_{ex}/f_{st}, Reynolds number, damping, virtual mass of the oscillating system, and some influencing parameters).

It is clear from the foregoing that, to sustain a self-excited, self-limiting resonant response, the motion of a freely vibrating VIV-body interferes at proper amplitudes and frequencies with the mutual interaction of the shear layers and becomes a dominant part of the overall instability mechanism: shear-layer interaction leading to vortex shedding, leading to alternating transverse and modulated in-line forces, leading to body motion, leading toward the common frequencies of the excitation and response in respective directions. Thus, the interaction of the body and the flow under proper conditions serves as the magnifier, organizer, and synchronizer of the phenomenon. It is not ever likely to happen at a *constant* amplitude and *frequency* for the freely oscillating bodies because of the significant amplitude/frequency/added-mass modulations. In the words of one of the pioneers (Koopmann 1967a), "when the wind velocity approaches the boundaries of the resonant region, the shedding frequencies are close enough to the natural frequency of the system to cause the system to respond in short bursts of periodic motion in the plane to the direction of the wind." If the motion during one of these bursts is large enough to correlate the vortex wake along the span, the cylinder jumps to a higher displacement amplitude than before and a steady-state oscillation follows. Once inside this instability boundary, the cylinder motion controls the wake frequency and the resulting oscillation takes place at the natural frequency of the system. "In addition, as the wind velocity is slowly increased in the resonant region, the displacement amplitude of the cylinder steadily increases until at some definite velocity, a peak displacement amplitude is reached."

Cheng and Moretti (1991) conducted a series of experiments with a circular cylinder subjected to forced transverse vibration in a uniform cross-flow at Reynolds numbers of 1500 and 1650. They measured the prevailing *vortex shedding frequency* f_{vs} with hot-film probes, placed $4D$ downstream and $2.5D$ across from the center of the cylinder. They varied f_{ex}/f_{st} (driving frequency/Strouhal frequency for a stationary cylinder) from zero to 4.5 and the amplitude A/D from zero to 0.75. The driving frequency was monotonically increased to avoid hysteresis effects. Figure 6.3 shows representative data identifying the relevant subharmonic, nonharmonic, and superharmonic wake frequencies. When f_{vs} is near f_{st}, on a horizontal line given by $f_{vs}/f_{st} = 1$, f_{vs} is unaffected by the excitation. However, when f_{vs} falls on a line from the origin with a slope of unity, the vortex structure in the wake locks onto the excitation frequency ($f_{ex}/f_{st} \approx 0.5$–1.6). A further observation is made that f_{vs} tends to decline with increasing excitation, possibly due to the widening of the wake.

A more recent set of the data of Fig. 6.4 by Krishnamoorthy *et al.* (2001) shows the wake frequencies as a function of the oscillation frequency,

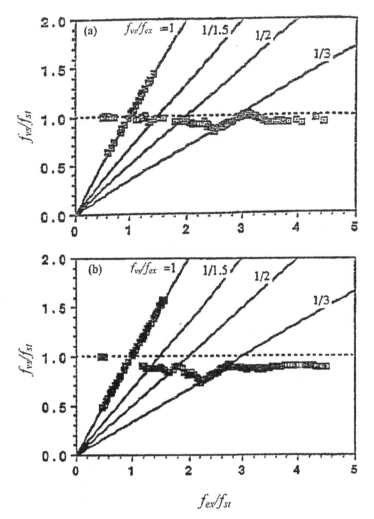

Figure 6.3. Variation of the shedding frequency with driving frequency of a single cylinder in uniform flow: (a) $A/D = 0.05$, $Re = 1500$; and (b) $A/D = 0.235$, $Re = 1500$ (Cheng and Moretti 1991).

vortex-shedding frequency, and remaining wake frequencies for $A/D = 0.22$ and $Re = 1500$. Cheng and Moretti (1991) found that the lock-in range exhibits an onion-shaped region, as shown in Fig. 6.5a. Figure 6.5b shows that C_L increases with A/D up to $A/D \approx 0.5$ and then decreases rapidly with increasing A/D. At small amplitudes, this dependence is essentially the same as that found by Koopmann (1967b). The center frequency of the lock-in is slanted toward lower frequencies, and the Reynolds number has a strong effect on the upper frequency lock-in boundary.

For long, rigid or flexible structures (e.g., a cable), the phenomenon is further complicated by the fact that the structure tends to respond at a *variety of frequencies over its entire length*. This, in turn, gives rise to additional

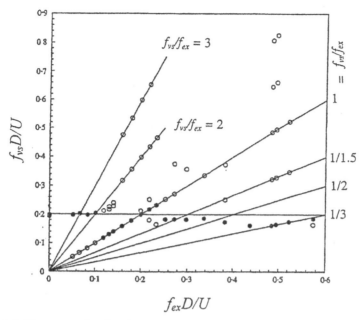

Figure 6.4. The normalized wake frequencies as functions of the normalized oscillation frequency, vortex-shedding frequency, and remaining wake frequencies for $A/D = 0.22$ and $Re = 1500$. (Krishnamoorthy *et al.* 2001).

and omnidirectional fluid forces whose prediction is at best approximate (see, e.g., Vandiver and Li 1994; Larsen and Halse 1995; Triantafyllou and Grosenbaugh 1995; Vikestad *et al.* 2000; Halse 2000; Dahl *et al.* 2006; Lie and Kaasen, 2006; Iranpour *et al.* 2008; Prastianto *et al.* 2008) and the references cited therein). When there is no synchronization (lock-in), the driving fluid force and the structure oscillate at their own frequencies. In field tests, a locked-in condition or a standing wave profile may not occur on long wires towed in the ocean (Alexander 1981). There is no total spanwise correlation along very long structures placed in the ocean environment partly because cooperating instabilities prevent such coherence, even in well-controlled laboratory experiments, and partly because larger amplitude disturbances and omnidirectional waves and currents surely prevent anything other than short coherence lengths (an extensive table is given by Pantazopoulos 1994). Only the relatively short test cylinders or cables result in well-separated modal frequencies, reducing the effects of modal interaction and enabling single-mode lock-in to be studied in some detail (Iwan and Jones 1987).

A fitting summary of the foregoing is given by Kim *et al.* (1986):

> Multimoded non-lock-in response did occur when the mean shedding frequency fell between natural frequencies. At these times, three or four modes were present in the cross-flow response. The in-line response would at the same time have several modes. Under lock-in conditions, the excitation

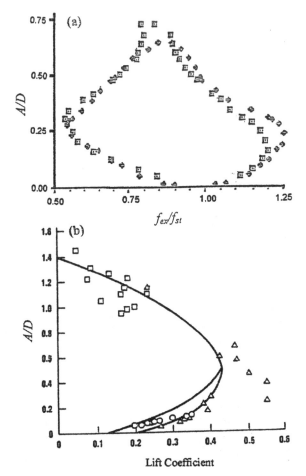

Figure 6.5. (a) The lock-in range exhibits an onion-shaped region ($A/D = 0.22$, $Re = 1500$). The center frequency of the lock-in is slanted toward lower frequencies (Cheng and Moretti 1991); (b) Lift coefficient for a pivoted rod at resonance: □, $L/D = 48$, △, $L/D = 15$, Vickery and Watkins (1962); ○, $L/D = 13.8$, Hartlen *et al.* (1968). Evidently, C_L increases with A/D up to $A/D \approx 0.5$ and then decreases rapidly with increasing A/D. More extensive C_L data for spring-supported cylinders and cantilevers are tabulated by Pantazopoulos (1994).

bandwidth is very narrow. Under non-lock-in conditions, even with very uniform flow, the excitation bandwidth broadens substantially. Under such circumstances, the lift force is best *characterized as a narrow band random process with sufficient bandwidth to excite two or more adjacent modes.* Lock-in occurs if and only if the separation of the natural frequencies of the cylinder are large compared to the bandwidth of vortex-induced forces.

Clearly, the vortex shedding, the character and timing of the vortices, and the amplification of the driving force are inextricably related. Also, self-excitation without lock-in is common but self-regulated lock-in without self-excitation is impossible.

If a cylinder is free to oscillate in both the transverse and the in-line directions, the common frequencies of the body and the driving forces in their respective directions may lead to lock-in and the axis of the body traces the path of figure **"eights"** (Moe and Wu 1990; Moe *et al.* 1994; Sarpkaya 1995). The figure-eight loop is caused by the considerable variation of drag force during large-amplitude oscillation. Dye (1978) tested a cantilevered cylinder in water and made the following observations: "During the swing of a cylinder from a dead-end, the wake is tilted and the projected drag in the velocity direction is reduced. Once the dead-end is reached, the wake realigns with the velocity and the drag force is increased. The cylinder is pushed back along the maximum amplitude line before the wake becomes tilted again." Zdravkovich (1990) noted that "the two-degree-of-freedom response is the interaction of the streamwise synchronization at $V_r = U/f_{ex}D \approx 2.5$ with the transverse synchronization at $V_r \approx 5$." It is a well-known fact that (see, e.g., Sarpkaya, 1979b) for $1.7 < V_r < 2.3$ (the so-called first instability region) oscillations occur in the in-line direction and the vortices are shed symmetrically. In the interval $2.8 < V_r < 3.2$ (the second instability region), the vortices are shed alternately. Chen and Jendrzejczyk (1979), working with a cantilevered tube in water, varied the reduced velocity from 1.92 to 4.92. They found a "beating" mode, typical of the presynchronization at $V_r = 4.53$, as shown in Fig. 6.6. A single-degree-of freedom system is inhibited from exhibiting these intricate variations.

The mobility of the separation points is important but not necessary for synchronization (e.g., for a square or rectangular cylinder). The separation points on a smooth *stationary* circular cylinder trace an arc whose magnitude depends on *Re* and the frequency and amplitude of the flow oscillation (Sarpkaya and Butterworth 1992; Sarpkaya 2002b). It is understood that the characteristics of turbulence of the approaching flow at each cycle depends on the energy stored in the fluid and is not independently controllable. In general, the point of separation on a cylinder depends not only on the pressure gradient but also on the type of unsteadiness of the ambient flow (e.g., sinusoidal), turbulence upstream of the separation points, roughness of the surface or other excrescencies, symmetry or the asymmetry of the cylinder, taper along the rod, presence of salient edges, and the mode of vortex (or vorticity) shedding. For a cylinder undergoing VIV, the actual instantaneous value of the wake angle is greater than that between the ambient flow velocity and the relative fluid velocity (Raudkivi and Small 1974) and the oscillation of a cylinder enhances both the strength of the vortices and the excursions of the separation points. Detailed discussions of the oscillation of bodies with salient edges may be found, for example, in Deniz and Staubli (1997, 1998), and in the references cited therein.

The lock-in also occurs at excitations that are super harmonics of the shedding frequency (see Figs. 6.3 and 6.4). Furthermore, the lock-in regions for the odd-number superharmonics appear to differ from those for the even-number superharmonics (Olinger and Sreenivasan 1988; Cheng and Moretti 1991; Rodriguez and Pruvost 2000). This is undoubtedly related to the nature of the shedding of the vortices.

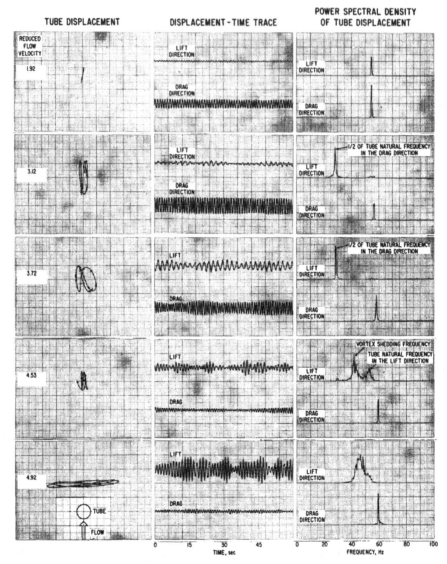

Figure 6.6. A cantilevered tube in water with reduced velocities from 1.92 (top) to 4.92 (at the bottom). The "beating" mode is typical of the presynchronization at $V_r = 4.53$. A single-degree-of-freedom system is inhibited from exhibiting these intricate variations (Chen and Jendrzejczyk 1979).

In the following section, we discuss a number of fundamental topics that are inextricably related to VIV and that are of major importance to the understanding of the intricate relationships between the governing parameters and the observed or predicted dynamics of the oscillating system. One of these parameters is the added mass or the added inertia, which has been discussed in detail in Sections 2.9–2.12. We will now revisit the added mass from a different perspective.

6.3 Added mass, numerical simulations, and VIV

At relatively small Reynolds numbers one does not need to deal with the added mass directly since the pressure and viscous contributions are explicitly incorporated into the numerical or theoretical solutions of the Navier–Stokes (NS) equations. The simulations at larger Reynolds numbers are complicated by the fact that VIV is not a small perturbation superimposed on a mean steady motion. It presents strong unsteady flow characteristics manifested by the existence of large-scale structures for which the use of standard turbulence models (RANS, LES, DVM, and others; see Section 6.13) is highly questionable. *Thus, one is forced to use empirical equations and approximate added mass coefficients.* As noted in Chapter 2, the cycle-averaged added-mass coefficient $C_a (= \Delta m / \rho_f V_b)$, where ρ_f is the density of fluid and V_b is a suitable reference volume, can be negative depending on the direction of *the flow of kinetic energy*, with far-reaching consequences for freely vibrating cylinders with low mass ratio $m^* (= \rho_m / \rho_f)$. This is because of the fact that $C_a / m^* = (\Delta m / m) = C_a \rho_f / \rho_s$ may approach –1 and the normalized virtual mass $(m + \Delta m)/m = (1 + C_a / m^*)$ may approach zero, resulting in an oscillating cylinder with *no apparent or virtual mass*. In reality, no mass is added to or subtracted from the body. The physical shape and mass of the body within the incompressible control volume remain invariant. What is imparted to the fluid (or the body) is positive or negative accelerations or inertia (per unit mass) or changes in kinetic energy due to the motion of the body, which can be negative or positive. In other words, the increase (or decrease) of the kinetic energy of the fluid within the control volume or the quotient of the additional force required to produce the accelerations throughout the fluid divided by the acceleration of the body manifest themselves as negative or positive added mass.

It has been noted, e.g., by Feng (1968), Bishop and Hassan (1963), and Sarpkaya (1979b) among others, that the transverse force needed to excite a cylinder to large-amplitude oscillations is far greater than that exerted by vortex shedding. It was not clear in the 1970s that *the virtual mass of the body may decrease to very small values.* The high-mass-ratio cylinders are obviously less affected by the added mass and its variations. In fact, the significance of several parameters stands out only at lower mass ratios. These issues will be discussed in more detail later.

It follows from the foregoing that in separated time-dependent flows (such as VIV), the common frequency (between the body and the vortices) at lock-in cannot remain constant throughout the synchronization range because the added mass is a function of time, the shape of the body and its surroundings, and the type of motion and orientation of the body through the fluid. For example, the added mass of a cylinder tracing the path of figure-eights is not the same as that of a cylinder constrained to move only in the transverse direction.

The negative added mass has been previously discussed a number of times (see, e.g., Keulegan and Carpenter 1958; Sarpkaya 1963, 1976b, 1976e, 1977a, 1986a, 1986b; Vandiver 1993; Vikestad *et al.* 2000). The negative added mass is not a mystery and occurs mostly in the approximate range of $0.5 < f_{ex}/f_{st} < 0.85$ where $f_{com}/f_{vac} > 1$, the phase angle and amplitudes are relatively large, and two pairs of vortices are shed per cycle (to be discussed later). A similar occurrence of negative added mass in transverse vortex shedding, from a cylinder undergoing sinusoidal oscillations in a fluid otherwise at rest, has been evaluated and discussed in great detail by Sarpkaya (1977a).

Subsequently, it will become clear that the fluctuations of the added mass hold the key to the understanding the similarities and differences between the forced and self-excited vibrations. The fluctuations of the amplitude of self-excited oscillations (about $\pm10\%$) impose higher-order harmonics on the cyclic variation of the added mass. This, in turn, leads to fluctuations in frequency. Their combined effect leads to further changes in amplitude, added mass, frequency, coherence length, phase angle and so on. Clearly, if one can eliminate the modulations superimposed on the cyclic variation of the added mass in self-excited vibrations, one will maintain its amplitude and frequency constant and thus make the motion behave more like a forced vibration. When no two cycles are alike, one should expect a large variety of wake states as the motion transitions from a *high-lift phase state* to a *low-lift phase state* (sometimes, mistakenly called "low-frequency" and "high-frequency" states, even through the frequency changes only slightly between the two states). What is of course remarkable in VIV is that a small change in frequency (say, f_{ex}/f_{st} or f_{com}/f_{st}) may cause rather large changes in phase, in wake structure, and thus in all the attendant consequences of VIV in or out of lock-in.

6.4 Governing and influencing parameters

6.4.1 *Parameter space*

A simple dimensional analysis shows that the parameters "controlling" the transverse vortex-induced oscillations of a cylinder are the density of fluid ρ_f, dynamic viscosity μ_f, velocity of the ambient flow U, diameter of the cylinder D, length of the cylinder L, spring constant k, mean roughness height of the cylinder k_s, *structural* damping factor ζ, mass of the body m (with no added mass), mean shear dU/dy, taper dD/dy, characteristic turbulence intensity

ε_t, and the integral length scales I_{ils} of the ambient flow, the force, and the uncertainty parameters S_p. Then the normalized amplitude may be written as

$$\frac{A}{D} = \mathbf{F} \left\{ \begin{array}{c} \zeta, \ \frac{\rho_f U D}{\mu_f}, \ \frac{L}{D}, \ \frac{4m}{\rho_f \pi L D^2}, \ \frac{D}{U} \left(\frac{k}{m}\right)^{1/2}, \ Re_{\mathrm{sl}}, Re_{\mathrm{cr}} \\ \frac{D}{U_0^2} \frac{dU}{dt}, \ \frac{D}{U} \frac{dU}{dy}, \ \frac{dD}{dy}, \ \varepsilon_t, \ \frac{k_s}{D}, \ \frac{I_{\mathrm{ils}}}{D}, \ \mathbf{S}_p, \ \frac{F_0}{\rho D^2 U^2} \end{array} \right\} \quad (6.4.1a)$$

and the normalized force may be written as

$$\frac{F_0}{\rho D^2 U^2} = \mathbf{F} \left\{ \begin{array}{c} \frac{A}{D}, \ \zeta, \ \frac{\rho_f U D}{\mu_f}, \ \frac{L}{D}, \ \frac{4m}{\rho_f \pi L D^2}, \ \frac{D}{U} \left(\frac{k}{m}\right)^{1/2}, \ Re_{\mathrm{sl}}, Re_{\mathrm{cr}} \\ \frac{F_0}{\rho D^2 U^2}, \ \frac{D}{U_0^2} \frac{dU}{dt}, \ \frac{D}{U} \frac{dU}{dy}, \ \frac{dD}{dy}, \ \varepsilon_t, \ \frac{k_s}{D}, \ \frac{I_{\mathrm{ils}}}{D}, \ \mathbf{S}_p \end{array} \right\} \quad (6.4.1b)$$

where Re_{sl} is the Reynolds number beyond which the transition eddies in the free shear layers disappear ($Re_{sl} \cong 2 \times 10^4$ for steady flow), and Re_{cr} is the critical Reynolds number beyond which the Strouhal number of the *vibrating* cylinder exhibits a *smooth transition* to a higher value (about 0.24), unlike its steady-flow counterpart. *The material damping ζ is composed of grain friction, dislocation friction (rather small), and the presence of vacancies (microscopic voids).* When a material is deformed, say by a VIV event, the structure moves against itself, causing the above phenomena (one or more depending on the structure of the material) to consume some of the energy of the motion: thus, damping. Clearly, there are no materials or structures without some damping and "the undamped natural frequency" does not exist.

The shear parameter $(D/U)(dU/dy)$ has been suggested by Vandiver (1993) and others in aeronautics after consideration of numerous possibilities. Humphries and Walker (1988) defined the shear parameter in terms of a characteristic velocity U_r (often as the midspan velocity) as $(D/U_r)(dU/dy)$. It is not unique and does not define all the dynamics of shear. However, the use of a larger number of shear parameters is not practicable. The current objective is to use as few parameters as possible, experimental data obtained from rigid cylinders, and strip theory in conjunction with approximate correlation-length models, as suggested by Triantafyllou *et al.* (2003). Clearly, when the shear parameter is small (say, less than 0.01), one would expect longer correlation lengths and cells. Additional parameters such as the number of excited modes, wave characteristics, cable properties, and axial-end conditions (i.e., free or constrained to move in the axial direction as the test pipe flexes) may have to be considered but are not included in Eq. (6.4.1b). This is primarily because of the fact that there are no programmable means to quantify some of them (or their interaction with other parameters) at any Reynolds number. The most important facts about the shear are as follows:

1. There should be no fear as far as safety is concerned (assuming no complications due to proximity effects to free-surface, bottom and/or other pipelines, stratification, and the strong nonlinearity of the shear);

2. It reduces and broadens the peak amplitude at all Reynolds numbers (from subcritical to supercritical); and

3. At high shear parameters (e.g., 0.03), the vortex excitation range usually extends over a larger reduced-velocity range, but at reduced peak amplitudes (Humphries and Walker 1988). These will be discussed in 6.11 in connection with VIV at high Reynolds numbers.

The ratio of the longitudinal integral length scale to the diameter, I_{ils}/D gives a good insight into the effect of turbulence. As shown by Basu (1985), the larger the I_{ils}/D, the smaller is the interaction between the free-stream turbulence and the cylinder boundary layer and wake. The intensity of turbulence of the ambient flow plays an important role (within reason) in establishing the critical and supercritical regimes (with or without the help of distributed roughness) and on the effectiveness (often degradation) of some singular VIV suppression devices at sufficiently large Reynolds numbers (Zdravkovich 1981).

The parameter $(D/U_0^2)(dU/dt)$, with U_0 as a reference ambient velocity (say, prior to the start of the acceleration or deceleration), or in more general terms, the parameter $(D^n/U^{n+1})(d^nU/dt^n)$, must be considered in assessing the significance of the local acceleration to the convective acceleration. If it is small (Sarpkaya 1991c, 1996b), the flow may be approximated by juxtaposition of steady states, i.e., by flows with negligible or no history effects. If large, it is not sufficient to define the changes in flow velocity with more or less arbitrary or qualitative measures such as "increasing velocity," "progressive change of velocity," "velocity with large steps," or "decreasing velocity."

It is a well-known fact that, depending on its magnitude, the rate of change of velocity (plus or minus), as defined by the subject parameters, gives rise to very interesting response characteristics, even for an isolated cylinder or cable (see, e.g., Sumer and Fredsøe 1988; Brika and Laneville 1993; and Frédéric and Laneville 2002). Thus, $(D/U_0^2)(dU/dt)$ must be quantified during both the acceleration and deceleration periods to assess the effect of the rate of change of velocity on the inception of the transient states (e.g., hysteresis, intermittent jumps) in both numerical and physical VIV experiments. Sarpkaya (1991c) has shown that the rate of deceleration at the end of the acceleration period is just as important as the acceleration period.

Obviously, it is not only the rate of change of velocity that can precipitate hysteresis effects. The rate of change of amplitude and the rate of change of frequency can produce equally interesting forms of hysteresis and can help to explain some puzzling observations. Thus, it is necessary to consider and quantify the following additional parameters in assessing the history effects: $(1/U)(\partial A/\partial t)$ and $(D/U)^2(\partial f_{ex}/\partial t)$. Such studies will help to resolve the consequences of a specific unsteady input of given type and duration on the subsequent stages of the fluid/structure interaction.

The parameter $(D/U)(k/m)^{1/2}$, or its inverse, may be written as $V_r = U/f_{vac}D$ using the "natural frequency" f_{vac} obtained in a vacuum because it is the *only natural frequency*. All other frequencies represent unsolved or partially solved fluid–structure interaction problems of the type pioneered by Stokes (1851) and Basset (1888). Nevertheless, a number of other reduced velocities have been used in the literature for a variety of reasons, including the need to emphasize the importance of the variation of one or the other parameter (added-mass, phase, in-phase, and out-of-phase components of the lift force): $V_r = U/f_{air}D$, using f_{air}, obtained from pluck tests in *still* air, or $V_r = U/f_{wtr}D$, using f_{wtr} obtained from pluck tests in a *still* test fluid (e.g., water), or $V_r = U/f_{com}D$, using the actual (or common) frequency f_{com} at which the lock-in occurs at a given velocity. Here a clear distinction must be made between forced and self-excited oscillations. In forced oscillations, $f_{com} = f_{ex} = 1/T$ (where T is the period of oscillation), and it was used by Sarpkaya (1978a) to define $V_r = UT/D$ against which the force-transfer coefficients were plotted. Thus, f_{com} or f_{ex} varies as the period of oscillation is varied. Furthermore, one is at liberty to change the frequency and/or the amplitude content of the oscillations.

6.4.2 *Uncertainties*

The uncertainty encompasses all the circumstances (influencing parameters) mostly beyond the capacity of the experimenter to control or to vary systematically (see, e.g., Schewe 1983a). They are unknowable facility-related constraints or a consequence of unsteady facility-interference coupling. Some of these are the end conditions, parameters controlling the mobility of the separation points on curved surfaces, 3D behavior of flow over 2D bodies, unpredictable (or difficult to predict/control) spanwise correlation and its possible dependence on the forced or free nature of the vibrations, consequences of restraining the in-line oscillations, the strong facility and amplitude dependence of the lift coefficient, spectral bandwidth of lift force, shedding frequency bandwidth, nonquantifiable growth of the various disturbances in the unsteady shear layers, effects on VIV of the distribution of the ambient velocity and turbulence across the test section, blockage ratio, size and shape of the end plates, nonuniformity of the surface roughness, yaw, body deformation, unwanted secondary vibrations of the body and its support system, noise, temperature gradients, and other parameters influencing a given experimental setup. Often a number of small (presumed innocuous) assumptions in analysis and experiments or both may make the interpretation of the results rather difficult. However, we must also heed Stokes' (1851) words of wisdom: "Such extreme precision in unimportant matters tends, I think, only to perplex the investigator, and prevent him from entering so readily into the spirit of an investigation." Obviously, all the dimensionless parameters cannot replace the physical insight that springs from experience and knowledge.

6.4.3 *Mass and structural damping*

Returning to the discussion of $m^*[= \rho_m/\rho_f]$ and ζ versus $m^*\zeta[= \zeta\rho_m/\rho_f]$, it is noted that there is no compelling reason to combine m^* with ζ. In fact, Sarpkaya (1978a, 1979b, 1989, 1995, 1996b, 1997) and Zdravkovich (1990) suggested over the years that they should not be combined to form a new parameter (or to eliminate an independent parameter). Nevertheless, it has almost become a common practice, at least until recent experiments with small m^* (large ρ_f and small ρ_m) to combine the two parameters into a so-called "mass-damping" parameter $m^*\zeta$ (with $\delta = 2\pi\zeta$). This will be discussed in more detail later in connection with the consequences of free or self-excited oscillations where the structural-damping-dependent response may not necessarily be a sinusoidal or stationary random process. Suffice it to note that m^* and ζ play very important roles in VIV. According to KW (1999), the range of synchronization is controlled primarily by m^* (when $m^*\zeta$ is constant), whereas the peak amplitudes are controlled principally by the product of $m^*\zeta$ in the range of Reynolds numbers $(3.5 \times 10^3 - 10^4)$ encountered in their experiments. However, the dependence of A/D on ζ, ρ_f/ρ_m, $U/f_{vac}D$, f_{vac}/f_{com}, the lift coefficient C_L, the phase angle φ, and the Reynolds number remain unresolved as shown below in Fig. 6.7.

The Skop and Balasubramanian data (1997) may be represented by

$$\frac{A}{\gamma D} = 1.12\,e^{-1.05 \cdot S_G} \tag{6.4.2}$$

The scatter in the data, particularly for S_G less than about 0.5 (i.e., mostly for experiments in water), is of practical as well as fundamental interest and is not entirely due to experimental errors. It is partly due to the use of relatively small Reynolds numbers (less than 10^4) for which the transition to turbulence on the free shear layers does not move sufficiently upstream, magnification of the uncertainties in the damping factors, the end conditions (i.e.,

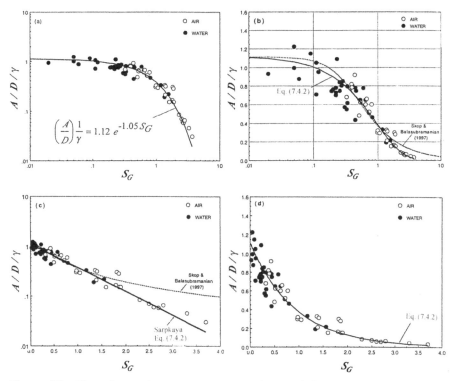

Figure 6.7. Experimental measurements of the modally normalized maximum amplitude versus the response parameter S_G : (a) a log–log plot, (b) a linear–log plot, (c) a log–linear plot, and (d) a linear–linear plot.

the restraint applied to the ends of the cables for or against *axial* motion), the history of the motion (e.g., whether a particular steady-flow velocity has been arrived at from a lower or higher value; Brika and Laneville 1993), and, most importantly, from the particular behavior of the wake and the lift coefficient C_L, leading to strongly nonlinear dependence of the force coefficients on A/D.

6.4.4 f_{vac}, f_{com}, *and added mass*

In 1995, we denoted f_{com} by f_{ny} (prevailing frequency) to include the self-excited oscillations. Moe and Wu (1990), working with self-excited as well as forced oscillations, introduced f_{true}, the frequency "at which the cylinder is actually vibrating." Here it is denoted by f_{com} to cover both the forced and self-excited oscillations. It does not remain constant throughout the synchronization range because the added mass coefficient C_a in viscous flow is not only a function of the shape, orientation, and the physical surroundings of the body, but is also a strong function of the resulting viscous fluid motion. In fact, f_{com} (either from experiments or from DNS) can be used together with f_{vac} to determine *a posteriori* the actual added mass,

$$\Delta m \left\{ = m[(f_{\text{vac}}/f_{\text{ex}})^2 - 1] \right\} \tag{6.4.3}$$

or the added-mass coefficient,

$$C_a = m^*[(f_{\text{vac}}/f_{\text{ex}})^2 - 1] \tag{6.4.4}$$

Clearly, the cycle-averaged value of Δm is zero for $f_{\text{com}} = f_{\text{vac}}$, and Δm (and C_a) are negative whenever $f_{\text{com}} > f_{\text{vac}}$, *regardless of the viscous medium in which the synchronization occurs at the common frequency f_{com}*. For experiments conducted in a dense fluid (say, water), the contribution of the air to the added mass may be neglected, i.e., $f_{\text{air}} \approx f_{\text{vac}}$. Then one has, for the added water mass, $\Delta m_{\text{wtr}} \approx m[(f_{\text{vac}}/f_{\text{com}})^2 - 1]$. In any case, C_a for a circular cylinder oscillating in water is not equal to its ideal value of unity. Rewriting the preceding expression as

$$C_a/m^* = \Delta m/m = (f_{\text{vac}}/f_{\text{ex}})^2 - 1 \tag{6.4.5}$$

it is seen that for a given m^*, C_a is strongly dependent on the accuracy of $f_{\text{vac}}/f_{\text{com}}$. Also, rewriting the normalized total mass or "virtual mass" as

$$(1 + C_a/m^*) = (1 + C_a\rho_f/\rho_m) = (f_{\text{vac}}/f_{\text{ex}})^2 \tag{6.4.6}$$

and noting that C_a may have *negative as well as large positive values* for a circular cylinder (see, e.g., Sarpkaya 1977c, 1978a; Gopalkrishnan 1993; Vikestad *et al.* 2000), one observes that the role of the added mass at high mass ratios is minimal or that f_{com} *approaches* f_{vac}. However, for small mass ratios, C_a becomes increasingly important as the mass ratio m^* becomes smaller, i.e., for a large (positive) C_a and small m^*, the *virtual mass* (added mass plus m) may acquire very large values. More importantly, however, for a relatively large *negative* C_a (say, –0.7) and small m^* (for example, 0.6), the *virtual mass becomes negative*, a result that is unacceptable. Thus, one must have $C_a/m^* > -1$ or $(m^* + C_a) > 0$. It is clear that bodies of very small m^* will have rather large in-vacuum natural frequencies (f_{vac}). Thus, under the circumstances in which C_a has a large negative value and the out-of-phase component of the lift force is positive (*energy transfer from the fluid to the body*), the system can develop rather large amplitudes. It must be emphasized that the positive as well as the negative values of the added mass strongly depend on A/D, as first shown by Sarpkaya (1977c, 1978a), and, to a lesser extent, on the Reynolds number, provided that the Reynolds number is larger than about 1.5×10^4 to 2×10^4.

Equation (6.4.6) may also be written as

$$\frac{m^*}{m^* + C_a} = \frac{(f_{\text{com}}/f_{st})^2}{(f_{\text{vac}}/f_{st})^2} \tag{6.4.7}$$

which emphasizes the role of f_{ex}/f_{st} and f_{vac}/f_{st} in determining the ratio of the normalized mass of the oscillating body to its virtual mass. Only when $C_a = 0$, one has $f_{\text{ex}} = f_{\text{vac}}$. For all values of $C_a < 0$, $f_{\text{ex}} > f_{\text{vac}}$, and, conversely, for all values of $C_a > 0$, $f_{\text{ex}} < f_{\text{vac}}$. Forced vibration experiments show that C_a for $A/D = 0.5$, for example, goes through zero at $f_{\text{ex}}/f_{st} \approx 0.85$ and abrupt changes occur in phase and the energy transferred to the cylinder from the fluid as f_{ex}/f_{st} decreases, say from, $f_{\text{ex}}/f_{st} \approx 0.88$ to $f_{\text{ex}}/f_{st} \approx 0.80$. For cylinders with relatively *large* m^*, $f_{\text{ex}}/f_{\text{vac}}$ remains close to unity, the shedding mode changes from a Karman-type of vortex shedding (two single vortices per cycle) to two pairs of vortices per cycle (to be discussed in more detail later). However, for cylinders with *small* m^*, the mode of vortex shedding remains essentially the same. The importance of these interesting phenomena, leading to a strong connection between the forces, masses, VIV suppression, and mode and phase changes will become apparent later.

6.5 Linearized equations of self-excited motion

The equation of motion for a body of single degree of freedom with linear springs and damping has been known for a long time and constitutes one of the classic examples in most vibration texts. Briefly, it may be written as

$$m\ddot{y} + 2m\,\zeta\,\omega_y\,\dot{y} + ky = \frac{1}{2}\rho DU^2 C_L \sin \omega_s t \tag{6.5.1}$$

in which m is mass per unit length of the cylinder and includes the added mass, ζ is *the* (*structural*) damping factor, k is the linear spring constant, ρ is the density of fluid (in mass units), D and L are the diameter and length of the cylinder, U is the "steady" ambient velocity, $\omega_y = 2\pi f_y$, and $F_L = 0.5\rho U^2 D C_L \sin(\omega_s t)$ where C_L is the strongly amplitude-dependent lift coefficient. Undoubtedly, its determination constitutes the weakest link in the analysis of VIV. This is partly due to the fact that events giving rise to lift are the least deterministic of all unsteady-separation events in fluid dynamics.

With the foregoing limitations on mind, we will proceed with the development of a "deterministic" model. Expecting a sinusoidal steady-state response with the amplitude A_y, and phase angle ϕ, we have

$$y = A_y \sin(\omega_s t + \phi) \tag{6.5.2a}$$

$$C_y = (C_L \cos \varphi) \sin(2\pi f_{ex} t) + (C_L \sin \varphi) \cos(2\pi f_{ex} t) \qquad (6.5.2b)$$

Inserting (6.5.2a) into (6.5.1) we have (Thompson 1988)

$$\frac{y}{D} = \frac{\rho U^2 C_L \sin(\omega_s t + \phi)}{2k\{[1 - (\omega_s/\omega_y)^2]^2 + (2\zeta \omega_s/\omega_y)^2\}^{1/2}} \qquad (6.5.3)$$

and the phase angle

$$\tan \phi = \frac{2\zeta \omega_s \omega_y}{\omega_s^2 - \omega_y^2} \qquad (6.5.4)$$

Obviously, the response reaches its maximum value when the shedding frequency and the natural frequency of the cylinder become nearly equal, *leading to a state of resonance.* We then have for $f_y = f_s$

$$(A_y/D) = C_L/4\pi S_t^2 \delta_r \qquad \text{with} \qquad \delta_r = 4m\pi\zeta/\rho D^2 \qquad (6.5.5)$$

Equation (6.5.3) is nonlinear because of the strong dependence of C_y on the amplitude of the cylinder displacement. However, at lock-in, the displacement y and the transverse fluid-force coefficient C_y are usually expressed, to an unknown order of approximation, by sinusoidal functions (known as the harmonic model approximation) in which the force leads the displacement by a phase angle φ,

$$C_y = (C_L \cos \varphi) \sin(2\pi f_s t) + (C_L \sin \varphi) \cos(2\pi f_s t) \qquad (6.5.6)$$

The displacement and acceleration are zero at $y = 0$ (the mean position) and the absolute values of the displacement and acceleration are maximum at $y = A$. The cylinder decelerates as it moves toward larger $|y|$ and accelerates as it moves toward smaller $|y|$ values.

It is a well-known fact that in *steady flow* the angle that the shear layer separates from a cylinder is affected by the Reynolds number. A shear layer on a body undergoing VIV is expected to be strongly affected by whether the relative flow past the cylinder is accelerating or decelerating. Thus, the variations of the total relative velocity about the cylinder and the alternating nature of the accelerations and decelerations of the cylinder could lead to highly complex excursions of the separation points and shear layer transitions. Krishnamoorthy *et al.* (2001) observed that "both shear layers develop abrupt double roll-ups as the cylinder accelerates from the top-dead-center to its mean position."

The parameter A/D is dependent on C_L, the phase angle φ, ζ, ρ_f/ρ_m, $(U/f_{vac}D)$, and $(1 + C_a\rho_f/\rho_m)^{1/2}$ or f_{vac}/f_{com}. There is, however, no mathematically derivable dependence on couple of parameters devised by KW (1999): $(m^* + C_A)\zeta$, where C_A is the potential flow added mass coefficient and $f^* = [(m^* + C_A)/(m^* + C_{EA})]^{1/2}$. These are purely empirical expressions, devoid of fluid physics.

Returning to the added mass and recalling that $C_a = m^*[(f_{vac}/f_{com})^2 - 1]$, one has for $f_{com} < f_{vac}$, i.e., for a positive added mass, $\tan \varphi > 0$ and for $f_{com} > f_{vac}$, i.e., for a *negative added mass*, $\tan \varphi < 0$. Thus, at $f_{com} = f_{vac}$, the phase angle shifts by 180°. Conversely, a 180-degree phase shift corresponds to a change in the sign of the added mass from positive to negative or vice versa, depending on the direction of the change of f_{com}. It must be emphasized that any change in C_a, and thus in phase, is not sudden. The reorganization of the wake occurs over a finite time or frequency range (see also Section 2.9).

6.6 Unsteady force decomposition

The lift coefficient in (6.5.1) must be determined analytically, numerically or experimentally if any progress is to be made toward solving industrially significant VIV problems. It is much easier and more reliable to quantify the lift coefficient using forced vibrations (because of the constancy of the desired amplitude and frequency) in spite of the fundamental differences between the self-excited and forced oscillations. It is a well-known fact that the force-transfer coefficients used in designs (say, that of risers) do not come from the free VIV tests or numerical simulations. They come from forced oscillation experiments (e.g., Sarpkaya 1977a, 1977c, 1986b, 1995; Moe and Wu 1990; Gopalkrishnan 1993, just to name a few) or from commercial test facilities enabling the use of large-scale models.

Sarpkaya (1978a), using $y = -A \sin \omega_{com}t$ and $U(t) = -U_m \cos \omega_{com}t$, expressed C_L as

$$C_L = C_{mh} \sin \omega_{com}t - C_{dh} \cos \omega_{com}t \qquad (6.6.1)$$

where $\omega_{com} = 2\pi f_{com}$ and the coefficients C_{mh} and C_{dh} are the Fourier averages, over many cycles of oscillation (about 100), of the transverse component of the normalized force acting on the cylinder. They are assumed to depend on A/D and the Reynolds number. Equation (6.6.1) may also be written as

$$C_L = -C_a \sin \omega_{com}t + C_d \cos \omega_{com}t \qquad (6.6.2)$$

Here, C_a is the added-mass coefficient and C_d is the drag coefficient. If the body were at rest and the fluid oscillated about it, C_a needs to be replaced by $1 + C_a$ to account for the effect of the imposed pressure gradient. The genesis of the above equations is the so-called "Morison–O'Brien–Johnson–Schaaf" equation or the "MOJS" equation (1950), according to which the time-dependent force exerted on a body moving with the velocity $U(t)$ in a fluid otherwise at rest is assumed to be a linear sum of an acceleration-dependent inertial force and a velocity-square-dependent drag force, i.e.,

$$F(t) = \frac{1}{2} \rho \, C_d D |U| U + \rho C_a \frac{\pi D^2}{4} \frac{dU}{dt} \qquad (6.6.3)$$

The coefficients of the two forces are determined experimentally by measuring the force and calculating their Fourier or least squares averages (Sarpkaya 1976b, 1976e, 1977a, 1986a, 1986b, 1987). More intricate formulations of the time-dependent force at the price of complexity turned out to be arguably not beneficial (Sarpkaya 1985, 2000, 2001b).

For a cylinder oscillating sinusoidally with $U = U_m \sin \omega_{\text{com}} t$, the first term on the right-hand side of the above equation has often been *linearized* (provided the circumstances be such that the nonlinearity of the square of the velocity may be neglected) to yield (6.6.2) (Sarpkaya 1978a). Rewriting (6.5.6),

$$C_y = (C_L \cos \varphi) \sin(2\pi f_{\text{com}} t) + (C_L \sin \varphi) \cos(2\pi f_{\text{com}} t) \qquad (6.5.6\text{-R})$$

and comparing it with (6.6.2), one has

$$C_a = -C_L \cos \varphi \qquad \text{and} \qquad C_d = C_L \sin \varphi \qquad (6.6.4)$$

from which one can find $C_L = \left(C_a^2 + C_d^2\right)^{1/2}$ and the phase angle $\varphi = \tan^{-1}(-C_d/C_a)$. The maximum of C_L, from (6.6.2), occurs at $\theta_m = \tan^{-1}(-C_a/C_d) = \pi/2 - \varphi$. In other words, the following holds true for the linear approximation of the sinusoidal VIV regardless of whether it is obtained from the assumptions leading to the sinusoidal displacement and a sinusoidal force with a phase of φ or from a *linearization* of the MOJS equation,

$$\varphi = \tan^{-1}\left(-\frac{C_d}{C_a}\right) = \frac{\pi}{2} - \tan^{-1}\left(-\frac{C_a}{C_d}\right) = \frac{\pi}{2} - \theta_m \qquad (6.6.5)$$

or, combining it with (6.5.4), one has

$$\varphi = \tan^{-1}\left(-\frac{C_d}{C_a}\right) = \tan^{-1}\left(\frac{2\zeta\, f_{\text{vac}}\, f_{\text{com}}}{f_{\text{vac}}^2 - f_{\text{com}}^2}\right) \qquad (6.6.6)$$

As far as the forced oscillations are concerned, the linearized version of (6.6.3) is no more or no less empirical than Eq. (6.5.2a). However, for *the self-excited vibrations*, the original nonlinearized form of the MOJS equation is certainly more representative of the prevailing state of the flow, particularly at industrially significant Reynolds numbers.

The linearization represents a considerable simplification for both the ocean and the laboratory environment. The transverse as well as the in-line force for more complex oscillations (e.g., beating phenomena, oscillations with irregular or modulated amplitudes) cannot be represented by a two-coefficient model and certainly cannot be analyzed on a cycle-by-cycle basis in search of a new phenomenon. One needs to determine the Fourier averages of the said coefficients, deduced from many cycles (say, 100) to obtain representative results. There will always be considerable uncertainty stemming from the ambient flow environment. This, in fact, is one of the fundamental reasons for the difficulty of dealing with fluid–structure interactions in a nonlaboratory environment and with self-excited vibrations in both nature and laboratory.

If the motion contains amplitude- as well as frequency-modulated fluctuations (as in the case of free or self-excited oscillations), one must perform a *complex demodulation analysis* to capture the basic features of the data (see, e.g., Bloomfield 2000). In such cases, the assumption of a mean frequency and/or amplitude may lead to less accurate force components and phase angles.

As noted earlier, the single most important parameter in VIV is φ and its dependence on all the parameters cited in (6.4.1). Clearly, φ is destined to exhibit large scatter, particularly in tests with self-excited oscillations and in numerical simulations employing, for example, LES. An orbicular leaf attached to a tree responds with exuberance to droplets of uniform shape, weight, and frequency falling on it from a pipette when the frequency and the phase angle relative to the motion of the leaf are set precisely. When the phase angle is a few degrees off, the leaf comes to a dead stop even if the frequency is kept the same.

6.6.1 *Lighthill's force decomposition*

Lighthill (1986) asserted that the viscous drag force and the inviscid inertia force acting on a bluff body (subjected to unsteady motion in a viscous fluid otherwise at rest) operate independently. This assertion is false. *Viscosity is not drainable or distilled from viscous fluids, except at temperatures near absolute zero* (about 2 K). Notwithstanding this fact, Lighthill expressed the MOJS equation, (6.6.3), as

$$F = C_a^* \rho \, (dU/dt) V_b + \frac{1}{2} \rho \, A_p U^2 C_d \tag{6.6.7}$$

where C_a^* is the ***ideal inviscid-flow value of the added-mass coefficient***. It is obvious from the form of (6.6.7) that Lighthill (1986) meant by "viscous drag force" a force that contains *no inertial force* that is due to the motion of vortices. Thus his second term on the right-hand side of (6.6.7) is only a *velocity-square-dependent drag* force.

Sarpkaya (2001b) has shown conclusively that the above assertion is invalid. This is rather obvious from the fact that the subtraction of the ideal inertial force from the total force leaves behind *a vortex-motion force* or, just simply, a *vortex force*, which necessarily contains both a "velocity-square-dependent" drag force and an "acceleration-dependent" inertial force. In other words, the remainder of the total force (the second term) cannot be expressed as a velocity-square-dependent force, with a simple drag coefficient. Lighthill (1986) stated, "I want to argue that, as we necessarily move to more refined methods of estimation, we can appropriately continue to separate hydrodynamic loadings (as Morison's equation does) into vortex-flow forces and potential-flow forces as Taylor (1928a) did." A careful reading of Taylor's papers (1928a, 1928b) shows that they deal with the use of distributed ideal sources and sinks to calculate the added mass of airfoils immersed in an *inviscid fluid* and the determination of the effect of *convective acceleration* on bodies immersed in converging or diverging wind/water tunnels (the so-called horizontal buoyancy) in *steady* inviscid flow. They do not deal with unsteady flows, vortex motion, or "separation of hydrodynamic loading into vortex-flow forces and potential-flow forces," contrary to Lighthill's (1986) assertions. Unfortunately, Lighthill's eminence prevented others from readily accepting the fact that only a Morison-type equation with two or more terms (drag, inertia, and, if further refinement is needed, the addition of one or more combined drag–inertia terms) can closely mimic the physics of viscous fluid loading in Marine hydrodynamics. The facts are clearly brought out in Stokes' solution (1851) of oscillating viscous flow about a sphere:

$$F(t) = \left(\frac{1}{2} + \frac{9}{2}(\pi\beta)^{-1/2} \right) \frac{\rho\pi \, D^3}{6} \frac{dU}{dt} + \left(1 + \frac{1}{2}(\pi\beta)^{1/2} \right) (3\pi\mu \, DU)$$

$$\tag{6.6.8}$$

Evidently, both the inertial and drag components of the force are modified by the effect of viscosity (so-called Basset's history terms, 1888) as seen from the following version of the preceding equation,

$$F(t) - \left(\frac{1}{2}\frac{\rho\pi}{6}\frac{D^3}{dt}\frac{dU}{dt}\right) = \frac{9}{2}(\pi\beta)^{-1/2}\frac{\rho\pi}{6}\frac{D^3}{dt}\frac{dU}{dt}$$

$$+ \left(1 + \frac{1}{2}(\pi\beta)^{1/2}\right)(3\pi\mu DU) \quad (6.6.9)$$

We have to emphasize that at higher Reynolds numbers the effects of separation, turbulence, cavitation, and the proximity of other bodies will quickly render exceedingly complex the understanding of "resistance in unsteady flow" in general and the resolution of flow-induced oscillations in particular.

6.7 Limitations of forced and free vibrations

6.7.1 General discussion

The motivation of this section is not to assess how the forced oscillations can be used to predict the "free" VIVs under otherwise similar conditions, but rather to examine in depth their similarities, differences, and limitations. Obviously, the two cases are not identical: one is driven internally by the wake at an *average* frequency f_{ex} (dictated by the past and the prevailing state of the motion and the periodic forcing arising from it) as the Reynolds number increases with increasing U in $U/f_{wtr}D$. Furthermore, some or all of the Reynolds numbers in the synchronization range may not necessarily be larger than Re_{sl}, making the response dependent on both Re (i.e., on the state of the shear layers) and $U/f_{wtr}D$. Forced oscillations are driven externally at an *exact* frequency f_{ex} at a desired amplitude A/D and Reynolds number throughout the range of f_{ex}/f_{st} or $V_r = U/f_{ex}D$. One is limited largely by the lock-in regions and the other to a large range of reduced velocities and amplitudes. Forced oscillations help to regularize and idealize almost every aspect of the vortex-induced oscillations, leading to nearly pure sinusoidal oscillations, forces, and almost repeatable wake states. Thus, one has an exact knowledge of the frequency of motion and of the magnitude and shape of the displacement (and hence, of the velocity and acceleration). Even then, however, one expects and finds differences in the structure of the wake from cycle to cycle, for example due to changes in the motion of the separation points and the coherence length.

The bodies undergoing VIV in nature are neither constrained (except by their supports) nor forced to oscillate at a constant amplitude and frequency. Their reduced amplitude (A/D) and Reynolds number vary with

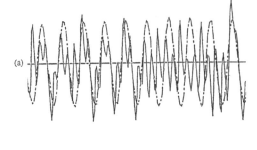

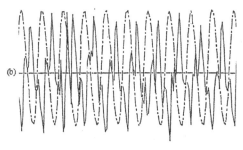

Figure 6.8. Force (continuous lines) and displacement (broken lines) time series. Sample data show that perfect synchronization is not "perfect" and 10% variation in peak amplitudes of force (in both forced and self-excited oscillations) is quite common. (a) Self-excited, in-line spring-supported, $V_r = 5.93$; (b) forced, in-line spring-supported, $V_r = 5.93$. (Moe and Wu 1990).

the reduced velocity and the accurate determination of their individual effects on the remaining governing parameters in general and on the lock-in phenomenon in particular becomes difficult particularly when some or all *Re* values are below Re_{sl}. The extraction and interpretation of reliable force information from free oscillations are more difficult. This is particularly true when the body responds very rapidly to changes in reduced velocity as in Feng's (1968) or in Brika and Laneville's (BL's) experiments (1993) in certain ranges of their reduced velocity $V_{rb} = (V_r/2\pi) = U/(2\pi f_{ex} D)$. Thus, neither the instantaneous center of gravity nor the cycle-averaged value of the virtual mass, nor the force acting on the body, nor the acceleration of the body is known, except for the fact that the virtual mass, the force, and the acceleration are nonlinearly related. This is analogous to deducing information from $F = $ (virtual mass)$\times$ acceleration without knowing *any one of the three elements of the most fundamental equation of motion*. In spite of this, it should be the ultimate objective of the VIV research to predict, to the extent possible, the kinematics and dynamics of self-excited vibrations from forced vibration (physical/numerical) experiments and, equally important, the dynamics of forced oscillations (say, e.g., drag, lift, and inertia coefficients and the phase angle) from the physical/numerical experiments with self-excited oscillations. However, perfect synchronization is not "perfect" and 10% variation in peak amplitudes of force (in both forced and self-excited oscillations) is quite common, as shown in Fig. 6.8 (Moe and Wu 1990).

Griffin (1972) carried out both self-excited and sinusoidally forced oscillations under nearly similar conditions. At the same average amplitude, the

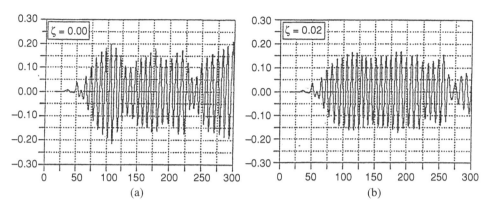

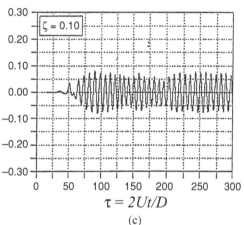

Figure 6.9. Example showing that the computed lift (after sufficient transition time) exhibits "quite large variability" (all for $m^* = 7.85$, $f_{vac} = 0.14$); (a) damping $\zeta = 0.0$; (b) $= 0.02$; (c) $\zeta = 0.10$ (Al-Jamal and Dalton 2004).

free oscillations exhibited amplitude (and, most certainly, phase) modulations, similar to those noted above. The average velocities in the wake were quite similar, but their instantaneous values exhibited the expected discrepancies. The numerical simulations of Al-Jamal and Dalton (2004) and Blackburn *et al.* (2001), among many others, have shown that the computed lift (after sufficient transition time) exhibits "quite large variability" as in Fig. 6.9, and the wave form exhibits interesting "double peaks" as in Fig. 6.10, "likely to be related to the shedding of four concentrated regions of vorticity per motion cycle," as noted by Blackburn *et al.* (2001).

6.7.2 Amplitude and phase modulations

The immediate ramifications of the facts noted above are that (a) the response amplitude given by (6.5.3) becomes increasingly more approximate as the amplitude and phase modulations increase; (b) the second-order fluctuations affect the frequency ratio and the amplitude differently, and (c) the appearance of m^* and ζ as $m^*\zeta$ in (6.5.3) does not mean that $m^*\zeta$ is a

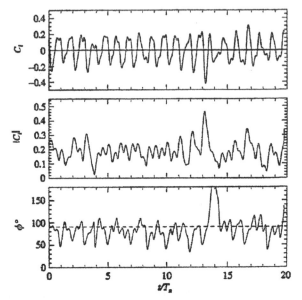

Figure 6.10. Time series of lift, its instantaneous magnitude, and phase angle in relation to cross-flow displacement for the 3D simulations ($Re = 1250$, $St \cdot V_r = 1.33$) (Blackburn *et al.* 2001).

universal parameter of VIV for all ranges of m^* and ζ for a number of reasons. The parameter $m^*\zeta$ is based on a set of linear equations whereas the oscillations are affected by the start-up effects, history effects, and transitional second-order changes, separation excursions, correlation length fluctuations, pressure distribution, to name just a few of the governing and influencing parameters. The imposed amplitude and frequency drive the forced oscillations whereas the free oscillations are driven by the past and the prevailing state of the motion and the forces arising from it. This is particularly significant when transient phenomena take place at certain frequencies in both the idealized experiments (at low Reynolds numbers) and in practical applications (pipes and cables) at larger Reynolds numbers undergoing both in-line and transverse vibrations.

The relative amplitude of the *forced sinusoidal* oscillations can be accurately represented by

$$y_r = (A/D)\sin 2\pi f_{\mathrm{ex}}t \qquad (6.5.2a)$$

where f_{ex} is the driving frequency. However, the representation of the fluid-force coefficient C_y by a linearized equation, such as

$$C_y = (C_L\cos\varphi)\sin(2\pi f_{\mathrm{ex}}t) + (C_L\sin\varphi)\cos(2\pi f_{\mathrm{ex}}t) \qquad (6.5.2b)$$

may be valid only for small A/D values for the following reasons. The so-called Keulegan–Carpenter number $K = 2\pi A/D$ for a cylinder subjected to oscillations transverse to a uniform stream varies in the approximate range of $1.25 < K < 10$. Extensive investigations of sinusoidal flow about a cylinder (Sarpkaya 1976e, 1977a, 1977c, 1986a, 1986b) have shown that the drag and inertia coefficients in the MOJS equation vary with both Re and K. In the range of K values noted above, the variation of the added-mass coefficient is particularly strong for K larger than about 7. This is primarily due to the changes in the structure of the wake on both sides of the cylinder. In the case of a circular cylinder subjected to transverse oscillations, there are additional complications and major differences: the direction of the ambient flow relative to the cylinder changes with time, the oscillations are in the cross-flow direction but the ambient flow is normal to it, and the wakes of the two cases are not even similar. However, one can easily surmise that both the added mass and velocity-square-dependent lift force are even more complex than that for the pure sinusoidal oscillations in a fluid otherwise at rest. This has already been shown by Sarpkaya (1978a) in his forced oscillation experiments. The larger the amplitude of VIV oscillations, *the more nonlinear is the dependence of the lift and inertial forces on A/D, particularly at mode changes, phase jumps, hysteresis, and intermittent switching.* In other words, Eqs. (6.5.2a) and (6.5.2b) must be replaced by a nonlinear force equation to enhance its accuracy if it is to be compared with self-excited vibrations. The only such equation (other than the addition of higher-order terms with either arbitrary or Fourier-averaged coefficients) is Morison's equation. It is easy to show that it reduces to (Keulegan and Carpenter 1958; Sarpkaya 1977a)

$$C_y = C_d \frac{3\pi}{8} |\cos \omega t| \cos \omega t - C_a \sin \omega t \qquad (6.6.12a)$$

The maximum force occurs at $\omega t = \theta_m$ where

$$\theta_m = \sin^{-1} \left(-\frac{4}{3\pi} \frac{C_a}{C_d} \right) \qquad (6.6.12b)$$

It is interesting to note that $|C_d/C_a|$ must be larger than $4/3\pi$. Using (6.6.12.b), one can determine the phase angle with respect to displacement, velocity, or acceleration.

As noted above, the relative amplitude of self-excited vibrations is not constant and the motion is not a pure sinusoidal oscillation. Separation points, pressure distributions, and correlation lengths are history dependent and thus the instantaneous states in forced and self-excited cases at the same amplitude and average frequency do not necessarily give rise to instantaneous similar correlation lengths, pressure distributions, and sectional or total forces. The character of every cycle is determined by the cumulative effect of the prevailing conditions (Basset effect) and by the uncertainty parameters. Every change in amplitude indicates a change in the lift force, which in turn is a reflection of a mismatch between the vortex-shedding frequency and the body frequency, leading to fluctuations in the shedding of vorticity. One can find better force and amplitude representations by accounting for the additional harmonics or by resigning to the use of time-averaged values. At present, there are no analytical models that would suggest the form of the nonlinearity likely to be contributing to the dynamics of VIV. In recent years new methods of multivariate and especially spacio-temporal time series analyses have been developed. Thus, the amplitude may be represented with sufficient accuracy through the use of one of these approximate methods such as (a) the *harmonic analysis;* (b) *proper orthogonal decomposition* (POD, also known as the Karhunen–Loève decomposition); (c) *force-state mapping* (Meskell *et al.* 2001); or (d) *complex demodulation analysis* in dealing with nonexact periodic series (Bloomfield 2000). The POD is an optimal expansion scheme to discretize a random process (Loève 1977) and has been used, e.g., by Lenarts *et al.* (2001); Sarkar and Païdoussis (2003); and Cohen *et al.* (2003). Evangelinos and Karniadakis (1999) have successfully used the *complex demodulation analysis* in connection with their work on the VIV of a cable. If none of the above methods is used, it is suggested that Fourier-averaged force-transfer coefficients be calculated using sufficiently large number of oscillations (experimental or numerical) and incorporated into (6.6.1) to create a model to predict the force as a function of time. Such a force equation should produce, for similarly normalized parameters, the lift and phase, or the drag and the added-mass coefficients obtained from the forced oscillations.

The question is not whether the two fairly idealized VIVs (forced and free, with minimum *Re* larger than about 15 000) are exactly alike or not, but rather whether they are sufficiently alike to extract reliable information from each for purposes of comparison toward a physics-based understanding of VIV. The final decision regarding the applicability of the fixed-body data to the prediction of the characteristics of a freely oscillating body will depend on many more experiments (forced and self-excited) with other rigid and flexible bodies at much higher Reynolds numbers.

The comparison of the flow kinematics (often of DPIV pictures at low *Re*) of two in-line constrained cylinders (one self-excited, one forced to oscillate) under highly idealized circumstances (in-line constrained rigid cylinder, small L/D, better correlation, small *Re*, uniform flow with relatively small turbulence) is far from sufficient to draw scientific and/or industrially significant conclusions regarding the dynamic similarity of self-excited and forced oscillations. The unconstrained rigid or elastic bodies (long cables or pipes with varying degrees of bending and *support* stiffness) often subjected to omnidirectional waves, currents, and shear at much higher Reynolds numbers present highly complex problems. What may be true for the physics of the simplest cases may not at all hold true for the more realistic circumstances.

6.8 Experiments with forced oscillations

6.8.1 *A brief summary of the existing contributions*

Following the pioneering experiments of Bishop and Hassan (1963) with cylinders subjected to forced oscillations in uniform flow, Protos *et al.* (1968), Toebes (1969), Jones *et al.* (1969), Stansby (1976), Sarpkaya (1978a, 1979b), Chen and Jendrzejczyk (1979), Staubli (1983), Moe and Wu (1990), Cheng and Moretti (1991), Gopalkrishnan (1993), Moe *et al.* (1994), Sarpkaya (1995), and Carberry (2002) conducted experiments using forced oscillations.

Bishop and Hassan (1963) reported their lift and drag forces in arbitrary units and assumed the added-mass coefficient to be equal to its ideal value of $C_a = 1$. This invalidated their force measurements and, for the same reason, those of Protos *et al.* (1968) and Toebes (1969). However, Protos *et al.* and Toebes were the first to point out the importance of the phase angle φ in the determination of the direction of the power transfer between the fluid and the cylinder. Bishop and Hassan have identified two critical frequencies ($f_{com}/f_{st} = 0.86$ and 0.95 at $Re = 6000$ and $A/D = 0.25$) delineating a hysteresis loop. The phase angle φ (between the displacement and the excitation) started with a negative value and increased gradually to a positive value of about 90° at the lower critical frequency. Then the motion became unstable and the phase angle jumped from 90° to about 180°, with no intermediate values. However, when f_{com}/f_{st} was increased, from the lower to higher values, the branch of $\varphi \cong 170°$ was followed until the upper critical frequency was reached. Then the phase angle changed abruptly by 180°. In other words, there was no gradual phase change in the interval $0.86 < f_{com}/f_{st} < 0.95$. Many years later, Krishnamoorthy *et al.* (2001) noted, "the phase switch does not occur abruptly. Instead, over several cycles of cylinder oscillations, both "in-phase" and "out-of-phase" vortex shedding occurs during the transition." Our high-speed photographic recording (at a rate of 500 frames/sec) of the forced oscillations of smooth as well as roughened cylinders have shown that the phase changes do in fact occur over several cycles.

Jones *et al.* (1969) forced a large cylinder to vibrate at small amplitudes, at very large $Re = 1.9 \times 10^7$. Their lift coefficients were very similar to those reported at relatively low Reynolds numbers. Stansby (1976) conducted

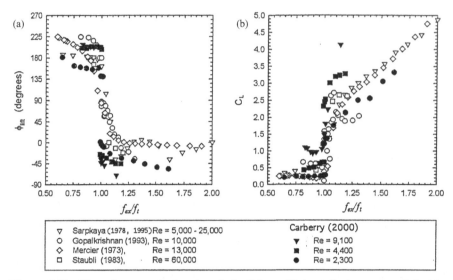

Figure 6.11. The phase angle (φ_{lift}) and the lift coefficient C_L in terms of $f_{\text{ex}}/f_{\text{tr}}$. Apparently, the data for *Re* less than about 10^4–1.5×10^4 differ most from the rest even in a plot pulled together by f_{ex}/f_t. This suggests that the average position of the transition from a disturbed laminar state to a turbulent state has not yet fully migrated upstream in the free shear layers (from Carberry 2002).

forced vibration experiments with a circular cylinder, and he too observed that the phase angle jumped about 180° in the lock-in range for decreasing f_{com}/f_{st} and increasing A/D. The switch occurred at $f_{\text{com}}/f_{st} = 0.86$ (as in the case of Bishop and Hassan) at $Re = 3600$ and $A/D = 0.25$. Stansby associated his observations with the wake width being greater for f_{com}/f_{st} below critical, to being smaller for f_{com}/f_{st} above critical. This reduced the question to what makes the relative wake width vary.

Carberry (2002) subjected circular cylinders to controlled sinusoidal oscillations transverse to a uniform flow at $Re < 10^4$. She has observed many of the well-known characteristics of the forces and transitions as the frequency of oscillation passes through the Strouhal frequency. She called them a "transition" between the "*low- and high-frequency states*" although the frequency changes only slightly during the said transition. As noted earlier, it is a "phase transition" where the "jump" in phase could indeed be very large. Carberry (2002), in an attempt to compare her data with those obtained previously, defined a "transition frequency" f_t at which sharp changes occur in the phase and amplitude of the lift force. Obviously, f_t differed from one data set to another. Thus, using f_{ex}/f_t rather than f_{ex}/f_{st}, she was able to bring into closer agreement the phase (φ_{lift}) and C_L data of Sarpkaya (1978a), Staubli (1983), Gopalkrishnan (1993), and Carberry (2002), as shown in Fig. 6.11. All the data have been obtained at subcritical Reynolds numbers,

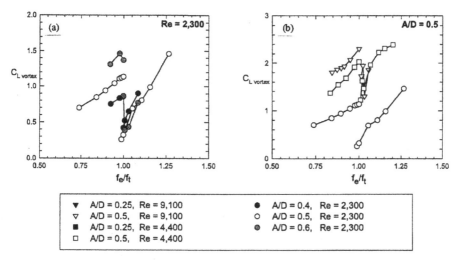

Figure 6.12. Carberry's (2002) data show that $C_{L\text{vortex}}$ increases with A/D for a given Re and with Re for a given A/D in the range of Reynolds numbers from 2300 to 9100.

ranging from as low as 2300 to as high as 6×10^4, over the past 30 years. The Re values of Carberry (2002) are within the said range ($Re = 2300, 4400,$ and 9100), but her data exhibit notable differences from the mean of the others even in a plot pulled closer by f_{ex}/f_t. This appears to be partly due to the fact that her *Reynolds numbers are considerably smaller* than that required for *the transition in the free shear layers* to mature and partly due to the fact that a "transition frequency" f_t does not uniquely characterize a VIV event. Figure 6.12 shows $C_{L\text{vortex}}$ data of Carberry (2002) for a range of A/D at $Re = 2300$ and for a range of Re (2300–9100) at $A/D = 0.5$ as a function of f_{ex}/f_t. Their additional data at $A/D = 0.25, 0.40,$ and 0.60 depict equally large variations with Re and A/D probably partly due to amplitude and frequency modulations, and the increase of the instability in the free shear layers with increasing Re, and partly due to the effects of a number of secondary parameters noted earlier. Carberry (2002) also noted that *small differences in the motion of the cylinder can result in significant changes in the energy transfer.*

Sarpkaya (1977a, 1977c, 1978a) carried out systematic measurements of forces acting on a rigid cylinder vibrating sinusoidally transverse to a uniform water flow ($Re = 6 \times 10^3$ to 3.5×10^4) and expressed the transverse-force coefficient in a manner consistent with the force decomposition discussed earlier [see Eqs. (6.6.4) and (6.6.5)]. For an ambient velocity of $U = U_m \cos \omega_{\text{com}} t$, it was reduced to

$$C_a = -C_L \cos \varphi \quad \text{and} \quad C_d = C_L \sin \varphi \qquad (6.6.4\text{-R})$$

and to

$$\varphi = \tan^{-1}\left(-C_d/C_a\right) = \tan^{-1}\left(2\zeta\,\frac{f_{\mathrm{vac}}\,f_{\mathrm{com}}}{f_{\mathrm{vac}}^2 - f_{\mathrm{com}}^2}\right) \qquad (6.6.6\text{-R})$$

Sarpkaya (1978a) presented the inertia or added-mass coefficients (component of lift in phase with the cylinder acceleration) and drag coefficient (component of lift in phase with the cylinder velocity) for various values of A/D in the range $6 \times 10^3 < Re < 3.5 \times 10^4$. He then used the data in a linear equation of motion to predict the amplitudes of oscillation of an elastically mounted self-excited cylinder. His predictions were in good agreement with the experimental data of Griffin and Koopmann (1977).

Staubli (1983) measured the fluid forces acting on a transversely oscillating circular cylinder in a towing tank. His work was essentially similar to that of Sarpkaya (1978a), but at a higher Reynolds number ($Re \approx 60\,000$). Staubli predicted the vibrations of a freely oscillating cylinder (by Feng 1968) using the results of his measurements. In general, he found good agreement with the experimental data. He has shown that hysteresis effects, which are observed in experiments with elastically mounted cylinders of certain damping and mass ratios, are caused by the nonlinear relation between the fluid force and the amplitude of oscillation. Staubli did not present drag and inertia coefficients. However, they can be derived from his lift force and phase data and vice versa as described in the present paper.

Moe and Wu (1990) made a major effort to conduct both free and forced oscillation experiments using the same apparatus. The cylinders were suspended such that they were (a) free in both the in-line and transverse directions; (b) clamped in the in-line direction, free in the transverse direction; (c) clamped in-line, forced in the transverse direction; and (d) free in the in-line direction and forced to vibrate in the transverse direction. Moe and Wu (1990) obtained a number of important results: (1) the use of the true or prevailing oscillation frequency (our f_{com}) results in very similar lock-in ranges for both the forced and free oscillations; (2) large self-excited cross-flow motions occur for a wider range of the reduced velocity if the cylinders are free to execute figure-eight motions than if they are restrained in the in-line direction; (3) the lift force is irregular for all test cases, and most so for the self-excited case; and (4) relatively large random effects exist in the lift force of the cylinder undergoing self-excited or forced vibrations as seen in Figs. 6.9a and 6.9b. These effects are *stronger* for the in-line fixed cases than for the in-line spring-supported ones.

Gopalkrishnan (1993) carried out *forced oscillation* experiments in a towing tank ($Re = 10^4$) and produced extensive plots of the lift force and phase.

He measured the reaction force only at one end of the test cylinder (suspended at both ends) but performed the data reduction by assuming a uniformly distributed load. Thus, the effects of three-dimensional conditions (e.g., the spanwise variations of the flow) in a nominally two-dimensional flow were not accounted for. His data, as well as those obtained by Sarpkaya (1978a) and Moe and Wu (1990), were compared by Sarpkaya (1995) by plotting the in-phase and out-of-phase components of the lift force (for $A/D = 0.25, 0.50$, and 0.75) as a function of $Vr St = (U/Df_{ex})(f_{st}D/U) = f_{st}/f_{ex}$ in order to account for the variations in Strouhal number (and the Reynolds number) among the three experiments. The level of agreement between the three sets of data suggested to Blevins (1999) that the drag and inertia coefficients obtained by Sarpkaya (1978a), Staubli (1983), Wu (1989), Deep Oil Technology (1992), and Gopalkrishnan (1993) may be combined in a single database and represented with semi-empirical correlations for design purposes only.

The test cylinders in the experiments previously noted were constrained to remain in the transverse plane. As noted earlier, Moe and Wu (1990) carried out additional, but limited, tests with cylinders not constrained in the in-line direction (using suitable springs). Interestingly enough, the random variations in the lift force turned out to be smaller for the in-line-unrestrained cases than for the in-line-restrained cases. Clearly, the data obtained with in-line-restrained rigid cylinders may be of limited use under certain circumstances, particularly when the natural frequency of the cylinder in the in-line direction is made or becomes increasingly larger than that in the transverse direction. Thus the exploration of the biharmonic motion of a cylinder elastically supported in both directions becomes an issue even more important than that of an in-line restrained cylinder subjected to amplitude-modulated beating motions. Gopalkrishnan *et al.* (1992) subjected rigid cylinders, restrained in the in-line direction, to amplitude-modulated (beating) motions and expressed the transverse force in terms of an "equivalent lift coefficient."

6.8.2 *Detailed discussion of more recent experiments*

Sarpkaya (2004) repeated his 1978 experiments in a new water tunnel (with no free surface) at $Re = 45\,000$ with two 50-mm-diameter cylinders with $L/D = 7$ and $A/D = 0.50$. One of the cylinders was polished to make it as smooth as possible and the other was specially modified for the exploration of VIV suppression. The velocity at the test section (356 mm × 510 mm) ranged from 0.04 m/sec to 0.92 m/sec and the ambient turbulence at the lowest Re was 0.8% and at the highest Re was 1.2%. The objectives of the rather ambitious program were (a) to obtain reliable data at Reynolds numbers higher than that encountered in the 1978 experiments; and, more importantly, (b) to determine the dependence of the key quantifiable parameters (phase angle and the in-phase and out-of phase components of the transverse force) on the Reynolds number at discrete values of Re in the range $2.5 \times 10^3 < Re < 45 \times 10^3$ using the same facility, experimental procedures, and test cylinder. Sample data are shown in Fig. 6.13. The new C_a, C_d, and phase data for $A/D = 0.5$ and $Re = 45 \times 10^3$ have confirmed the earlier findings (within the respective error bands of 4%) and substantiated many of the observations made earlier by Sarpkaya (1978a, 1979b, 1995) and others since then (e.g., Staubli 1983; Gopalkrishnan 1993). Evidently, important variations occur in both C_a and C_d in the range $0.80 < f_{ex}/f_{st} < 0.90$ and the positive values of C_d are the regions of primary and secondary synchronizations, which signify power transfer *from the fluid to the cylinder*.

The phase angle decreases rapidly from a value of about 180° to little above zero at $f_{ex}/f_{st} \approx 1.5$. In addition, important variations occur in all three parameters outside the primary f_{ex}/f_{st} range noted above. Figure 6.14 shows the same data as a function of $V_r = U/Df_{ex}$ where, for a given U and D, the forcing frequency decreases along the V_r axis but the Reynolds number *remains constant* at $Re = 45 \times 10^3$. The data reveal that C_a decreases sharply from about 3.9 to about –0.6, as the reduced velocity increases from $V_r \approx 3.5$ to 5.90. C_a then rises slowly to about -0.4 as V_r increases toward 10. This phenomenon occurs for a wide range of A/D values with different C_a values (see Sarpkaya 1978a). The changes in C_a may be interpreted in a number of interesting ways using (6.4.6) or (6.4.7), i.e., $(1 + C_a/m^*) = (f_{vac}/f_{ex})^2$ or the virtual $mass = (m^* + C_a) = m^*(f_{vac}/f_{ex})^2$. The decrease of C_a from large positive values toward zero, as V_r approaches 5.85, shows that f_{ex} is rising toward f_{vac} (i.e., the virtual mass of the body is decreasing). At $f_{vac}/f_{ex} = 1$, $C_a = 0$, $V_r \approx 5.85$ and the virtual mass of the body becomes equal to the actual mass of the body. Subsequently, C_a becomes negative and acquires its minimum value of about –0.6 (for $A/D = 0.5$) at about $V_r \approx 5.90$. At that point, the body's apparent mass is the smallest it will ever be and f_{ex} is the largest it will ever get for this particular amplitude ($A/D = 0.5$). In short, f_{ex} increases up to about $V_r = 5.90$ and then decreases gradually while remaining larger than f_{vac}.

Figure 6.14 also shows that the drag coefficient C_d (the normalized out-of-phase component of the total instantaneous transverse force) rises sharply for V_r values from about 5 to 5.90, i.e., the drag is in phase with the direction of motion of the cylinder and helps to magnify the oscillations (positive energy transfer). It is positive also in the ranges $3.2 < V_r < 4.4$ and $5.90 < V_r < 10$. The foregoing substantiates the fact already noted by Sarpkaya (1978a) that "synchronization or lock-in is manifested by a rapid decrease in inertial force and a rapid increase in the absolute value of the

Figure 6.13. Inertia and drag coefficients (or the in-phase and out-of phase components of the lift force) and the phase angle as a function of f_{ex}/f_{st} ($A/D = 0.50$, $Re = 42\,500$, $L/D = 7$, smooth cylinder). Perfect synchronization is seen to occur at $f_{ex}/f_{st} \approx 0.85$, accompanied by rapid changes in phase and force-transfer coefficients (Sarpkaya 2004).

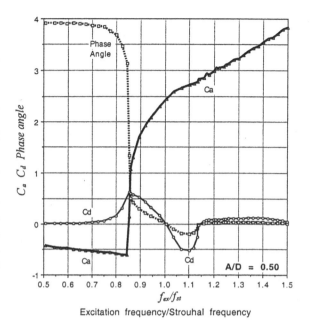

Excitation frequency/Strouhal frequency

Figure 6.14. Inertia and drag coefficients (or the in-phase and out-of phase components of the lift force) and the phase angle as a function of $V_r = U/f_{ex}D$ ($A/D = 0.50$, $Re = 42\,500$, $L/D = 7$ for a smooth cylinder). Perfect synchronization is seen to occur at $V_r = 5.80$–5.85, accompanied by rapid changes in phase and force-transfer coefficients (same as Fig. 6.13 except that the horizontal axis is changed to V_r) (Sarpkaya 2004).

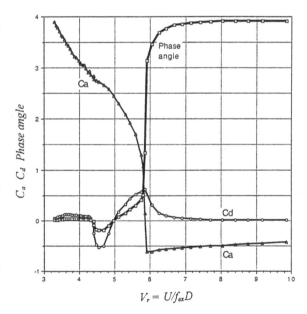

drag force" and that "lock-in is a phase transformer." The data also show that the use of an inertia or added-mass coefficient equal to unity, as determined by oscillating the cylinder in a fluid otherwise at rest, is not correct for modeling the vortex-induced oscillations. Obviously, this fact has nothing to do with the shifting of the ideal added mass to the left side of the equation of motion and dealing only with the so-called "vortex-induced forces," as long as the *actual added mass* (Δm), and *not its ideal value, is used in calculating the prevailing body frequency*, i.e., $f_{ex} = (1/2\pi)[k/(m + \Delta m)]^{1/2}$. Clearly, high m^* values diminish the contribution of C_a/m^* or $\Delta m/m$ to the virtual mass. Conversely, very small values of m^* magnify the contribution of C_a/m^* (for positive values of C_a) and lead to lock-in at very small excitation frequencies, as per (6.4.6). Furthermore, the range of synchronization increases because the variation of f_{vac}/f_{ex} with respect to C_a/m^* decreases according to $(1 + C_a/m^*)^{-1/2}$ and f_{vac}/f_{ex} loses its ability to delineate sharper frequency boundaries.

It has been noted a number of times that the added-mass coefficient C_a can be negative with far-reaching consequences for freely vibrating cylinders. At first it appears paradoxical. As described in great detail in Chapter 2, ***no mass is added to or subtracted from the body.*** The physical shape and mass of the body within the *incompressible control volume* remain invariant. *What is imparted to the fluid (or the body) is positive or negative accelerations or inertia (per unit mass) or changes in kinetic energy due to the motion of the body, which can be negative or positive.* In other words, the *increase* (*or* decrease) of *the kinetic energy* of the fluid within the control volume or the *quotient of the additional force* required to produce the accelerations throughout the fluid divided by the acceleration of the body manifests itself as "added mass." Furthermore, C_a is often the cycle-averaged value of the sum of the masses transported during the periods of acceleration (cylinder moving toward the mean position, i.e., $y \to 0$) and deceleration (cylinder moving toward its maximum amplitude, i.e., $y \to |A|$). Thus, (cycle-averaged) negative added mass means that the drift mass during the deceleration periods is larger than that during the acceleration periods. The basic difference between the kinematics of the two cases is that, during the deceleration periods, the velocity vector relative to the cylinder is oriented toward the axis of the wake, whereas during the periods of acceleration the said net velocity is toward the shear layer from which the cylinder is coming. Undoubtedly, this is a consequence of the particular behavior of the wake vortices during the said periods and the forces they exert on the cylinder (awaiting a direct numerical simulation of VIV at high Reynolds numbers, say $Re > 15\,000$). Similar "negative" added mass has been previously discussed by Keulegan and Carpenter 1958) and by Sarpkaya (1976b, 1976e) in connection with the sinusoidal motion of flow relative to smooth and roughened cylinders. Subsequently, it has been discussed by numerous investigators (see, e.g., Vandiver 1993; Gopalkrishnan 1993; Vikestad *et al.* 2000). In self-excited oscillations, the occurrence of amplitude and frequency modulations is accompanied by corresponding changes in the added mass. For example, a change of $\Delta(f_{vac}/f_{com})$ leads to a change of

$\Delta(C_a/m^*) = 2(f_{vac}/f_{com})\Delta(f_{vac}/f_{com})$, which could be very large when f_{com} is relatively small as in the case of small m^*.

The second objective of these experiments was to determine the dependence of the phase angle and the in-phase and out-of-phase components of the transverse force on the Reynolds number at seven discrete values of Re in the range $2.5 \times 10^3 < Re < 45 \times 10^3$, using the same facility, experimental procedures, and test cylinder. This was prompted by the well-known fact that the Gerrard–Bloor transition waves in the free shear layers disappear in the range of Reynolds numbers from about $Re \cong 2 \times 10^4$ to 5×10^4 in steady flow about a stationary cylinder. Bloor and Gerrard (1966) measured the frequency of transition waves within $1.3 \times 10^3 < Re < 4.5 \times 10^4$ (see also Gerrard 1978; Wei and Smith 1986; Ahmed and Wagner 2003; Zdravkovich 1997, 2003; Norberg 1994, 2003). When the Reynolds number reaches about 2×10^4, the near wake becomes highly three-dimensional, and the eddy formation length does not move closer to the separation point. Unfortunately, there are no comparable *direct* observations and measurements with cylinders subjected to VIV. Thus, the critical value of Re_{sl}, at which the cycle-averaged eddy formation length remains nearly stationary, is not yet known. Furthermore, direct measurements to gather such data appear to be prohibitively difficult. Owing to these facts, we have chosen the three parameters noted above (C_a, C_d, φ) to deduce *indirect* information about the effect of the evolution of the unsteady free shear layers, knowing fully well that the variations observed from one Re to another at the same A/D are not exclusively due to the variations in the shear layers. However, it is not too difficult to assume that the shear layers will certainly have a predominant effect on what happens to the generation of vorticity, shedding of vortices, phase angle, and the force components.

The data obtained at different weeks at each Reynolds number were compared with each other as well as with those at consecutive Reynolds numbers (2.5×10^3, 7.5×10^3, 12.5×10^3, 15×10^3, 20×10^3, and 45×10^3, all with $A/D = 0.5$). For smaller values of Re (2.5×10^3 and 7.5×10^3), the differences between the force coefficients were indeed as large as those encountered by Carberry with $A/D = 0.5$ and $Re = 2.3 \times 10^3$, 4.4×10^3, and 9.1×10^3, as shown in Fig. 6.12. However, when the Reynolds number was increased to 12.5×10^3, the variations in data relative to $Re = 7.5 \times 10^3$ became smaller but certainly larger than the usual scatter in the data. The phase angle exhibited the largest difference, particularly for $f_{ex}/f_{st} > 0.9$. When Re was increased to 15×10^3 and then to 20×10^3, the differences between the tracing parameters reduced to the level of the scatter observed in each experiment. The comparison of the data for $Re = 20 \times 10^3$ and $Re = 45 \times 10^3$ confirmed the conclusion that all three "identification parameters" stabilize and do not materially depend on the Reynolds number, at least in the range $15 \times 10^3 < Re < 45 \times 10^3$. It is tempting to assume that this conclusion will remain true for all Re in the lower subcritical range of Reynolds numbers for smooth uniform flows about smooth circular cylinders, at least for $A/D = 0.5$. Obviously, experiments at larger A/D values and Reynolds numbers are desirable.

6.9 The wake and the VIV

Enormous progress in computers, and in flow visualization and measurement devices, led to the description and classification of the wake states of the simplest bodies, albeit at relatively small Reynolds numbers, and reduced the fundamental questions of VIV from (a) under what circumstances do the various types of VIV occur, and (b) what are the controlling and influencing parameters in all ranges of $m^*\zeta$, to (1) why do the vortices in the wake shed as two vortex pairs per cycle under certain conditions and as two single vortices in a cycle (as Karman vortices) under other circumstances, and (2) what is the role of vortices in the cause-and-effect relationship between their motion, hysteresis, and the controlling and influencing parameters?

Angrilli *et al.* (1972) pioneered a number of experiments to establish a relationship between the vortex shedding and the cylinder displacement. They were the first to introduce the concept of "time of vortex origin" and to suggest that VIV should be called a "self-controlled" or "self-regulated" phenomenon rather than a "self-excited" one since the alternating lift force is not initiated and maintained by the body motion to exist or to persist like galloping. They have carried out experiments with a self-excited cylinder in the range of Reynolds numbers from 2500 to 7000. Figure 6.15 shows the successive positions of the vortices. Angrilli *et al.* (1972) have found that "as long as the oscillation is small," as in Fig. 6.15a, "vortex trails are not very different from those produced by a stationary bluff body, but for larger oscillations," as in Figs. 6.15b and 6.15c, the two vortex trajectories must cross each other twice to reach a stable configuration in the wake." Figure 6.15d shows an unstable configuration.

Zdravkovich (1982, see also 1990, 1996a) was the first to compare the flow visualizations of others about circular cylinders subjected to forced or free vibrations. He suggested that the phase change (about π) in the unsteady lift force near synchronization is connected to a switch in the timing of vortex shedding. This has subsequently been confirmed by Lu and Dalton (1996) through numerical simulations (Fig. 6.16). Based on the flow visualization studies of Angrilli *et al.* (1972), Zdravkovich (1982) pointed out that two

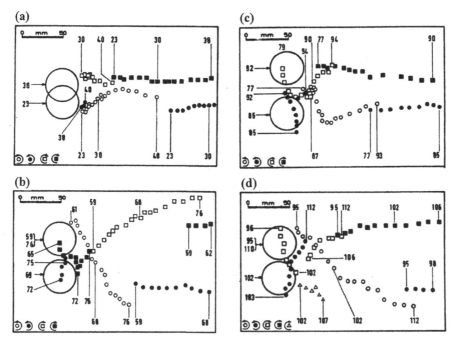

Figure 6.15. Correlation between the vortex patterns and the oscillation of an elastically mounted cylinder in the range of Reynolds numbers from 2500 to 7000. The symbols represent the successive positions of the apparent center of vortices: (a) for $f_{ex}/f_{wtr} < 1$, the vortex trails are not very different from those produced by a stationary cylinder; (b) at $f_{ex}/f_{wtr} = 1$, the vortex trajectories cross each other (twice) to reach a stable configuration in the wake; (c) at $f_{ex}/f_{wtr} \approx 1$, the amplitude becomes maximum; and (d) for $f_{ex}/f_{wtr} > 1$, the vortex configuration becomes unstable (Angrilli *et al.* 1972).

modes of drastically different vortex shedding occur in the lock-in range. At the beginning of the said range, the vortex formed on one side of the cylinder is shed when the cylinder is near the maximum displacement on the opposite side. However, toward the end of the lock-in range, the vortex is shed when the cylinder is near its maximum displacement on the same side. Zdravkovich (1982) also observed that the limit between these two modes is at the reduced velocity at which the maximum amplitude occurs.

Williamson and Roshko (1988, hereafter referred to as WR) drove a vertical cylinder along a sinusoidal path ($300 < Re < 1000$) in a fluid otherwise at rest and photographed the path of suspended aluminum particles on the free surface of the tank. They asserted that the dynamics of the vortices in the near wake is basically inviscid over the said Reynolds number range. As noted earlier, the shear layers could not have yet developed instabilities in the said *Re* range toward transition and turbulence. Thus, the observations, however ingenious, *may not be relevant to higher* Reynolds *numbers*.

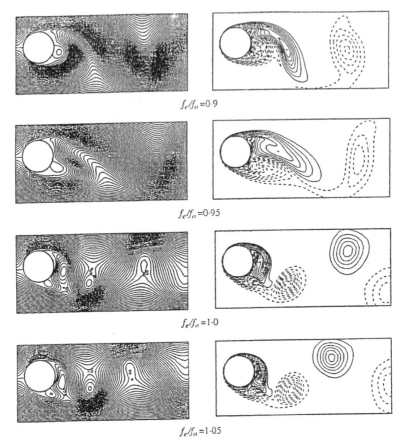

$f_e/f_o = 0.9$

$f_e/f_o = 0.95$

$f_e/f_o = 1.0$

$f_e/f_o = 1.05$

Figure 6.16. Instantaneous streamlines (left) and vorticity contours (right) for various f_{ex}/f_{st} ($A/D = 0.4$, $Re = 1000$). In all cases, the location of the cylinder is at its extreme upper position. At $f_{ex}/f_{st} = 0.95$, a new vortex is being shed from the upper surface. However, at $f_{ex}/f_{st} = 1.0$, the shedding has switched to the lower surface (Lu and Dalton 1996).

WR (1988) described, in the range ($300 < Re < 1000$), the emergence of various regimes (see Fig. 6.17) in a map of A/D versus f_{st}/f_{ex} (as well as $\lambda/D = U/f_{ex}D$), where λ is the wavelength. For $U/f_{ex}D$ (or f_{ex}/f_{st}) larger than a critical value, the wake manifests itself *as long attached shear layers*. Two counter-rotating vortex pairs are shed per cycle in the so-called 2P mode, as shown in Fig. 6.18a (from Brika and Laneville 1993). The 2P mode is associated with the splitting of a region of vorticity in each half-cycle. The DPIV measurements of GW (2000) have shown that the vortices of the 2P mode "convect laterally outwards from the wake centerline, causing a downstream oriented jet type flow close to the cylinder, which in turn results in a "double-wake" type velocity profile." The 2S mode, depicted in Fig. 6.18b, represents the alternate shedding of vortices in the classical Karman mode. Figure 6.19 is a plot by KW (1999) of their data ($Re \approx 3700$) on the map

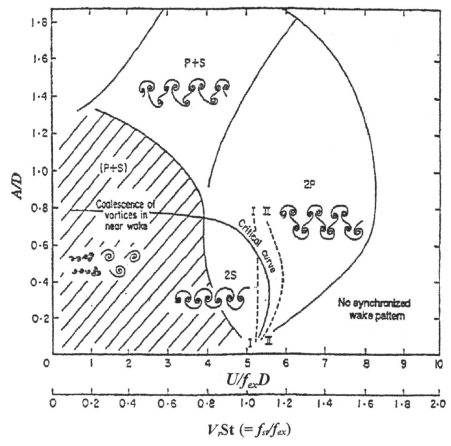

Figure 6.17. Map of vortex synchronization patterns near the fundamental lock-in region. The critical curve marks the transition from one mode of vortex shedding to another (Williamson and Roshko 1988).

shown in Fig. 6.17 showing that the lower-branch regime (see Fig. 6.2 for definitions) collapses well when plotted against f_{st}/f_{ex}.

The region delineated by $0 < A/D < 0.8$ and $0.40 < f_{ex}/f_{st} < 1.80$ in Fig. 6.17 has been replotted in Fig. 6.20b in terms of increasing excitation frequency to relate the various modes relative to the changes in the added mass, the out-of-phase component of the force, and the phase angle in our data (see Figs. 6.14 and 6.20), notwithstanding the large differences between the Reynolds numbers of WR (1988), (300 < Re < 1000), and ours, $Re = 45\,000$.

Near the left boundary of the 2P mode, in the WR map, the out-of-phase component of the lift force (energy input to the cylinder) is small and the phase angle is large (about π) as seen in Fig. 6.20a. As one approaches the critical curve along a constant A/D line (say, 0.5), the phase angle decreases rapidly, the out-of-phase component of the lift increases toward its maximum (i.e., more energy transfer from the fluid to the cylinder), and the

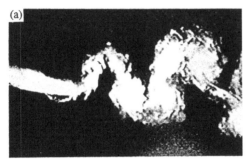

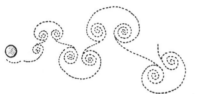

Figure 6.18. Photographs and sketches showing the two near-wake patterns responsible for the hysteresis loop ($V_{rb} = 0.93$, $Re = 7350$): (a) 2P mode ($A/D = 0.40$); (b) 2S mode ($A/D = 0.27$). Both photographs are taken at maximum negative displacement of the cylinder (Brika and Laneville 1993).

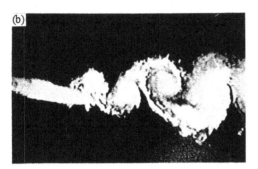

added-mass coefficient approaches its minimum value while remaining negative (i.e., the virtual mass of the oscillating system achieves its minimum value for the specific amplitude). On the critical curve, the 2P mode changes rather abruptly to the 2S mode, shown in Fig. 6.20b, but only for A/D values smaller than about 0.75. The 2S mode is characterized by the shedding of a single vortex in each half-cycle, like Karman vortex shedding, and by rapid increases in $C_L \sin\varphi$, phase angle, and the added-mass coefficient. In fact, the well-known recirculation bubble of the mean velocity field of the wake of a *stationary* cylinder manifests its presence in the 2S mode, showing that the 2S-mode of shedding, coupled with the cylinder motion, may lead to even more dynamic and organized motion. In particular, the coefficient of lift, enhanced by improved correlation, becomes *even larger than that for a fixed circular cylinder in steady flow*, emphasizing the fact that the use of all

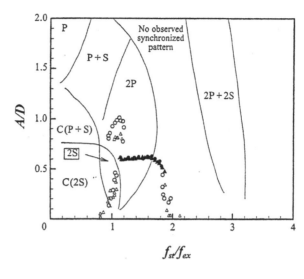

Figure 6.19. A/D for $Re \approx$ 3700 and two m^* values plotted in the WR (1988) map. Circles, $m^* = 1.19$ and $(m^* + C_A) = 0.011$; triangles, $m^* = 8.63$ and $(m^* + C_A) = 0.0145$. Solid symbols indicate the lower-branch regimes (Khalak and Williamson 1999; and Govardhan and Williamson 2000).

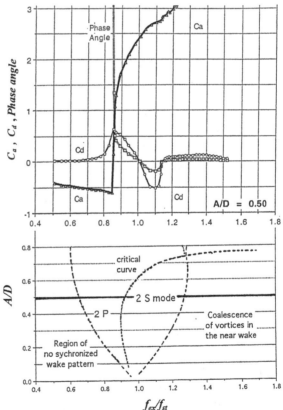

Figure 6.20. Top half: Sarpkaya's data shown in Fig. 6.13, $Re = 42\,500$; bottom half: a replot of the Williamson–Roshko (1988) map, $Re = 300$–1000.

active and passive means to prevent the wake from shifting to the 2S mode is most desirable for all practical purposes.

The rate at which the velocity step is increased changes the vorticity generated on the cylinder boundary. The increased vortex strength might precipitate an earlier hysteretic jump in A/D. For small increments, the vorticity generated is small and diffuses faster without making an impact on the strength of vortices or on the energy transferred to the cylinder. The critical value of f_{ex}/f_{st} is an important example. For $A/D = 0.50$, it is seen to be about 0.85 in Figs. 6.14 and 6.20a. According to Carberry (2002), the transition occurs in the range of $f_{ex}/f_{st} = 0.84 - 0.87$ (for $Re = 2300$). However, according to WR (1988), the critical value is at $f_{ex}/f_{st} \approx 0.94$ for $A/D = 0.50$, as seen in Figs. 6.17 and 6.20b. The difference between this value and those noted above may be due to the use of relatively low Reynolds numbers (300 $< Re < 1000$), as noted above. Even though in the said Re range, and up to about $Re = 10^4$, the Strouhal number for a stationary cylinder remains nearly constant, the all-the-more important shear layer transition to turbulence for VIV (coherence length and separation angle) occurs at Reynolds numbers well above 10^3. Apparently, on the basis of what has been said so far, the critical line in Fig. 6.20b needs to be shifted to the left (to the right in the original plot of WR, Figs. 6.17 and 6.20b) by an amount $\Delta(f_{ex}/f_{st}) \approx 0.10$ at the level of $A/D = 0.5$. Other values of A/D and Re may require different amounts of shifts. This will improve the agreement between their map and the data by BL (1993), KW (1999), and GW (2000). Finally, it is worth noting that accurate numerical solutions at relatively low Reynolds numbers (e.g., Blackburn and Henderson 1999 at $Re = 500$) do not find the 2P mode. Evangelinos and Karniadakis (1999) at $Re = 1000$ found multiple vorticity concentrations and transient mixtures of (P + S) and 2P modes in the near wake and general wake instability further downstream. The repeat of the WR (1988) experiments in a closed system at Reynolds numbers larger than 1.5×10^4, without the use of surfactants, is most desirable in both the forced and self-excited oscillations.

Rodriguez and Pruvost (2000) used a vertical tank, a slightly heated metal cylinder, spanning the width of the test section, and the schlieren technique (based on variations of the liquid-refraction index with temperature). Experiments were conducted at a Reynolds number of 700, at eight different amplitudes ranging from $A/D = 0.125$ to $A/D = 1$. The vortex-shedding phase was defined with respect to the cylinder position y/A at the time of shedding. A vortex is considered shed when it is cut from its feeding sheet in a manner similar to that introduced by Sarpkaya and Shoaff (1979a) in their discrete vortex simulation of flow-induced vibrations. Rodriguez and Pruvost (2000) spanned the synchronization region by varying the cylinder oscillation frequency (f_{ex}) in small steps while holding the amplitude A/D constant at one of the desired values. They have measured the prevailing vortex-shedding frequency f_{vs} *in the wake,* as previously done by Cheng and Moretti (1991). They have presented extensive data and photographs (see Fig. 6.21) for a large number of subharmonics. Their methods of identifying

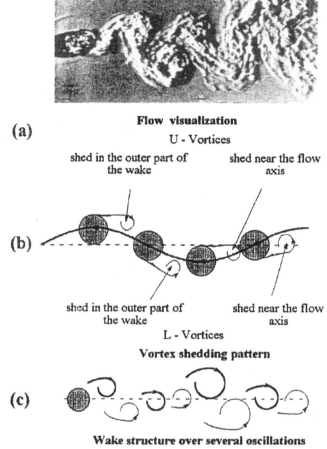

Figure 6.21. Wake-vortex structure over several oscillations of a heated cylinder for $f_{ex}D/U = 0.109$, $f_{ex}/f_{vs} = 1/2$, and $A/D = 0.237$; (a) shows the flow past the transversely oscillating cylinder; (b) depicts the motion of the cylinder relative to the ambient flow and the shedding of the upper U-vortices (at $y/A = \pm1$) and lower L-vortices ($y/A = 0$); and (c) shows that the alternate shedding of the L- and U-vortices, coupled with the direction of the half-cycle, creates a wake with two vortices convected on either side of the axis, followed by two vortices convected along the axis for $f_{ex}/f_{vs} = 1/2$ (Rodriguez and Pruvost 2000).

the various phenomena and their use of significantly different notations (e.g., L and U for the vortices shed from the lower and upper sides of the cylinder) made their results somewhat difficult to interpret. Nevertheless, Rodriguez and Pruvost (2000) have substantiated the fact previously noted by Blevins (1990) and Blevins and Coughran (2008) that the maximum lift coincides with the shedding of the L-vortices and the minimum lift with that of the U-vortices. This is because the shedding of an L-vortex causes the separation point on the upper surface to move *downstream*. This, in turn, accelerates the flow on the top surface and increases the lift. Rodriguez and Pruvost

(2000) have also noted that the motion of a cable (their primary objective) is directly related to the nature of the coupling of the numerous synchronization ranges and therefore it is important to assess whether such coupling can occur in the range of $0 < A/D < 1$, and how and when the transition from one mode to another occurs if the velocity in the far field varies with time.

It has been known since the 1960s that the transition, the wake states on either side of the transition, and the way in which they occur depend on the governing parameters and the uncertainty parameters given by (6.4.1). The singling out of a frame or two in the high- or low-phase regions of the oscillation is not too meaningful because the inertial forces called into action prevent an abrupt rupture of the prevailing conditions where there can be almost seamlessly incremental transitions from cycle to cycle. As noted earlier, Krishnamoorthy *et al.* (2001) have observed that "the phase switch does not occur abruptly. Instead, over several cycles of cylinder oscillations, both "in-phase" and "out-of-phase" vortex shedding occur during the transition." Other works dealing with the comparison of the wake states of free and forced oscillations is deferred to the next section for reasons, which will become clear later.

6.10 Self-excited vibrations

Numerous contributions have been made toward the understanding of the kinematics and dynamics of self-excited vibrations of mostly circular cylinders: Marris and Brown (1963); Koopmann (1967b); Feng (1968); Angrilli *et al.* (1972); Griffin (1972); Stansby (1976); Griffin and Koopmann (1977); Zdravkovich (1982, 1990, 1996); Williamson and Roshko (1988); Brika and Laneville (1993); Vandiver (1993); Moe *et al.* (1994); Sarpkaya (1995); Skop and Balasubramanian (1997); Balasubramanian *et al.* (2000); Hover *et al.* (1997, 1998); Atsavapranee *et al.* (1998); Zhou *et al.* (1999); Khalak and Williamson (1999); Govardhan and Williamson (2000); Davis *et al.* (2001); Laguë and Laneville (2002); Voorhees and Wei (2002); and on numerical simulations of long flexible cables by Newman and Karniadakis (1997); Bartran *et al.* (1999); Evangelinos and Karniadakis (1999); Triantafyllou *et al.* (2003), Laneville (2006); Huarte *et al.* (2006); Brankovic and Bearman (2006); Jhingran and Vandiver (2007); Iranpour *et al.* (2008); and others. Here only a few of these contributions will be discussed in some detail.

As noted earlier, the most interesting phenomena in VIV are hysteresis and lock-in/lock-out. For a given system, the occurrence of hysteresis depends on the approach to the resonance range – the rate of change of the velocity (with small or large increments or decrements) from a low or from a high velocity. The jump condition (double amplitude response) originates in the fluid system, and not, as once thought, in the cylinder elastic system.

Marris and Brown (1963) were interested with the elastic response of Pitot tubes under normal operating conditions. Each tube was cantilevered (in its plane of motion) to one of the walls of a water channel and subjected to uniform flow in the range ($1000 < Re < 2000$). Marris and Brown were able to change the length of the tube and hence its frequency. They have determined two lengths (and hence two frequencies), one of which was 35% larger than f_{st} and the other about 35% lower than f_{st}. Thus, they were able to bracket the synchronization range of the tubes. To the best of our knowledge, no one has ever varied the frequency of the test beam by changing its length (and, of course, mass and damping ratios), except in industrial applications to avoid synchronization.

As noted earlier, Feng (1968), inspired by Professor G. Parkinson, made one of the more widely known contributions to VIV. Experiments were carried out in a wind tunnel with a single-degree-of freedom flexible cylinder with $m^* = 248$, $\zeta = 0.00103$, and $m^*\zeta = 0.255$, as shown in Figs. 6.1 and 6.2. Feng measured f_{ex}, f_{st}, A/D, and the phase angle φ for three types of experiments: (a) the cylinder started from rest at a prescribed velocity U; (b) the velocity was increased by small increments while the cylinder was oscillating at a steady-state amplitude; and (c) the velocity was decreased by small increments while the cylinder oscillated at a steady-state amplitude. His phase and A/D data were presented as a function of $V_r = U/f_{vac}D$. For the first type of experiments, Feng found that for V_r less than about 5, A/D is very small (the inception phase of oscillations) and the frequency of excitation of the cylinder, f_{ex}, is smaller than f_{vac} and f_{vs}. For V_r larger than about 5, f_{vs} and f_{ex} became one and the same or, what we prefer to call, the common frequency f_{com}, i.e., the lock-in occurred. The amplitude smoothly rose to a maximum value of $A/D = 0.32$ at about $V_r = 6$. However, at $V_r \approx 7$, lock-out occurred, i.e., the vortices returned to their Strouhal relationship and the cylinder to a frequency very close to f_{vac}, and, with further increases in velocity, the amplitude dropped to negligible values near $V_r = 8.6$.

For the second type of experiments (velocity increased by small increments while the cylinder was oscillating), the amplitude reached a much higher value ($A/D = 0.53$) at about the same $V_r(\approx 6)$ where the maximum of ($A/D = 0.32$) occurred in the previous case. However, when V_r reached a value of about 6.4, A/D dropped sharply to the value that was reached from rest at the same wind speed. This drop was accompanied by a change of about 35° in phase and, most certainly, by a change in the wake structure. When the speed was decreased, the amplitude data followed the "steady-flow" case up to a special V_r value of 5.9 at which a second "smaller" jump back to the higher values occurred. This resulted in a *clockwise oscillation-hysteresis loop*, as shown in Fig. 6.1. This jump was accompanied by a phase change of about 60° and a *counterclockwise phase-hysteresis loop*, signaling very interesting changes in the shedding mode of vortices.

There are a number of similarities as well as differences (e.g., the synchronization and hysteresis covered by the velocity range, jumps in the phase) between the data obtained by Bishop and Hassan (1963), Feng (1968), and BL (1993), and for a flexible cylinder by Wu (1989) and Saltara *et al.* (1998), probably due to the differences in $A/D, Re, m^*, \zeta$, the test bodies (cable and rigid cylinders), and the uncertainty parameters.

BL (1993) performed a series of very thorough and well-documented experiments in a wind tunnel with a *flexible circular tube* ($m^* = 2054$) in the range of relatively low Reynolds numbers (from 3.4×10^3 to 11.84×10^3) and small damping ratios (from $\zeta = 0.83 \times 10^{-4}$ for small amplitudes to $\zeta = 2 \times 10^{-4}$ for higher amplitudes, up to a maximum of $A/D = 0.52$). A flexible cylinder was to simulate half of the wavelength of a vibrating cable and to eliminate the end effects. BL (1993) have performed tests in two regimes: PR, the "progressive regime" in which the velocity of air was either (a) increased or (b) decreased *at small increments* while the cylinder was oscillating at its steady-state amplitude; and IR, the "impulsive regime" in which the velocity of air was fixed and the cylinder was either (a) released from rest or (b) externally excited by a shaker, at an amplitude A/D of the order of 0.85, and then released. Furthermore, as an additional experiment, the regime of PR-a was repeated using a velocity step *twice as large* as the original small step. Figure 6.22 shows a comparison of the data obtained in the impulsive regimes with those obtained in the progressive regime as a function of $V_{rb} = (V_r/2\pi) = U/2\pi f_{ex} D$. It should be noted that BL (1993) have plotted their data using f_{ex} in defining V_r rather than f_{vac} (the so-called undamped natural frequency). However, the two frequencies in their in-air experiments do not differ much.

The obvious and striking features of the data shown in Fig. 6.22 are their similarity to those obtained by Feng (1968). The hysteresis emphasizes once again the fact that "one can go a little further if one goes slower and then jumps to where one should be" or "one jumps sooner, a little higher, if one goes faster, and then ends up at the same A/D." The phase angle (not shown here) follows the jumps in A/D quite faithfully. In the regime PR-a (small increments in velocity), A/D increases and reaches 0.53 at $V_{rb} = 1$. The next step (at $V_r \approx \times 1.01$) is into an abyss that reduces A/D to 0.38 (i.e., onto the data for PR-b, the small-decrement line) and increases φ by about 70°. Subsequently, A/D decreases to very small values at $V_r \approx 1.31$. If the regime PR-a is repeated with a larger step, as noted above, the amplitude data follow the PR-a (small-increment data) up to $A/D = 0.13(V_r \approx 0.87)$ and then jump

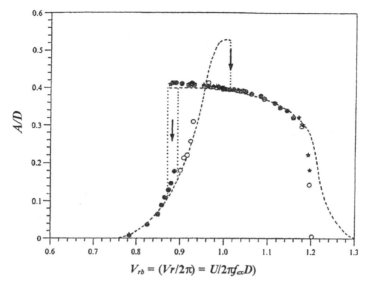

Fig. 6.22. The steady-state vibration amplitude A/D of the impulsive regimes as a function of $V_{rb} = (Vr/2\pi) = I/(2\pi f_{ex}D)$ is compared with the results of the progressive regime: $*$, from rest; $\circ$, from a preexcited amplitude; - - -, progressive regime (BL, 1993). The similarity of the data to those obtained by Feng (1968) is quite striking.

to $A/D = 0.40$ and then continue to decrease as in the case of PR-a (small increment). Obviously, it takes a sufficiently large velocity increment (at the right time) to force the vortices to change to a new configuration sooner. In the absence of a stronger impulse, a stronger velocity, arrived at gradually, serves the same purpose. However, a closer look at the data and the amplitude traces (not reproduced here) reveal that there are subtle differences between the two modes. For example, in the narrow range of $V_r = 0.88$ and $V_r = 0.95$, there are two possible steady-state amplitudes for a given velocity as far as the preexcitation procedure (IR-b) is concerned. BL (1993) have pointed out that for velocities larger than the synchronization onset velocity ($V_{rb} = 0.78$) and smaller than the lower critical velocity (LCV), the recording of the amplitude build-up from rest show intriguing behavior. While the system first tends toward an *unavailable final state* on an imaginary extension of the lower branch ($V_r < 0.88$), it suddenly departs, at a point defined by a break in the envelope curve, toward a second and available steady state of the upper branch. Such bifurcations in the envelope curves are accompanied by an abrupt change in the phase angle and in the wake-flow regime.

BL (1993) ascertained that the phase remains constant along the cable (a flexible tube) within $\pm 5°$ and concluded that the flow mode is not affected by the variation of the vibration amplitude along the cable and that their results should be comparable with those obtained with uniform rigid cylinders. Figure 6.18, presented earlier, shows their photographs as well as sketches of the two near-wake vortex patterns responsible for the hysteresis loop ($V_{rb} = 0.93$, $Re = 7350$) when the cable is at $y = -A$. The upper half shows the 2P mode at $A/D = 0.40$ and the lower half the 2S mode at $A/D = 0.27$. According to Fig. 6.17, $V_r = 5.84$ ($V_{rb} = 0.93$) does not intersect the critical curve and the vortex-shedding mode should be 2P for both cases. However, the large differences in the Reynolds numbers used by WR (1988) and BL (1993), amplitude variations along the cable used by BL, and the small or large increments or decrements of velocity imposed on the ambient flow, or the release of the cable from a preexcited state, may lead to different wake states. It appears that 2P → 2S mode jump is most sensitive to hysteresis, as intimated by BL (1993). This may, in part, be because the 2P mode is the most precarious, and the 2S mode is the most robust of all the known modes.

Clearly, many ways exist (some yet untried) to lead a body to excitation with unimaginable jumps and hysteretic behavior, even without the effect of neighboring bodies (multiple tubes or cables). These may exhibit distinctly different excited modal shapes and hysteresis. The foregoing confirms the fact that the dynamics of the wake is very responsive to changes imposed on the kinematics of the flow and the fact (stated earlier) that $(D/U_0^2)(dU/dt)$ must be quantified during both the acceleration and deceleration periods for one to understand the effect of the rate of change of velocity on the inception of the transient states in both the numerical and physical experiments. Only then can one go beyond the identification of the vortex positions and types of shedding for a given experiment to further the state of the art toward prediction.

BL (1993) compared their work with the contributions of Feng (1968) and Bishop and Hassan (1963), as shown in Fig. 6.23.

The similarities as well as the differences (the synchronization and hysteresis covered by the velocity range, jumps in phase) between the three studies are probably due to the differences in A/D, Re, mass parameter, damping, the test bodies (cable and rigid cylinders), end conditions, and the uncertainty parameters. These need not be discussed here in further detail.

Sarpkaya (1995) discussed the significance of two-directional or biharmonic free oscillations (in both the in-line and transverse directions) in light of experiments undertaken for that purpose, to simulate more closely the true nature of flow-induced vibrations. His data, reproduced here as Fig. 6.24, have shown that the variation of A/D with V_r for the case of "*the same natural frequency in both directions*" yields about 20% larger amplitudes over a 20% larger $V_r St = f_{st}/f_{ex}$ range for a Reynolds number of about 35×10^3.

However, the variations of A/D for other natural frequency ratios (in-line versus transverse) are considerably more difficult, indicating dramatic changes in the wake. Moe and Wu (1990) have noted that the separation

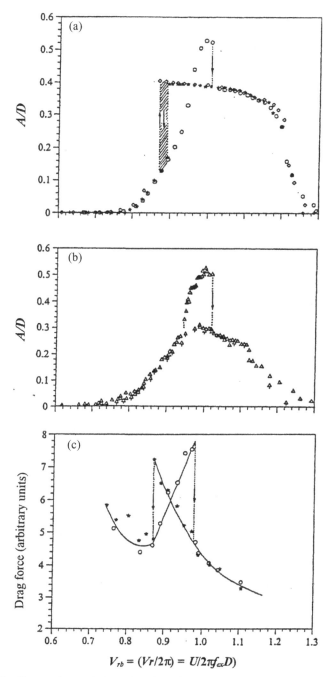

Figure 6.23. Comparison of BL's data (1993) with those of Feng (1968) and Bishop and Hassan (1963). For symbols see Fig. 6.22 and BL (1993).

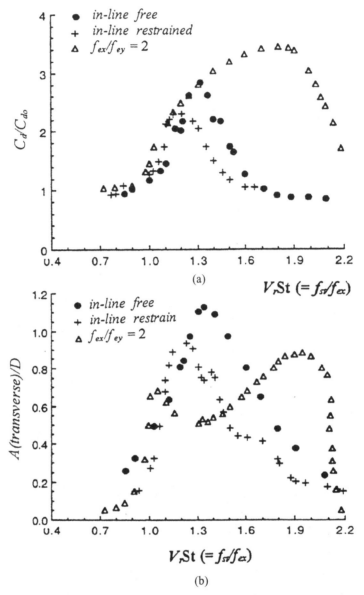

Figure 6.24. (a) the normalized drag coefficient; and (b) A/D are shown for three cases for $Re = 35\,000$: (1) in-line free, (2) in-line restricted, and (3) $f_{ex}/f_{ey} = 2$. It is seen that both C_d/C_{do} and A/D for the case of "*the same natural frequency in both directions*" yield about 20% larger values over a 20% larger $V_r St(= f_{st}/f_{ex})$ range. However, the variations of A/D for other natural frequency ratios (in-line versus transverse) are considerably more complex (Sarpkaya 1995).

points and pressure distribution are strongly affected by the previous history of the motion and the average forces differ from the forces at the average amplitude due to the strong nonlinearity of the in-phase and out-of-phase components of the transverse force. Jong and Vandiver (1985) and, subsequently, Vandiver and Jong (1987) studied the identification of the quadratic system relating cross-flow and in-line vortex-induced vibrations and the relationship between the in-line and cross-flow vortex-induced oscillations of cylinders. They have concluded that a strong quadratic relationship exists between the in-line and cross-flow motions under both lock-in and non-lock-in conditions and that the well-known frequency-doubling phenomena in the in-line response is a consequence of such a quadratic correlation. In other words, the motions in two directions are *not independent* of each other. As noted in Section 4, added mass is a function of the type of motion of the body. Thus, the added mass of a cylinder undergoing 2-DOF oscillations (tracing the path of figure-*eights*) is not the same as that of a cylinder constrained to move only in the transverse direction. The issue is further complicated by the differences between their dynamics (correlation length, kinetic energy imparted to the cylinder, and the phase angle). Marcollo and Hinwood (2002) found, in connection with their work on "the cross-flow and in-line responses of a long flexible cylinder subjected to uniform flow," that "The in-line vibration is found to have a *strong dependency on the cross-flow vibration* and is forced at frequencies very different to that which would be predicted *a priori*." It is obvious from the foregoing and from a more careful perusal of our data shown in Fig. 6.24 that the interaction of cross-flow and in-line oscillations leads to *substantially different results from those found at smaller Reynolds numbers*. The field data at large Reynolds numbers are in conformity with our findings. Figure 6.25 shows that that there is not sufficient data over a large range of Reynolds numbers and mass damping parameters to conclude that the inertia coefficient remains constant along the lower branch. In fact, one of the major objectives to be pursued in the years to come is to carry out experiments of flow-induced vibrations for which the shear layers have reached a stable state at sufficiently high Reynolds numbers (additional recommendations are cited in Sarpkaya 2004).

Triantafyllou *et al.* (2003) have carried out experiments with a single rigid cylinder (1-DOF) and a single flexible cylinder (pinned beam, 2-DOF) at the 2-DOF-Carriage research facility of the Massachusetts Institute of Technology, hereafter referred to as MIT. The experimental conditions were as follows: $Re = 3 \times 10^4$, aspect ratio $= 26, m^* = 3.0, \zeta = 0.035$, and $m^*\zeta = 0.105$. The results are shown in Fig. 6.26. The amplitude A/D (1/10 highest average, treating the response as a random process) of the 1-DOF rigid cylinder reaches a maximum just a little over 1.0 at $V_r \approx 5.6$ and reduces to very small values after $V_r \approx 11$. What is perhaps most important is the absence, rather than the presence, of some observations made at lower Reynolds numbers with low $m^*\zeta$: a hysteretic jump from the initial excitation branch to the upper branch (as in Fig. 6.2). In fact, the variation of A/D,

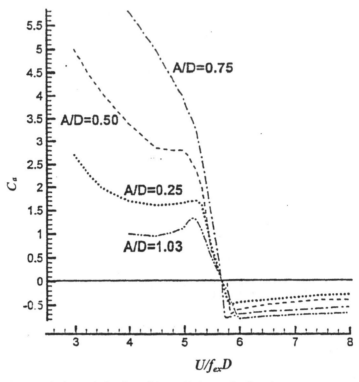

Figure 6.25. Variation of the inertia coefficient C_a (in-phase component of the transverse force) with $V_r = U/f_{ex}D$ for representative values of A/D (from Sarpkaya 1978; V_r is multiplied by 1.13 to match the Strouhal numbers at critical transitions).

in Fig. 6.26a, from its inception to its maximum proceeds rather smoothly. This may be either due to the fact that in the experiments of Triantafyllou *et al.* (2003), $m^*\zeta(= 0.105)$ is not small enough or the Reynolds number in GW's (2000) experiments is not high enough for the Gerrard-Bloor transition waves in the free shear layers to disappear and the shear-layer transition to reach its maximum upstream temporal mean position. In other words, the initial branches observed at low $m^*\zeta$ and at low Re may be a consequence of incomplete transition in the shear layers or, in other words, *a low-Reynolds number effect.* Experiments are needed with $m^*\zeta < 0.01$ where the minimum Re at $V_r = 3$ is larger than about 15000 to resolve the existence or absence of various regimes in the A/D versus V_r plot at industrially significant Reynolds numbers.

Figure 6.26b, for the flexible cylinder (a pinned beam with 2-DOF), shows essentially the same overall behavior as that for the rigid cylinder as far as the absence of various "branches" is concerned. However, the position of the maximum A/D is shifted to larger V_r (about 7) and to larger A/D

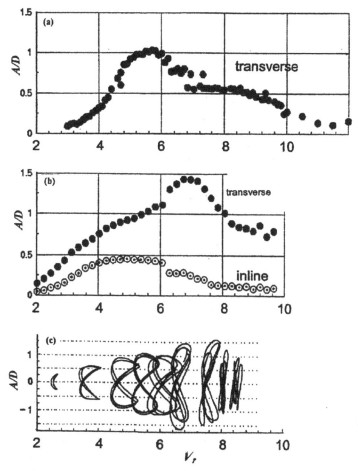

Figure 6.26. (a) One-degree-of-freedom rigid cylinder ($Re = 30\,000$, $L/D = 26$, $m^* = 3.0$, $\varsigma = 0.035$, and $m^*\varsigma = 0.105$). The amplitude A/D (1/10 highest average) reaches a maximum just a little over 1.0 at $V_r \approx 5.6$. There is no upper branch, but only a smooth increase in A/D; (b) a flexible cylinder (a pinned beam with 2-DOF) exhibits essentially the same overall behavior. However, the position of the maximum A/D is shifted to larger V_r (about 7) and to larger A/D (about 1.5); (c) the maximums of the in-line and transverse motion trajectories do not occur simultaneously (the former precedes the latter); see also the forces in 6.26b. (Triantafyllou *et al.* 2003).

(about 1.5). Figure 6.26c shows the motion trajectories of the flexible cylinder. It is evident from the in-line force coefficient as well as from Fig. 6.26c that the largest excursions in the drag direction occur at $V_r \approx 5.5$, not at $V_r \approx 7$ where the maximum amplitudes occur in the transverse direction. Clearly, the maxima of the in-line and transverse oscillations do not occur simultaneously; as seen in Figs. 6.26b and 6.26c, the former precedes the latter. Also the larger in-line amplitudes in the range of $4 < V_r < 6$ suggest improved and sustained correlations conducive to in-line forces.

6.11 Discussion of facts and numerical models

In the foregoing, we have presented mostly experimental facts in the range of Reynolds numbers from about 500 to 6×10^4. These were obtained through small- and medium-scale laboratory experiments under controlled conditions. Many relevant parameters such as the correlation length, pressure distribution, separation points, etc. were not measured and/or reported.

Experiments with forced and self-excited cylinders in nonsheared flows have shown that the amplitude ratio A/D, the phase angle φ, and the in-phase and out-of-phase components of the lift force depend (mostly) on the reduced velocity V_r or its various versions. These parameters are needed for the iterative design of relatively simple cables or pipes subjected to uniform flows.

However, the ocean environment is considerably more complex: lack of data, sufficient insight, and practical experience on the occurrence of VIV at critical to supercritical Reynolds numbers in omnidirectional waves and currents, uniform and/or nonuniform shear, stratification, ambient turbulence, with various types of excrescencies and possible multimodal response of structures, force the designers to continue to use relatively high safety factors. The matter is further complicated by the fact that data obtained either in the large basins around the world or in the oceans by the industry are often case-specific and proprietary. The wide dissemination of the existing ocean data would have helped to resolve many of the issues cited above for the betterment of all concerned. We are grateful to a few who shared some information with us during the past few years so that we could glean some facts from them.

As we have noted previously, in connection with the discussion of the "governing and influencing parameters," Humphries and Walker (1988, hereafter referred to as HW) carried out extensive experiments with a cylinder ($D = 0.168$ m, $L/D = 33$, with a "smooth external finish," $m^* = 1.98$, $\zeta = 0.0143$, $m^*\zeta = 0.0283$, and $f_{\text{wtr}} = 1.23$ Hz) in a deep flume in the range of Reynolds numbers from about $50\,000$ to 4×10^5 in both the nominally uniform flow, $(U_{\max}/U_{\min}) = 1{:}1$, and in linear (positive) shear with 1:1.5, 1:2, and 1:3, with the corresponding shear parameters of 0.0, 0.012. 0.02, and 0.03, respectively. The cylinder was effectively a *pinned–pinned* beam. As they have noted, "The top end of the 'pin' mounting was connected to a hydraulic ram to allow vertical movements as the vertical cylinder flexed. This prevented the introduction of axial tension, and hence changes in natural frequency, due to the model deflection."

The results of HW are shown in Figs. 6.27 and 6.28 for the global drag coefficient and the cross-flow amplitude response, respectively. Figure 6.27 shows that the drag coefficient for the uniform flow ($U_{\max}/U_{\min} = 1{:}1$) is

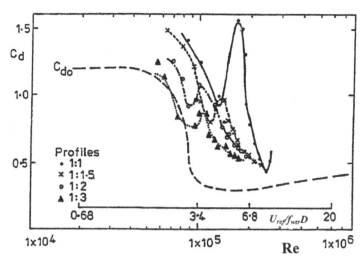

Figure 6.27. The nature of the variation of C_d is like that of a smooth cylinder entering the critical transition, with C_d dropping sharply. The VIV (with 2-DOF) first arrests the drop in C_d and then increases it sharply, proving that VIV occurred within the zone of critical transition. The drag for the uniform flow ($U_{max}/U_{min} = 1:1$) is dramatically amplified with respect to the rigid cylinder (at rest). As to the effects of shear, the maximums of C_d decrease with increasing shear and occur at smaller $U_{ref}/f_{wtr}D$ (Humphries and Walker 1988).

dramatically amplified with respect to the rigid cylinder (at rest) and that VIV in uniform flow produces the largest drag (after an initial anticipated sharp drop) in the critical regime. This highlights an extreme sensitivity to VIV in the critical regime, and perhaps a good "flagship" example of the state of affairs in *Re* scaling, at least for an effectively *pinned–pinned* beam, as noted above.

The nature of the variation of C_d is very much like that of a cylinder entering the critical transition, with C_d dropping sharply. However, when lock-in occurs (with 2-DOF), C_d increases sharply, proving that VIV occurred within the zone of critical transition. Furthermore, even in the critical *Re* region, the occurrence of maximum C_d (at about $U_{ref}/f_{wtr}D \approx 5.6$) in Fig. 6.27 precedes the occurrence of maximum A/D (at $U_{ref}/f_{wtr}D \approx 6.2$) in Fig. 6.28, as in the case shown in Figs. 6.26a and 6.26b, for a subcritical flow. Furthermore, the maximum A/D for the no-shear data (about 1.5 at $U_{ref}/f_{wtr}D \approx 6.2$) corresponds quite well with that in Fig. 6.26b (about 1.5 at $U_{ref}/f_{wtr}D \approx 7$). As to the effect of shear, the data under consideration clearly show that C_d decreases with increasing shear and the lock-in occurs with smaller peaks at smaller $U_{ref}/f_{wtr}D$. Likewise, A/D maxima decrease toward unity. It is further noted that shear extends the range of the lock-in region. However, these observations and the results shown in Figs. 6.27 and 6.28 are of a global nature only since they are based on the reference velocity

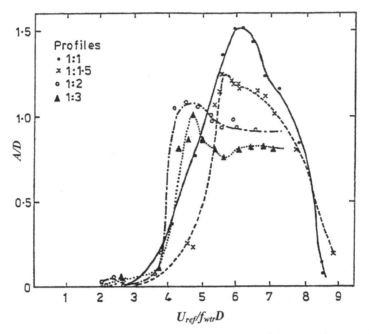

Figure. 6.28. The maximum A/D for the no-shear case (about 1.5 at $V_{\text{ref}}/f_{\text{wtr}}D \approx$ 6.2) corresponds quite well with that in Fig. 6.26 (about 1.5 at $U_{\text{ref}}/f_{\text{wtr}}D \approx 7$). With shear, the maximums of A/D occur with reduced peaks at smaller $U_{\text{ref}}/f_{\text{wtr}}D$ and decrease toward unity. It appears that shear extends the range of the lock-in region. Even in the critical Re region, the occurrence of maximum C_d (at about $U_{\text{ref}}/f_{\text{wtr}}D \approx 5.6$) precedes the occurrence of maximum A/D (at $U_{\text{ref}}/f_{\text{wtr}}D \approx 6.2$), as in Figs. 6.26a and 6.26b for the subcritical flow (HW 1988).

at midspan. With higher shear, the character of the flow (supercritical flow, high-intensity turbulence) at the top of the pipe may be significantly different from that at the bottom (subcritical flow, lower turbulence), and there is no simple way to average out or to explain the difficult-to-quantify occurrences and transitions along the pipe. In spite of all these, shear should not necessarily instill a "fear of the unknown" in most designs.

6.12 Suppression devices

The suppression of vortex-induced oscillations can be accomplished either mechanically or fluid dynamically. The system can be detuned mechanically by increasing the structural stiffness or using mechanical dampers. This detuning generally tries to ensure that the natural frequency of the structure is separated by at least an order of magnitude from the vortex-shedding frequency. The installation of dampers or the stiffening of the structure is not always technically or economically possible and a fluid-dynamic solution would be needed.

Fluid dynamically, the oscillations of a cylinder can be substantially reduced by introducing disturbances on or near the surface of the cylinder that interact with the vortex-shedding mechanism. This interaction can affect the shedding mechanism in four different ways: (a) minimizing the adverse pressure gradient by influencing the point of separation; (b) interfering with the vortex interaction in the near wake: (c) disrupting the vortex-formation length in the wake; and (d) disrupting the coherence of the vortex shedding or the spanwise coherence.

Numerous devices have been proposed in the past 20 years. Their use requires sound judgment, experience, and additional experiments. An excellent review of the rigid and soft suppression devices is presented by Hafen *et al.* (1976). They have noted that streamlined fairings, vortex generators, and studs have been used to influence the boundary-layer separation; splitter plates have been used to prevent vortex interaction; "hair" fairings, "fringe" fairings, and ribbons have been used to disrupt the vortex-formation length; helical strakes, "hair," ribbons, herringbone, and twisted pairs of cable have been used to disrupt the spanwise coherence. Figure 6.29 shows some of the representative suppression devices. Clearly, it is not possible to compare the advantages and disadvantages of these various devices. In general they tend to increase the in-line drag. Furthermore, the use of cables and pipes covered with suppression devices may introduce operational difficulties.

Splitter plates increase the base pressure (reduced drag) and can entirely eliminate the vortex formation. The omnidirectionality of the ocean waves and currents preclude the use of fixed splitter plates. The use and maintenance of self-aligning splitter plates (weather vane fairings) are rather difficult and costly. Doolittle (1974) investigated the effect of trailing ribbons, helical wraps, fairing bodies, masses, rings, and collars. He found that the trailing ribbons can reduce strumming as much as 99%, depending on their length and width, and a single helical wrap, reversing its spiral at midspan, as much as 66%.

Fabula and Bedore (1974) carried out extensive experiments with various suppression devices. Their results show a wide range of strum reduction effectiveness dependent on speed and flow angle. Heavy flags with adequate width and leeside tie-on hair had outstanding performance. Single helical ridges were mostly moderately effective but ineffective for certain conditions. Lighter flags and various ribbon treatments were moderately effective. Boundary-layer trips were only slightly effective and mainly made the vibration more tonal. The hydrodynamic load coefficients varied appreciably with the type of strum reduction device. It is important to note that these results have been obtained with steady flow about cables. Information about the effectiveness of practically every suppression device in time-dependent flows is not available. Thus, special care must be exercised in

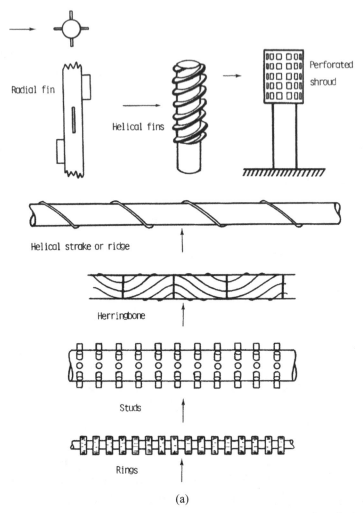

Radial fin

Helical fins

Perforated shroud

Helical strake or ridge

Herringbone

Studs

Rings

(a)

Figure 6.29. (a) Representative rigid strumming suppression devices (Hafen *et al.* 1976).

interpreting the results obtained in steady flows and in using the devices in wavy flows.

Strakes are among the mostly widely used devices. Usually, a three-start helical fin of 5 diameters pitch, each helix protruding about 0.1 diameter from the cylinder surface, prevents in-line and transverse oscillations. The drag coefficient of the straked cylinder in steady flow is nearly independent of Reynolds number and has a value of 1.3 (based on the cylinder diameter). Shrouds consist of an outer shell, separated from the cylinder by a gap of about 0.1 diameter, with many small rectangular holes. Circular holes are not found to be as effective. Shrouds as well as other devices may be used to cover only part of the pipe. However, final design must be based on a careful experimental investigation through the use of sound modeling techniques.

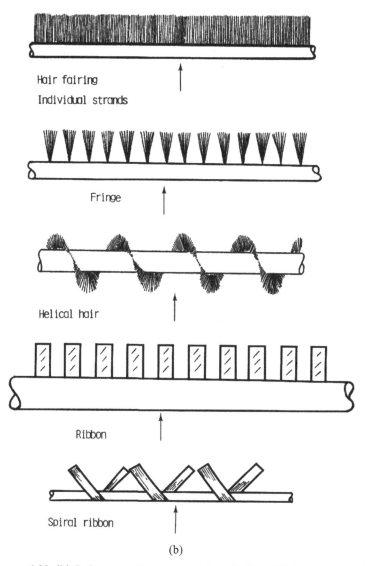

Hair fairing

Individual strands

Fringe

Helical hair

Ribbon

Spiral ribbon

(b)

Figure 6.29 (b) Soft strumming suppression devices (Hafen *et al.* 1976).

6.13 Evolution of numerical models

In recent years, several models have been developed for the prediction of VIV of slender marine structures such as risers and cables. A method for predicting approximately the static and lift responses of a flexible cylinder in a unidirectional sheared current was outlined by Patrikalakis and Chryssostomidis (1986). Their approach, based on the experimental data obtained by Sarpkaya (1978a), represents the multifrequency lift response of a flexible cylinder in a sheared current by predicting a number of independently determined, monochromatic, multimode dynamic solutions. A numerical example assuming bimodal solutions is included to illustrate the method for the geometry of a single-tube marine riser.

Larsen and Halse (1995) provided a direct comparison and comprehensive discussion of the most commonly used models (Larsen and Bech 1986; Lyons and Patel 1989; Skomedal *et al.* 1989; Vandiver 1993; Nedergaard *et al.* 1994; Triantafyllou *et al.* 1994; Vandiver and Li 1994). Only one model was based on computational fluid dynamics (Skomedal *et al.* 1989) and only one model considered a stochastic load process (Vandiver 1993).

Larsen and Halse (1995) concluded that large discrepancies exist between the predictions of the models compared and that *"all aspects of vortex-induced vibrations are still not understood."* These include, but are not limited to, description of the spatial attenuation, definition of the excitation zones, the stochastic nature of vortex-shedding process in time and space, the need to use experimentally determined coefficients, lack of correspondence between the experimental conditions and the application, and the use of different parameters in various databases. Individually, each modeling group tracks the forecast skill of their model.

Future efforts may be directed toward the development of models based on 3D unsteady CFD codes, full-scale and model-scale experimental data, polyspectral methods, and nonlinear control. The objective of the second effort is to generate a database that will be used to develop and calibrate finite-degree-of-freedom models for design and analysis of VIV problems.

A numerical simulation tool was developed by Dalheim (1999) at Det Norske Veritas for the prediction of VIV of flexible risers in sheared currents with some promising results. The studies of large-scale model testing of deep-sea risers (90 m long) in a shear current (Huse *et al.* 1998) have produced a number of important results: (a) vortex-induced vibrations may cause resonant *axial* vibrations in a deep-sea steel riser whether it is pinned at both ends or free at one end and pinned at the other; (b) such axial vibrations may lead to excessive stresses in the risers, significantly larger than the bending fatigue stresses, normally considered to be the main problem of VIV; and (c) reducing or eliminating the lateral as well as the axial excitation by employing the most suitable VIV suppression devices goes a long way toward reducing the high axial stresses.

It is evident from the foregoing that the end conditions of the test pipe, that is, the freedom allowed to the ends to move cyclically in the axial direction, emerges as an important parameter, particularly in the field tests. If both ends of a pipe (say, pinned–pinned) are constrained to extensional motion, the A/D will be smaller than if one or both ends were free to move in the axial direction, as in the case of tests by Humphries and Walker (1988). In a long pipeline, every span is elastically connected to each other. This serves as a strong damper and increases the stiffness of the line, which, in turn, reduces the amplitude of VIV. The amount of reduction in A/D is not, however, easy to quantify because of its dependence on several other parameters (e.g., m^*, ζ, L_{span}/D, stiffness, shearing strains, temperature and its gradients, friction, time-dependent roughness, etc.). It is also clear from the foregoing, as well as from other sources, that high-Re data are not significantly different from those seen at subcritical Reynolds numbers above 20000.

Bruschi *et al.* (1982) described two experimental investigations at high Reynolds numbers. The first was conducted in a wind tunnel with a model cylinder to quantify the wall-proximity effects. The second was carried out with a full-scale pipeline span (with various types of roughness) immersed in a tidal current ($1.7 \times 10^5 < Re < 2.2 \times 10^5$). Bruschi *et al.* (1982) have not presented any graphical or numerical results but made the following observations: (a) the oscillations at the center of the span reached $0.8D$ at $Re = 1.7 \times 10^5$ for a pipe of $D = 0.15$ m, and $A = 2D$ at $Re = 2.4 \times 10^5$ for a pipe of $D = 0.51$ m; and (b) the flow was in the critical transition region at the lower Re and in the supercritical region at the upper Re. They have conducted engineering tests with helical ropes and other "damping" devices (see also Stappenbelt and Lalji 2008).

Allen and Henning (1997, 2003) performed experiments with two flexible as well as nonflexible circular cylinders in the critical/supercritical Reynolds number range ($D_1 \approx 0.088$m, $k/D_1 = 1.37 \times 10^{-4}$, in the Re range of $2 \times 10^5 < Re < 6 \times 10^5$; and $D_2 \approx 0.14$ m, $k/D_2 = 9.94 \times 10^{-5}$, in the Re range of $6 \times 10^5 < Re < 1.5 \times 10^6$). Varying degrees of difficulties were encountered with the end conditions, alignments, Froude numbers exceeding unity (in the test basin), and the duration of the data-acquisition period. Their stationary nonflexible-cylinder experiments tended to agree with the well-known results of Roshko (1961), Schewe (1983a), and Shih *et al.* (1992) at similar k/D values. *The results have once again illustrated the fact that the case of a smooth cylinder is a pathological case and the lightly roughened cylinder becomes the canonical bluff body, particularly at supercritical Reynolds numbers, for both stationary and vibrating cylinders.*

For the smaller cylinder undergoing VIV, the drag coefficient decreased from about 1.2 to 0.6, the in-line rms displacement remained nearly constant at about $A/D = 0.1$, and the amplitude of the transverse oscillations leveled off at $A/D = 0.4$ for Re larger than about 3.2×10^5. For the larger cylinder undergoing VIV, the drag coefficient increased from 0.45 (at $Re = 7 \times 10^5$) to a maximum of 0.7 (at $Re = 1.3 \times 10^6$) and then decreased to about 0.6 at $Re = 1.5 \times 10^6$. The in-line rms displacement remained nearly constant at about $A/D = 0.05$, except at a single point ($Re = 1.05 \times 10^6$) at which it jumped to about 0.13. It was not clear whether this singular event was a Reynolds number effect or, more likely, the effect of the Froude number exceeding unity. The rms amplitude of the transverse oscillations increased almost linearly from zero (at $Re = 7 \times 10^5$) to $A/D = 0.44$ (at $Re = 1.3 \times 1.3 \times 10^6$) and then decreased linearly to about 0.3 at the highest Reynolds number encountered ($Re = 1.5 \times 10^6$).

It is a well-known fact that as the Reynolds number increases from about 4×10^5 to about 10^6, the Strouhal number is indefinable for a steady flow past a "smooth" cylinder. However, for a cylinder subjected to VIV, the *Strouhal number becomes definable in the said Re range* and smoothly transitions from about 0.18 to 0.24, due to the enhanced correlation. The wake width decreases from about D to 0.7 D, and the drag coefficient decreases accordingly. The limited data show that the case of a "smooth" cylinder is a pathological case and the lightly roughened cylinder becomes the canonical bluff body at high Reynolds numbers (see, e.g., Achenbach 1971; Schewe 1983a; Shih *et al.* 1992; Okajima *et al.* 1999). Cylinders of relatively small roughness exhibit relatively small A/D in the usual range of reduced velocities. However, with a roughness of about 1 mm (for instance, over a cylinder of 1 ft in diameter), one may expect a maximum A/D ratio of about 0.95. The same cylinder subjected to forced oscillations yields results very similar to those obtained with self-excited oscillations. The foregoing results are mostly for cylinders constrained in the in-line direction.

Experiments with 2-DOF (both in-line and transverse oscillations) at very high Reynolds numbers show that the results are indeed nearly identical to those reported by Sarpkaya (1995) (see Fig. 6.24), as we have noted earlier. The 2-DOF case produced 10 to 20% larger A/D values reaching about $A/D = 1$ over a wider range of reduced velocities (from about 4 to 13). The maximum A/D values occur in the range of reduced velocities from 8 to 10. In the case of the constrained or 1-DOF cylinders, the maximum A/D is somewhat smaller (about 0.85) in the range of reduced velocities from about 5 to 6, as would be expected. These results require further confirmation, hopefully, in the very near future.

The effect of shear is more difficult to account for in terms of various transitions and axial variations in the flow states. However, the experiments of Humphries and Walker (1988) provide strong guidance for design. Currently, strip theory, based on forced cylinder data and correlation models, is used in spite of its obvious shortcomings (e.g., phase and amplitude differences between the segments, directional changes in velocity and/or shear).

None of the high Reynolds number experiments ($Re > 20000$) show such phenomena as "initial branch" (seen only at Reynolds numbers smaller than about 5000). The vortex modes of 2P, 2S, and others have been mapped only at Reynolds numbers below 1000. Blackburn and Henderson (1999) did not find the 2P mode at $Re = 500$. Evangelinos and Karniadakis (1999) at $Re = 1000$ found multiple vorticity concentrations and transient mixtures of

(P + S) and 2P modes in the near wake and general wake instability further downstream. Brika and Laneville (1993) found 2P and 2S modes in the range of Reynolds numbers from 3.4×10^3 to 11.8×10^3. As we have noted earlier, the mean position of the line of transition to turbulence does not reach upstream enough for Re less than about 15×10^3 to 20×10^3. Lastly, it should be noted that at Re larger than about 20000, it might not be possible to photograph coherent vortex structures in the wake.

In free oscillations at sufficiently large Re, the oscillations are not sinusoidal, as evidenced by many experiments and numerical simulations. Consequently, the flow does not become fully established, say "periodic," because the amplitude, added mass, frequency, phase angle, vortex structures, and shear layers never become fully established. Each cycle is affected by the character of the previous cycle. Consequently, the ever-changing topology of the flow prevents it from exhibiting sharp changes (as branches) in the A/D versus V_r plots at high Re.

Guilmineau and Queutey (2001) used a 2D finite-volume analysis in conjunction with Reynolds-averaged Navier–Stokes solutions (RANS) and a K–ω model for turbulence to simulate the 1-DOF response of a cylinder in the range of Re from 900 to 15×10^3 and compared their predictions with those of KW (1999) at $Re \approx 3700$. Three initial conditions were used in the simulations: a cylinder starting from a state of rest, increasing velocity, and decreasing velocity. For the conditions of "from rest" and "decreasing velocity" they have predicted only the lower branch. With "increasing" velocity, on the other hand, the maximum amplitude corresponded to the experimental value, but the upper branch did not match the experiments (Fig. 6.30). The use of a fully developed turbulence model, (K–ω) for a 2D finite-volume analysis, in the range of *transitional* Reynolds numbers (for the shear layers) makes it difficult to discern the reasons for the differences between the experiments and their numerical simulations.

It may be stated with little reservation that high-Re data (A/D and force-transfer coefficients, such as lift and added mass) with smooth and reasonably roughened, nontapered cylinders exhibit essentially the same results as those the research community has produced in the laboratories during the past 15 years in the range of Reynolds numbers from about 20000 to 60000 (except, as noted above, with regard to the smooth transition in Strouhal numbers and drag coefficients in the critical to supercritical regimes). Thus the issue is not the discovery of great surprises at high Re, but rather how to refine the high-Re data to reduce the error to less than 10%, how to deal better with shear, how to quantify and to translate the effects of the end conditions (axial motion) for single-span laboratory experiments versus continuous pipes, how to account for and quantify multimodes and mode interference, how to suppress VIV at any Re without drag penalty, and how to

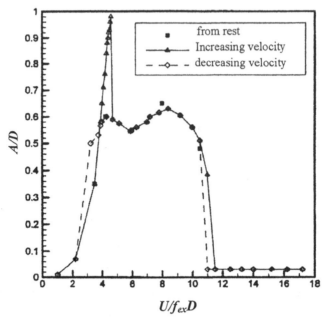

Figure 6.30. A 2D finite-volume analysis in conjunction with RANS equations and a K–ω model for turbulence to simulate the 1-DOF response of a cylinder in the range of Re from 900 to 15000 with $m^* = 2.4$ and $m^*\zeta = 0.013$. In the preceding analysis three initial conditions were used: cylinder starting from a state of rest, increasing velocity, and decreasing velocity. For the conditions of "from rest" and "decreasing velocity," only the lower branch was predicted. With "increasing" velocity, on the other hand, the maximum amplitude corresponded to the experimental value, but the upper branch did not match the experiments (Guilminaeu and Queutey 2001).

reduce the safety factors. The matter is further complicated by the fact that data obtained either in the large basins around the world or in the oceans by the industry is often case-specific and proprietary. The wide dissemination of the existing ocean data would help to resolve many of the issues cited above for the betterment of all concerned.

Notwithstanding the concerns expressed by Grinstein and Karniadakis (2002), LES has been used by a number of investigators. Saltara *et al.* (1998) used DVM and LES for a 1-DOF VIV at $Re = 1000$. Tutar and Holdo (2000) used LES in conjunction with the finite-element method at $Re = 2.4 \times 10^4$ for a cylinder subjected to forced oscillations and found that 3D representation was necessary to obtain accurate enough results. Zhang and Dalton (1996) performed a 2D LES study for a transversely oscillating cylinder at $Re = 13000$. Their results exhibited the same trends as the experimental results of Feng (1968). Lu and Dalton (1996) examined the VIV problem for a 2D viscous flow at $Re = 200$ and 2D turbulent flow at $Re = 855$. However, the lift calculations of Lecointe and Piquet (1989) were about 35% larger at

$Re = 855$ than those of Zhang and Dalton because they had not used a turbulence model. More recently, Al-Jamal and Dalton (2004) have performed 2D LES calculations of the self-excited response of a cylinder at $Re = 8000$ for a range of damping ratios and natural frequencies. In spite of the shortcomings of a 2D simplification, their results predicted the expected vibratory response in the range $0.72 < f_{com}/f_{st} < 1.26$. Decreasing material damping increased the lock-in range. A beating behavior was observed in the oscillations, which decreased with increasing damping (see Fig. 6.9). Clearly, one does not expect the 2D LES to represent the inception and growth of the instabilities accurately, let alone the transition to turbulence, in unsteady shear layers. It appears that the question of the 3D numerical simulation of VIV even for a smooth circular cylinder in the range of Reynolds numbers from a few thousand to over a million will remain unresolved indefinitely. Higher-order modes caused by secondary instabilities resulting from the changes in phase and amplitude in self-excited oscillations will certainly complicate the simulations with any turbulence model in any indirect numerical simulation.

At present, none of the methods discussed above ensures sufficient generality in a large parameter space as far as VIV is concerned. Large-scale benchmark experiments at large Reynolds numbers, coupled with 3D numerical simulations using RANS and LES, may allow one to develop industrial codes that may be used (after many calibrations) for design purposes in small domains of the controlling parameters. From a scientific point of view, the real purpose of the numerical models is to uncover the physics of the motion of the continuum.

It has been demonstrated for a number of different flows that the shear stress and strain tensors involved in subfilter eddy-viscosity models (LES: large-eddy simulations) appeared to be a compromise. In fact, Grinstein and Karniadakis (2002) noted recently, "After more than 30 years of intense research on large-eddy simulations (LES) of turbulent flows based on eddy-viscosity subfilter models (Deardorff 1970), there is now consensus that such an approach is subject to fundamental limitations." Nevertheless, Michelassi *et al.* (2003) expressed a more optimistic point of view in connection with their work on flow around low-pressure turbine blades at Reynolds numbers 5.18×10^4 and 2×10^5: "Direct numerical simulation and the large-eddy simulation are able to provide a much deeper insight in the wake–boundary-layer interaction mechanism as compared to two-dimensional unsteady Reynolds averaged Navier-Stokes simulations."

In the light of the foregoing, we will briefly discuss some of the current efforts and the progress made toward the enhancement of computational methods in marine hydrodynamics. Undoubtedly, they too will be upgraded in due course.

6.14 Experiments with advanced models

The more familiar finite-difference method (FDM) has been used for a long time. It was followed by the finite-volume method, which conserves mass, momentum, and energy. However, it requires fairly accurate distributions of pressure and velocity in the immediate vicinity of the boundary to arrive at reasonable lift and drag forces (Versteeg and Malalasekera 2007). The next major step has been the use of direct numerical simulation (DNS), as noted above. It limits the size of the achievable Reynolds number and one cannot imitate a continuum without enough grid points to achieve reasonable accuracy. Even then, a non-uniform distribution of the grid elements is necessary to realize reasonable accuracy in the gradients of the velocity and pressure. In other words, DNS without some sort of turbulence modeling may not be for the marine hydrodynamics.

This brings us to new and inevitable challenges of modeling of subcritical and supercritical flows with separation. So far, it has not been possible to compute the details of separation and the fluctuations of the separation points (in the vicinity of the separation zones) even on a smooth circular cylinder at Reynolds numbers as small as 10000. It appears that the discretization of a turbulent continuum for computational purposes is much harder than expected. Thus, to arrive at reasonable engineering solutions one has to resort to LES, an ingenious compromise based on "multilayer" filtering (like stacks of graded sand filters), Reynolds-averaged NS equations, and DES (detached eddy simulation), which emerged to alleviate the difficulties that are inherent in both LES and RANS solutions (see, e.g., Mittal and Moin 1997, and Breuer 1998). Other models, such as DESIDER (detached eddy simulation for industrial aerodynamics), UFAST (unsteady effects in shock wave-induced separation), POD (proper orthogonal decomposition), unsteady-RANS, and others are relatively more recent entries into the art of modeling. However, there is no assurance that these and many more models will contribute to the improvement of the modeling methodologies in the higher Reynolds number range.

Most of the marine hydrodynamics occurs in the range of supercritical Reynolds numbers. Thus, the DNS is not the computational answer to industrial applications (at least not for now). We may combine LES and RANS to obtain the DES. The large-scale part is the spatially filtered form of the instantaneous velocity. It represents the value of the velocity down to the grid size, and it can be calculated directly. An improved subgrid scale (SGS) description is called the dynamic model (DSGS) in which the modeling constant is not prescribed *a priori*. It is calculated as a part of the solution at each time step, based on the energy content of the smallest resolved scale.

The latest DNS of Dong and Karniadakis (2005), flow past a circular cylinder at $Re = 10000$, has encountered separation-point problems. In fact, their results produced the correct flow structures and statistics (away from the immediate separation regions), but provided no information about the details of the flow at or near the separation regions (see, e.g., Sarpkaya 2004). It appears that all other methods have encountered similar problems with the treatment of the time-dependent separation zone, presumably due to the difficulty of dealing with adverse velocity and pressure gradients with random fluctuations of magnitude and frequency during part of the cycle.

At $Re = 10000$, their calculated CD (1.143) agreed well with the experimental value (1.14); the rms lift coefficient, CL' (0.448), agreed well with the average CL' (0.458), varying from 0.384 to 0.532, and the calculated base pressure coefficient (-1.129) was somewhat over the experimental value of about -1.06. In any case, the accuracy of the experimental values is more questionable. The quantities such as base-pressure CL' have been debated for the past 100 years! We do not expect numerical predictions of such accuracy in the midst of natures' vagaries, especially in marine hydrodynamics. Evidently, only relatively stationary parameters (e.g., drag and Strouhal number) can be reliably calculated as long as Re is low enough. *One must also be mindful of the fact that the current safety factors in the design of offshore structures run over 20!*

A few other LES simulations may be mentioned: Lu *et al.* (1997), 3D steady flow past a fixed circular cylinder (Re up to $Re = 44200$). Breuer (2000), 3D LES, $Re = 140000$ (finite-volume calculation, and Smagorinsky and dynamic subgrid scale models). A comprehensive study at supercritical Reynolds numbers by Catalano *et al.* (2003), reducing the computational effort using a wall model, i.e., the turbulent boundary layer (at least in part), was modeled, not calculated. There are many more models such as FLUENT, where the eddy viscosity was obtained from the standard k-epsilon model of Launder and Spalding (1972). We must also note that there are a number of commercial codes such as FLUENT and AcuSolve, (a finite-element code) with RANS and DES capabilities, recommended for supercritical Re. These are described in Constantinidis *et al.* (2005) and Constantinidis and Oakley (2006).

Kim and Moin (1985) performed an LES calculation using FLUENT at Reynolds numbers of 1.4×10^5 and 1×10^6 for flow past a smooth cylinder. Their calculations at $Re = 1.4 \times 10^5$ agreed well with experimental results. However, their and Breuer's calculated separation points were well over $98°$ (much larger than the expected value of $80°$). *This is due to the high-frequency fluctuations imposed on the mean fluctuations of the separation points and sharp and random pressure gradients, imposed on the mean separation angle* (a commonly overpredicted value). Evidently, one will encounter greater difficulty in dealing with rectangular cylinders (with large-scale separation) with any code. Oakley obtained good agreement with measurements at $Re = 2 \times 10^5, 4 \times 10^5$, and 6×10^5. As expected, the results for the smooth cylinders had no surprises. Constantinidis and Oakley (1996) have examined the straked-cylinder VIV for $3.9 \times 10^5 < Re < 8.2 \times 10^6$ and $3.7 < VR < 22$. The drag coefficient, as expected, was 3 to 4 times larger than that of the unstraked cylinder.

It is evident that CFD has the capability to provide significant input to the determination of the forces acting on a cylinder subjected to an approach flow. The ultimate goal is to be able to predict VIV adequately and to predict the suppression of VIV (in relatively regular waves and uniform currents) using strakes or fairings.

At this point in time, however, CFD has the capability to produce reasonable agreement with experimental data for the stationary parts of the parameters (drag, etc.) with stationary cylinders at least through the subcritical range of Reynolds numbers. In the supercritical range, there remain a number of unresolved issues. The numerical approaches, LES, RANS, DES, and many others, appear to offer some short-range potential. Future progress will benefit from advances in electronic computing and from the formulation of high-Re turbulence modeling. While the efforts to enhance "modeling" for all needs is commendable, one must bear in mind that the meeting of man and nature in a stormy environment will always be stochastic and expensive (large safety factors), to say the least. For valuable insight into the attributes of many commercial codes, the reader is referred to Spalart and Alamaris (1992) Constantinidis *et al.* (2005) and Constantinidis and Oakley (2006), among many others. It is hoped that robust models/codes will increase in number and enhance their predictive capabilities within the limits of our understanding of turbulence.

7

Hydrodynamic Damping

7.1 Key concepts

Damping is a fundamental as well as practical problem in fluid dynamics. It deals with small amplitude oscillations of a body (e.g., a cable in the ocean environment). The classical solutions of Stokes (1851) and Wang (1968), valid only for $K \ll 1$ and $\beta \gg 1$, have shown that the oscillatory boundary layer gives rise to skin friction and normal pressure and, hence, to damping force, in anti-phase with velocity. Its prediction for large as well as small components of offshore structures is rather difficult. It gives rise to fascinating flow phenomena, and opens new questions on flow instabilities: Honji instability and the symmetry breaking in an ever decreasing K and increasing β environment render the analytical as well as the experimental understanding of damping a challenge worth pursuing.

7.2 Introduction

Often, a distinction is made between bodies subjected to vortex-shedding excitation and those (e.g., a TLP) that undergo simple sinusoidal motion at very small amplitudes and high frequencies in water, presumably due to excitation that comes from outside, i.e., not caused by the flow itself. The latter is called "hydrodynamic damping" or "viscous damping," meaning the decrease of the amplitude of an oscillation by forces in anti-phase with velocity. There are two reasons for such a distinction: The classical solution of Stokes (see also Wang), valid for $K \ll 1$ and $\beta \gg 1$, has shown that the oscillatory boundary layer gives rise to skin friction and normal pressure, and hence, to a damping force, in anti-phase with the velocity. The reduced damping δ_r is then expressed as $\delta_r = 2m(2\pi\zeta)/\rho D^2$ where m is the mass per unit length (including the added mass) and ζ is the damping factor.

Hydrodynamic "damping" is the decrease of the amplitude of oscillation of a body (for example, a tension leg platform or a simple circular cylinder) from cycle to cycle. It is a measurable but not easily predictable quantity. The very thin layer near the wall and its complex instabilities (coherent and quasi-coherent structures) are at the heart of the phenomenon but cannot yet be predicted analytically. The use of experimentally measured quantities suffice for engineering purposes but remain unsatisfying if further progress is to be made toward the understanding of the physics of the near-wall instabilities. The question would not have been too challenging if it were not due to the fact that *coherent and incoherent structures have profound effects on the wall friction and on the understanding of the evolution of the stability/instability of wall flows.* We must hasten to note that we have not yet introduced the effects of often non-uniform shape, size, distribution, etc. of the ever-present roughness elements. Clearly, the subject is far more challenging than the determination of a seemingly simple damping coefficient. Thus, it is clear that there would not have been any need for the "fluid damping coefficient" if one were able to predict all the fluid forces resulting from the fluid–structure interaction, whether the excitation is imparted by vortex shedding or comes from outside (i.e., not caused by the flow itself). Obviously, we have not yet touched upon the effects of wall roughness.

It is of considerable importance for structures undergoing dynamic excitation in the range of flow parameters defined by very small Keulegan–Carpenter numbers (K) and very large frequency parameters (defined earlier as $\beta = Re/K$). The reason for this is that typical K values for the tension leg platforms are in the range of 0.005 to 0.02. The large values of β mean large rates of diffusion of vorticity, presumably dictated by the prevailing coherent and/or incoherent flow structures to be discussed later.

At present, the prediction of time-dependent flows (unsteady ambient flow and/or body response) is far from developed in the range of parameters of practical significance. The turbulent diffusion cannot be simulated without ad hoc assumptions and there are no reliable numerical models at a level that could be used for industrial explorations. Thus, the question reduces to the *experimental determination of the hydrodynamic damping coefficients in the range of governing parameters of industrial significance.*

The damping is affected by everything and anything that one can possibly imagine: is the test tank open or covered, is the tank in the sun or shade, is the cover soft or stiff, are there "small" apertures around the tank, is the cylinder barely stiff, is the cylinder surface smooth enough, are any people or machines adjacent to the tank causing vibrations (only barely), is the free surface in the tank left free, do you get the same damping coefficient the next day, have the bubbles been removed, were there dirt or bubble accumulations on the test cylinder, are there mosquitoes deposits on the cylinder, are your repeat measurements yielding higher damping coefficients, what is the range of your β values, how many times have you repeated the experiments under the initial set-up conditions (except, of course, temperature changes, exposure to the sun, rise of bubbles to the surface or cover, etc.), is there traffic in the vicinity of the test equipment, and many more questions, limited only by one's imagination.

In view of the foregoing, perhaps a more meaningful questions might be, what is the maximum damping one gets without going to extreme cautions? Since such an upper limit is bound to be more stable (and more likely to prevail), it is always wise to know not how small is the damping in your measurements, but rather how high is your damping after much professional care.

7.3 Elements of damping

The equation of motion for a vertically mounted cylinder is given by

$$m\ddot{x} + c\dot{x} + kx = F(t) \tag{7.3.1}$$

where x, is the displacement; m, the mass; k, the stiffness; c, the damping; and $F(t)$ is the total force acting on the cylinder. The precision of its solution depends on the accurate determination of the mass matrix, damping matrix, stiffness matrix, and the force vector. However, every term in (7.3.1) is subject to varying degrees of approximations. Usually, the damping is expressed as a coefficient ζ where $\zeta = c/(4\pi m f_0)$. The response is then expressed as

$$x/D = f\left(K, \beta, \frac{4m}{\pi \rho D^2}, \zeta, f_0/f_w\right) \tag{7.3.2}$$

where m *includes the added mass.*

As described in great detail in Chapter 2, the correct value of the added mass is a time-dependent quantity and cannot be evaluated precisely partly because it becomes part of the complex vibration problem. However, under most circumstances (e.g., mildly separated or unseparated flow about cylinders) *it may be assumed to be equal to its still-water value.* There are numerous other circumstances (like air – water interface, surface waves, body proximity, amplitude variation with depth, and roughness effects, noted earlier) that make the assumption somewhat tolerable.

7.3.1 *Stokes canonical solutions*

Stokes (1851) presented the solutions for both a sphere and a cylinder oscillating in a liquid with the velocity $U = -A\omega \cos \omega t$, assuming that the amplitude A of the oscillations is small and the flow about the bodies is laminar, unseparated, and stable. For a sphere, his solution may be written as

$$C_d = \frac{24}{Re}\left(1 + \frac{1}{2}\sqrt{\pi\frac{Re}{K}}\right) = \frac{24}{Re}\left(1 + \frac{1}{2}\sqrt{\pi\beta}\right) \tag{7.3.3}$$

and

$$C_m = \frac{3}{2} + \frac{9}{2}\sqrt{\frac{1}{\pi}\frac{K}{Re}} = \frac{3}{2} + \frac{9}{2}\sqrt{\frac{1}{\pi\beta}} \tag{7.3.4}$$

which shows that both the drag and inertial force are modified by β, expressing the rate of diffusion. In other words, it is impossible to decompose $F(t)$, for the flow under consideration, into an inviscid inertial force and a viscous force. Both are affected by the diffusion of vorticity in which resides the memory of viscous fluids. In other words the presence and the consequences of *viscosity cannot be extracted from a fluid in motion*. If diffusion has sufficient time to adjust to the unsteady conditions imposed on the flow, it may be said that the motion is a juxtaposition of steady states or a slowly varying unsteady flow.

Stokes' (1851) classical solutions formed the basis of many empirical and numerical models where the oscillations are presumed to be small enough to allow convective accelerations to be ignored. Stokes' solution for a *circular cylinder*, later extended to higher terms by Wang (1968, *with minor corrections to higher-order terms by* J. R. Chaplin), may be decomposed into "in-phase" and "out-of-phase" components as

$$C_a = 1 + 4(\pi\beta)^{-1/2} + 24(\pi\beta)^{-3/2} + \cdots \tag{7.3.5}$$

$$C_d = \frac{3\pi^3}{2K}\left[(\pi\beta)^{-1/2} + (\pi\beta)^{-1} - 6(\pi\beta)^{-3/2} + \cdots\right] \tag{7.3.6}$$

which shows the dependence of C_a and C_d on β and the fact that

$$\frac{C_a - 1}{KC_d} \cong \frac{8}{3\pi^3} \tag{7.3.7}$$

For sufficiently large values of β, (7.3.7) may be approximated to

$$\{KC_d\sqrt{\beta}\}_{S-W} = 26.24 \tag{7.3.8}$$

which shows that both $d(KC_d)/d\beta$ and KC_d approach zero asymptotically as β increases. However, the experiments reported herein show that KC_d *approaches* about 0.04 for the largest β *values* encountered.

The foregoing equations are valid only for sinusoidally oscillating, unseparated, stable, laminar flow about a vertical smooth cylinder. Hall (1984) carried out a stability analysis of the unsteady attached boundary layer on a cylinder oscillating transversely in a viscous fluid in both linear and weakly linear regimes. In order to simplify the problem, Hall further assumed that the oscillation frequency is large. This led to a critical K given by

$$K_{cr} = 5.78\beta^{-1/4}\left(1 + 0.21\beta^{-1/4} + \cdots\right) \tag{7.3.9}$$

Equation (7.3.9) is entirely consistent with the experiments of Honji (1981) within the range of comparison (Sarpkaya 1986a, 1993a, 2001b, 2002a, 2006a).

An extensive review of the existing force models has shown that the degree of empiricism increases with increasing Reynolds number, separation, convective accelerations, and the three-dimensionality of the wake. The practice of expressing the time-dependent fluid force as a function of the prevailing velocities and accelerations has subsequently been extended to nonlinear motions through the use of numerous approximations and numerical models. However, the issue is bound to remain unresolved as long as separation and turbulence remain ununderstood.

For sufficiently large values of β, Eq. (7.3.9) may be reduced to

$$\{KC_d\sqrt{\beta}\}_{S-W} = 26.24 \tag{7.3.10}$$

which is valid only within the limits noted above. The precision of its solution depends on the accurate determination of the mass matrix, damping matrix, stiffness matrix, and the force vector. However, every term in (7.3.10) is subject to varying degrees of approximations. In particular, the specification of the damping and force vector requires numerous assumptions and careful experiments. Here we will address only the damping issue.

Ordinarily, [c] in (7.3.1) would include only the structural damping (often assumed to be linear viscous), nonlinear interface damping at the joints, and, if any, the foundation radiation and foundation damping. These are not easy to determine and often recourse is made to field measurements and past experience (field measurements naturally include the fluid effects).

At present, the prediction of time-dependent flows (unsteady ambient flow and/or body response) is far from developed in the range of parameters of practical significance. The turbulent diffusion cannot be simulated without ad hoc assumptions and there are no reliable numerical models at a level that could be used for industrial design types of explorations. Thus, the question reduces to the experimental determination of the hydrodynamic damping coefficients in the range of governing parameters of industrial significance.

Obviously, the state of the art cannot offer anything better for the solution of (7.3.1), i.e., for the safe and economical interaction of nature and man-made structures. The hope is that, in the course of large-scale experiments, one will be able to accumulate enough empirical data to model the instabilities and the turbulent diffusion near the surface of a circular cylinder undergoing damping motions. The results may be of some use in the calculation of the heave and pitch motions of the tendons of tension leg platforms if one assumes that all other flow effects, such as waves and currents, are absent and the springing is taking place in a quiescent fluid, with relatively small amplitudes and periods.

Some of the techniques used to determine the damping levels in structural forms responding dynamically in their loading environment are discussed in detail by Haritos (1989). These are the "peak transfer function" method, the "half-power" bandwidth approach, "maximum entropy" methods, and the "equivalent-area" method. The "pluck" test is used from time to time to determine δf from the decay of the amplitude of oscillations. It is easy to show that even a small current in any direction relative to the pluck-induced vibration changes the damping experienced by the cylinder. Pluck tests may yield sufficiently accurate reduced damping for special cases, provided that the starting value of the vibration has $K \ll 1$.

7.4 Previous investigations

Considerable experimental work has been carried out in recent years on hydrodynamic damping of circular cylinders in oscillatory flow: Sarpkaya (1986a), in the range of $1035 < \beta < 11\,240$; Otter (1990), for $\beta = 61\,400$; Troesch and Kim (1991), for $\beta = 23\,200$ and $48\,600$; Bearman and Macwood (1992), in the range $14\,371 < \beta < 30\,163$; Bearman and Russell (1996), for $\beta = 60\,000$; Chaplin and Subbiah (1998), for $\beta < 166\,900$; Chaplin (2000), for $\beta = 670000$ and 1277000; and Sarpkaya (2000, 2001b), for $\beta = 748000$ and 1365000 for smooth, sand-roughened ($k/D = 1/100$), and porous pipes (30% porosity with uniformly distributed circular holes).

The results have shown that drag coefficients are indeed inversely proportional to K in the region of no vortex shedding, but the Λ_K values exceeded unity by varying amounts, except those of Sarpkaya (1986a) for $\beta = 1035$

and 1380 and Otter (1990) for $\beta = 61\,400$. The most recent free-decay tests of Chaplin (2000) have shown that Λ_K remains at about 2 even for β values larger than 10^6. Sarpkaya's (2000) data yielded Λ_K values somewhat larger than 2. The differences between various data are primarily due to the difficulty of measuring damping and the sensitivity of the damping to the environmental conditions.

The published data are summarized either in terms of $KC_d\beta^{1/2}$ or, better, in terms of another parameter Λ_K defined as

$$\Lambda_K = [\{KC_d\sqrt{\beta}\}_{\mathrm{exp}} / \{KC_d\sqrt{\beta}\}_{S-W}]_K \qquad (7.4.1)$$

where $\{KC_d\sqrt{\beta}\}_{S-W} \approx 26.24$ is the classical 2D laminar flow solution of Stokes and Wang. It should be noted that Λ_K is meaningful only at very small K values. Every attempt has been made to choose the smallest K at each investigation. The data for small K may be represented either by multiplying the constant 26.24 by Λ_K or by multiplying ν in β by $(\Lambda_K)^2$. In the latter case, $(\Lambda_K)^2$ transforms the kinematic viscosity to a virtual or eddy kinematic viscosity (à la Boussinesq).

The relative roughness of the free surface of the cylinder is given by k/D. Other parameters of interest are the relative thickness of the Stokes layer,

$$\delta_{St}/D = 2.82\beta^{-1/2} \qquad (7.4.2)$$

the oscillatory Reynolds number Re_δ defined by

$$Re_\delta = U_{\mathrm{max}}\delta_{St}/\nu = 2.82\,Re\,\beta^{-1/2} \qquad (7.4.3)$$

and the customary Reynolds number $Re = U_{\mathrm{max}}D/\nu$.

Some of the differences between various data may be attributed to the methods of experimentation, the use of pluck or forced oscillations, the range of β, damping of the supporting elements, free-surface and wall-proximity effects, cylinder surface roughness and aspect ratio, presence of air bubbles, etc. However, the consistency of the near doubling of the Λ_K at relatively large β can neither be denied nor explained away in terms of errors in the otherwise carefully conducted experiments. *This necessitated an extensive visual and photographic investigation of the structures in near-wall flow.* Schlichting (1932), speaking of "the wall layer," noted that "it is *so small that the eddy inside it cannot be seen in flow visualization pictures!*"

Sarpkaya's experiments with more advanced instruments in the 1990–2004 period were somewhat more successful in rendering them visible and measurable. Experiments were conducted in a U-shaped water tunnel, where the flow oscillated about cylinders at a constant frequency and in a rectangular basin where the test cylinder was subjected to forced oscillations.

The system was so designed that either pluck tests (free decay) or resonant-force-vibration tests could be performed at frequencies as large as 6.0 Hz (in water) in the range of K values from about 0.001 to 0.55. The data (displacements, accelerations, forces, and strains) were sampled at a rate of 2500 Hz. A comprehensive uncertainty analysis (95% confidence interval) has shown that the uncertainty for the drag coefficients was less than 4.8%.

The nominal size of the cylinders had larger L/D ratios. The most recent experiments reported herein were carried out with two sets of cylinders: 35.56-cm and 49.50-cm smooth nonperforated cylinders and a 49.50-cm perforated cylinder (porosity $= 30\%$). The solid cylinders were subsequently sand-roughened ($k/D = 1/100$, where D is the bare cylinder diameter here and elsewhere in this chapter). The massive static and dynamic data from the linear-variable displacement transducers, accelerometers, force transducers, and strain gages were sufficient to determine the system stiffness, added mass, and the net hydrodynamic damping (here the logarithmic damping) of the submerged parts of the pendulum and the structural damping (the latter was indeed very low). Then, the drag coefficients were deduced from the well-known expression (see Sarpkaya 1978c)

$$C_d = 3\pi m\delta/(2\rho D^2 K) \qquad (7.4.4)$$

where C_d is the drag coefficient, m is the effective mass of the cylinder, and δ is the logarithmic decrement ($\approx 2\pi\varsigma$). For large values of β, *the combination of Eqs.* (7.3.7) and (7.4.4) yields

$$\varsigma = \frac{1}{2}(C_a - 1)\frac{M_d}{m} \qquad (7.4.5)$$

$$KC_d = \frac{3\pi^3}{4}\varsigma\frac{m}{M_d} \qquad (7.4.6)$$

where $M_d =$ is the displaced mass of the cylinder per unit length. Equations (7.4.4), (7.4.5), and (7.4.6) show the extreme difficulty of the determination of KC_d for small values of K and large values of β. The concerns expressed above are equally valid for the forced-vibration experiments also.

7.5 Representative data

7.5.1 *Solid cylinders*

Preliminary plots of the drag coefficients as a function of K for two solid cylinders ($\beta = 748000$ and 1365000) have shown that there is no measurable

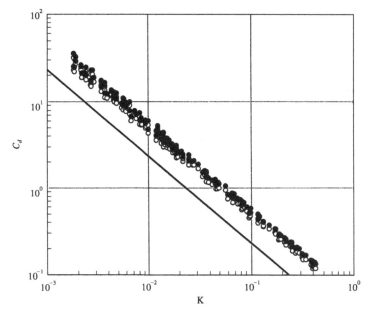

Figure 7.1. Drag coefficient versus Keulegan–Carpenter number for smooth solid cylinders at $\beta = 1365000$: $\circ$, pluck tests: $\bullet$, forced oscillation: ____, theory (Stokes 1851; Wang 1968).

difference between them. If anything, the difference is well within the scatter of the data. It is for this reason that the drag coefficients are presented herein only for $\beta = 1365000$.

Figure 7.1 shows the C_d for pluck and forced-vibration experiments for the above β. Evidently, there is some scatter in the data. The scatter is more apparent in Fig. 7.2 and the mean lines through the data appear to have a small upward slope with increasing K. The drag coefficient is larger than that predicted by the Stokes–Wang (unseparated laminar flow) analysis in both the pre-Honji ($K_{cr} < 0.169$) and post-Honji regimes, and that the forced vibration tests yield somewhat larger C_d values. The average Λ values (see Sarpkaya 2001a) are about 2.2 and 2.4 for the pluck and forced-vibration tests, respectively. Figure 7.2 shows C_d for the rough cylinder ($k/D = 1/100$) for $\beta = 1365000$. Its characteristics are similar to those shown in Fig. 7.1 except that the effect of the roughness increases Λ_k from 2.2 (for the smooth cylinder) to 3.2 for the rough cylinder. Forced vibration of the rough cylinders (not shown here) yielded an average Λ_k value of about 3.4, with equally larger scatter. As will be amplified later, the scatter of the data is not entirely due to the shortcomings of equipment and data evaluation. Some of the reasons are buried in the interstices of the roughness elements and the interaction between the roughness elements and the incoherent structures.

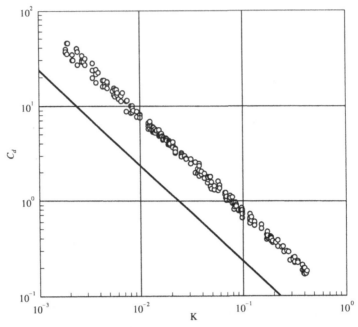

Figure 7.2. Drag coefficient versus Keulegan–Carpenter number for rough solid cylinders ($k/D = 1/100$) at $\beta = 1\,365\,000$: ○, pluck tests; ——, theory (Stokes 1851; Wang 1968).

7.5.2 *Perforated cylinders*

A perforated body is a rigid hollow shell whose surface is pierced by a distribution of small apertures, which allow the near-free passage of fluid. The essential characteristics of flow through a perforated surface depend on a number of parameters: the Reynolds number, roughness number, the relative spacing of the perforations, angle of the aperture to the incident flow, the ratio of the total open area to the total surface area of the body, whether or not the incident flow is laminar or turbulent, and on the Reynolds and Keulegan–Carpenter numbers of the cylindrical body.

The effect of flow unsteadiness in general and the added mass in particular has not been subjected to extensive theoretical, numerical, and experimental work. The existing works (e.g., Howe 1979 and Molin 1992) dealt with highly specialized cases and include the effect of perforations rather indirectly. The unsteady flow results show (Howe 1979) that perforations have a profound effect on the added mass and damping of the body. At very small amplitudes of oscillation or unidirectional surging, perforations reduce the added mass significantly since the body becomes transparent to the fluid motion. For porosities below about 30%, the added mass increases

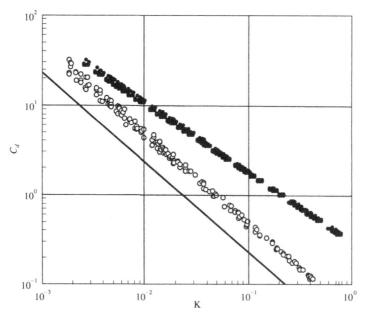

Figure 7.3. Drag coefficient versus Keulegan–Carpenter number: ○, pluck tests of smooth solid cylinders at $\beta = 1\,365\,000$ (see Fig. 7.1); ●, pluck tests of the porous cylinder (30% porosity) at $\beta = 1\,365\,000$; ——, theory (Stokes 1851; Wang 1968).

rapidly. Equally important is the fact that added mass also increases rapidly with increasing amplitude of oscillation for a given frequency and porosity.

Figure 7.3 shows the drag coefficient for the perforated cylinder (porosity $= 30\%$, hole size $= 1.168$ mm, center-to-center spacing $= 1.984$ mm, and $\beta = 1\,365\,000$), together with that for the solid cylinder. The perforations considerably increase the drag (and hence the damping). The drag data were not expected to fall on a line parallel to the Stokes–Wang line for many reasons, the primary one being the fact that near the wall the flow is very complex (part of flow goes through the cylinder and part around the cylinder). This is also the reason for the relatively small inertia coefficient shown in Fig. 7.4.

The mean line through the perforated-cylinder data shown in Fig. 7.3 may be represented by $KC_d^{1.3} = 0.21$.

Clearly, the perforated cylinders, alone or in conjunction with a coaxial pipe or cable, can provide very large damping without increasing the added mass. This is an important fact for design purposes.

7.5.3 Three-dimensional instabilities

The low-speed streaks and quasi-coherent vortical structures are now considered to be ubiquitous features of a turbulent boundary layer near a solid wall, even though their origin still remains unresolved in both steady and

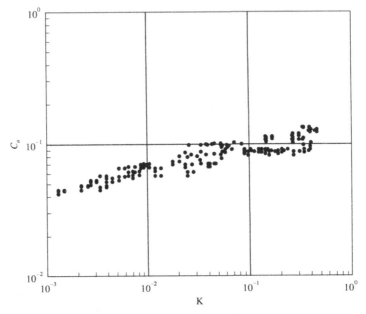

Figure 7.4. Inertia coefficient versus Keulegan–Carpenter number: •, pluck tests of the porous cylinder (30% porosity) at $\beta = 1\,365\,000$.

periodic flows (Sarpkaya 1993a). Incisive as well as eloquent reviews of the state of the art, deduced from 40 years of experimentation and about 20 years of numerical simulations of low-Reynolds-number (canonical) flows, have been given by Robinson and Kline (1990), Smith *et al.* (1991), and Robinson (1991). The studies on the instability of external oscillatory flows on curved walls are relatively new and deal mostly with periodic sand ripples (see, e.g., Hara and Mei 1990a, 1990b; Scandura *et al.* 2000, and the references cited therein). For understandable reasons, these studies deal with relatively small Reynolds numbers and prove once again that the simulation of more realistic three-dimensional flows is beyond the present computing power.

As noted earlier, Honji (1981) visualized the flow around a transversely oscillating cylinder in a fluid otherwise at rest and observed nearly regularly spaced, mushroom shaped, three-dimensional vortices. Subsequently, Sarpkaya (1986a) named them "the Honji instability" and extended the range of the observations to higher β values (about 5500 to 1 350 000).

The use of the direct numerical simulation at higher β values of interest is not yet feasible. In fact, only two efforts (to the best of our knowledge) were made to simulate the Honji instability: Zhang and Dalton (1999) at $\beta = 196$, using the primitive-variables form of the Navier–Stokes equations, and Suthon (2009) who carried out numerical investigations of the 3D

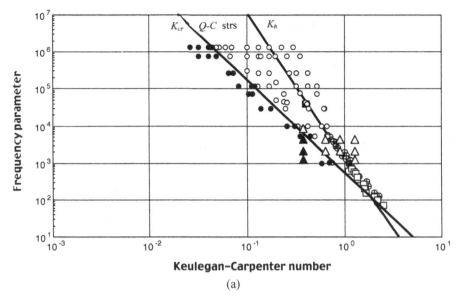

Figure 7.5. (a) This figure shows the data contributed by Sarpkaya (1986a), in the range of $1035 < \beta < 11240$; Otter (1990), for $\beta = 61400$; Troesch and Kim (1991), for $\beta = 23200$ and 48600; Bearman and Macwood (1992), in the range $14371 < \beta < 30163$; Bearman and Russell (1996), for $\beta = 60000$; Chaplin and Subbiah (1998), for $\beta < 166900$; Chaplin (2000), for $\beta = 670000$ and 1277000; and Sarpkaya (2000, 2001a, 2004), for $\beta = 748000$ and 1365000, and Johanning (2003), all for smooth cylinders. It must be noted that the shape, shading, and size of the data points are not related to the stability or the instability of the flow in the K–β plane. They are simply the identifiers of the contributors whose names are given in the preceding list, but not specifically identified in the figures in order not to clutter the already-crowded figure space.

SOF about a cylinder using the primitive-variables form of the incompressible Navier–Stokes equations in the cylindrical coordinate system. Three classes of computational experiments were conducted using $\beta = 1035, 6815$, and 9956, while K was increased until the stability and the Hall lines were reached, as expected by Sarpkaya (1993b, 2001a, 2006b). In view of the foregoing, it was imperative that the K–β plane be explored to the maximum extent possible.

Flow visualization experiments were conducted using, with few minor exceptions, almost exactly the same instruments and procedures described in detail in Sarpkaya (1986). Each experiment began at a point defined by (K_h, β) in Fig. 7.5a. Experiments were carried out either by maintaining β constant (i.e., the frequency of oscillation) and decreasing K from an initial value of $K > K_{cr}$ down to K values smaller than K_{cr}, or by maintaining K constant and increasing β (i.e., the frequency f). Each change in either K or β is followed by a long rest and "refueling" (new dye introduction) period. In the following, the inception of the instabilities near $K = K_{cr}$, the growth and

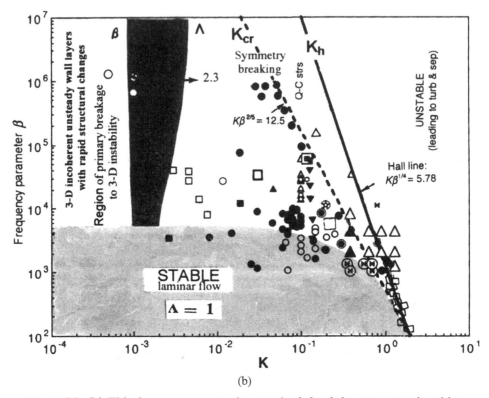

Figure 7.5. (b) This figure represents data to the left of the symmetry-breaking region plus the data in the stable region near the K axis. Further to the left, higher orders of symmetry breaking occur. The shaded STABLE region is defined on the basis of the data available (the structural change of the instabilities is depicted in Fig. 7.6).

decay of the instabilities in the region $K_{cr} < K < K_h$, larger-scale instabilities in the region K, K_h, and, finally, the qualification of the stability boundary and the wavelength s/D will be discussed.

First, it is necessary to describe the nature of small finite disturbances or the smallest observable structures that appear only during the periods of high ambient velocity (U larger than about 0.5 U_m). The structures always grow at a wave crest. The number of crests along the wave spanning the crown of the cylinder depends on K and β. For a wave with multiple crests, the wavelength decreases with increasing β for a given K. As noted earlier, these disappear quickly as $U \to 0$. The line, labeled K_h, after Hall (1984), represents Eq. (7.3.9). Each K on the Hall line corresponds to a critical K_h at which the mushroom-shaped structures of the Honji instability occur.

Experiments were carried out either by holding β constant (e.g., frequency of oscillation) and decreasing K from an initial value $K > K_{cr}$ down to K values smaller than K_{cr} or by maintaining K constant and increasing β.

Each change in either K or β is followed by a long rest and "refueling" (new dye introduction) period. The results are presented in two parts to avoid confusion.

First and foremost, it must be noted that all regions above the "stable region" (identified with $\Lambda = 1$) are *unconditionally unstable*. Obviously, the type and character of the instabilities depend on their position on the map. Instabilities may be in the form of quasi-coherent structures of various forms (varying with distance from the boundary) and various orders of symmetry-breaking instabilities discussed by Elston *et al.* (2006). The wavelengths decrease with increasing β and the structures become less "Honji" like (Sarpkaya 2001a, 2002a, 2006b). The dashed line (marked K_{cr}) separates the region of the primary and secondary symmetry-breaking instabilities (above the "stable" region on the left where β exceeds about 4500). It is the lower limit of the points defined by (β, K) where either no quasi-coherent structures (QCSs) are created during the *entire cycle* or those created (mostly during the high-velocity periods) barely survived the low-velocity period. Part of the subjectivity, aside from the human interpretation of the definition of the life of a QCS, comes from the fact that there cannot be a single line separating the stable region from the unstable region due to the statistical nature of the intermittency of the structures. In fact, the difficulty of the determination of the (fluctuating) stability line cannot be adequately emphasized. It depends not only on the parameters that can be controlled but also on those that are, for all intents and purposes, beyond the capacity of the experimenter to control (e.g., temperature gradients, residual background turbulence, very small air bubbles, higher-order harmonics of the vibrations, nonlinear interaction of various types of perturbations, and many other uncertainties have already been pointed out). Ironically enough, we are looking for the reasons as to why the flow does not become unstable at smaller K values in the region just above the "stable region" (where $\Lambda = 1$). The region between the lines K_{cr} and K_h is where there always are QCSs. The Hall line is where the "mushrooms" appear. In the region to the right of K_h one encounters, with increasing K, first Honji–Hall types of instabilities, then QCSs, and eventually, separation and/or turbulence followed by vortex shedding. All of the foregoing instabilities are shown in Fig. 7.6.

It is apparent from the foregoing that the vorticity acquired by each vortex blob is sensitive enough to the instantaneous distribution of vorticity (as expected in a highly time-dependent flow) but not sensitive enough to change the basic character of the formation and motion of the Honji blobs. *The motion picture attached herewith (see the Web) clearly shows the evolution and motion of the vortex tubes over several cycles.* The remarkable fact is that the vortex tubes *roughly return to their starting shapes and positions in spite of the relatively violent motions they undergo during a given cycle.*

7.5.4 *Closing remarks*

An effort was made to measure the drag of circular cylinders (smooth, rough, perforated) subjected to sinusoidally oscillation motion, particularly at large values of β, to gain some insight into the magnitude of their damping and, in particular, into the reasons for the deviation of measured values from the Stokes–Wang laminar-flow solution.

The second and probably the most important part of this effort yielded results significantly different from expectations. For smooth cylinders, the results have shown that there is a stable region ($\beta < 5000$) in which no discernable flow structures exist near the crown of the cylinder. On the Hall line, these structures take the form of Honji instability even at the highest

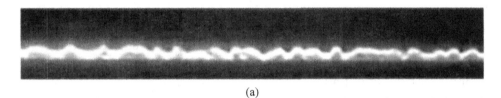

(a)

Figure 7.6. (a) The crests of a large number of waves spanning the width of the light sheet are shown at $U = U_m$ for $K = 0.045$ and $\beta = 1365 \times 10^6$ (a point slightly to the right of S). They disappear as U approaches zero. The appearance of these waves and the growth of structures from their crests are entirely consistent with the centrifugal nature of the instability leading to coherent structures on the Hall line ($f = 5.86\,\mathrm{Hz}$, $K = 0.045$, $\beta = 1365 \times 10^6$). (b) The evolution of wave crests and streamwise vertical structures is shown as samples from different runs as U approaches U_m. These disappear quickly as U decreases to zero. Except for the position and direction of the streamwise single vertical structures, the flow state repeats indefinitely ($f = 1\,\mathrm{Hz}$, $K = 0.098$, $\beta = 1.2 \times 10^5$). (c) The evolution of structures at a larger Stokes number at: (i) $U = 0$, (ii) $U = U_m/2$, and (iii) $U = U_m$ in a half-cycle ($f = 321\,\mathrm{Hz}$, $K = 0.14$. $\beta = 7.48 \times 10^5$). (d) These structures are at the highest β achievable in the experiments on the Hall line. The characteristic features of the mushroom-shaped structures are still apparent but not as precise as those of at lower β. Some of the mushrooms occasionally rise above the others and then continuously evolve during the cycle. The average relative spacing between them is based on over 20 cycles of observations ($f = 5.86\,\mathrm{Hz}$, $K = 0.169$, $\beta = 1.365 \times 10^6$). (e) Representative mushrooms recorded during part of a single cycle (at 32-ms intervals). The structures, regardless of the number of cycles, do not necessarily acquire the same size and shape at the same time. Some grow larger and others grow to the same size at a later time. This reinforces the notion that the mushrooms are inherently unstable even on the H line ($f = 0.623\,\mathrm{Hz}$, $K = 0.35$, $\beta = 7.4 \times 10^4$). (f) A single, nearly perfect (but never absolutely symmetrical), Honji instability on the Hall line.

See the movie noted in **OMAE 2006 Keynote Address, Hamburg. It is on the Web (www.omae2006.com) and may be viewed as desired.** *Please try a couple of times if the images do not respond the first time.*

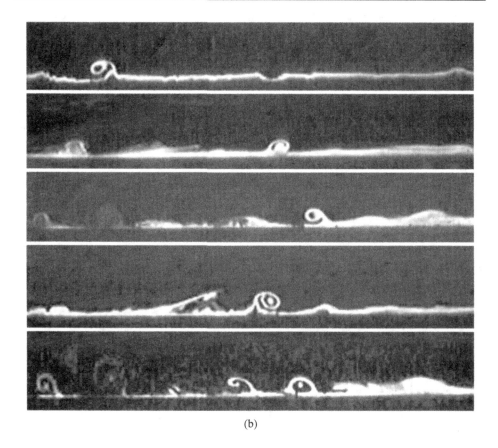

(b)

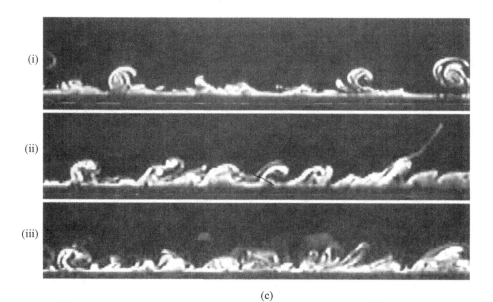

(c)

Figure 7.6 (*continued*)

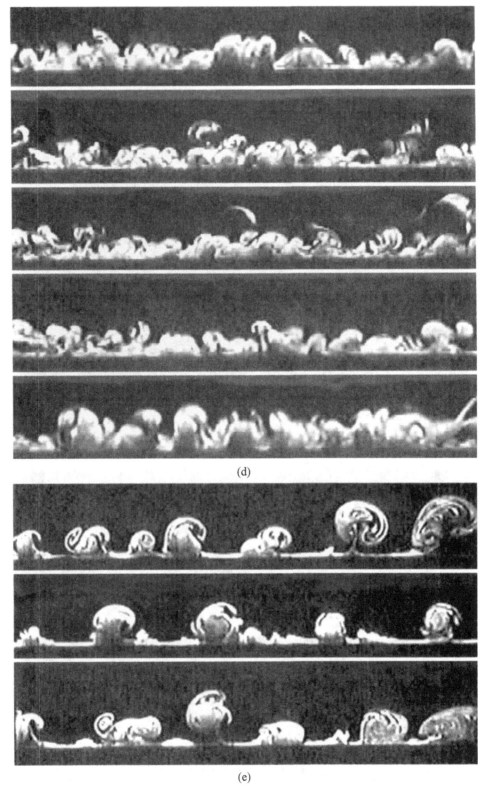

(d)

(e)

Figure 7.6 (*continued*)

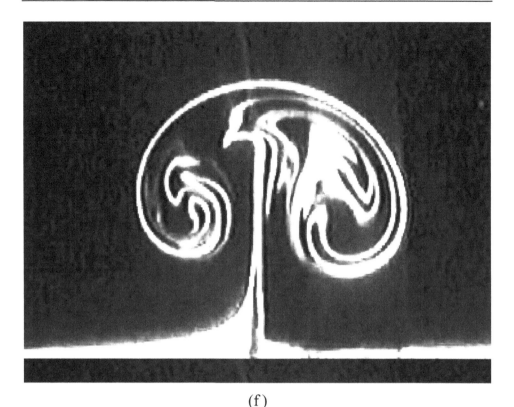

(f)

Figure 7.6 (*continued*)

β encountered in our experiments. To the right of the Hall line, instability takes many forms and, with increasing K, falls under the influence of separation and all of its attendant consequences. Beyond a second threshold the flow becomes turbulent.

The existence of a region between the stability line S and the Hall line H can only partly explain the measured increase in the drag coefficient relative to the Stokes–Wang analysis. However, there remains a large range of K values (say between 0.0003 and the stability line) in which there is no observable instability. Even though this does not preclude the existence of evolving instabilities, it is hard to imagine that they would be large enough to nearly double the drag coefficient.

The use of direct numerical simulations at β values of interest is not yet feasible. However, the tracking and quantification of similar coherent and quasi-coherent structures in experimental and simulated flows, at any point in the $K-\beta$ plane, through the use of, for example, the discriminant of the characteristic equation of the velocity gradient tensor, might help us to understand not only the relationship between damping, roughness, and porosity, but also the evolution of turbulence itself.

References

Abernathy, F. H. and Kronauer, R. E. 1962. The formation of vortex streets. *J. Fluid Mech.* **13**, 1–20.

Abramowitz, M. and Stegun, I. A. 1965. *Handbook of Mathematical Functions.* Dover, New York.

Achenbach, E. 1968. Distribution of local pressure and skin friction around a circular cylinder in cross-flow up to $Re - 5 \times 10^6$. *J. Fluid Mech.* **34**, pt. 4, 625–639.

Achenbach, E. 1971. Influence of surface roughness and the cross-flow around a circular cylinder. *J. Fluid Mech.* **46**, pt. 2, 321–335.

Ackermann, N. L. and Arbhabhirama, A. 1964. Viscous and boundary effects on virtual mass. *J. Eng. Mech. Div. ASCE* **90**, EM4, 123–130.

Ahmed, N. A. and Wagner, D. J. 2003. Vortex shedding and transition frequencies associated with flow around a circular cylinder. *AIAA J.* **41**, 542–544.

Alexander, C. M. 1981. The complex vibration and implied drag of a long oceanographic wire in cross-flow. *Ocean Eng.* **8**, 379–406.

Ali, N. and Narayanan, R. 1986. Forces on cylinders oscillating near a plane boundary. In *Proceedings of the 5th International Offshore Mechanical Arctic Engineering OMAE Sympsium*, Vol. III, pp. 613–619.

Al-Jamal, H. and Dalton, C. 2004. Vortex induced vibrations using large eddy simulation at a moderate Reynolds number. *J. Fluids Struct.* **19**, 73–92.

Al-Kazily, M. F. 1972. Forces on submerged pipelines induced by water waves. Ph.D. dissertation, College of Engineering, University of California, Berkeley.

Allam, Md. Mahbub and Zhou, Y. 2007. Strouhal numbers, forces and flow structures around two tandem cylinders of different diameters. *J. Fluids Struct.* **24**, 505–526.

Allen, D. W. and Henning, D. L. 1997. Vortex-induced vibration tests of a flexible smooth cylinder at supercritical Reynolds numbers. In *Proceedings of the ISOPE Conference*, Vol. III, Honolulu, pp. 680–685.

Allen, D. W. and Henning, D. L. 2003. Vortex-induced vibration current tank tests of two equal-diameter cylinders in tandem. *J. Fluids Struct.* **17**, 767–781.

Allen, H. J. and Perkins, E. W. 1951. A study of effects of viscosity on flow over slender inclined bodies of revolution. NACA Tech. Rep. 1048.

Allender, J. H. and Petrauskas, C. 1987. Measured and predicted wave plus current loading on a laboratory-scale, space frame structure. In *Proceedings* of the Offshore Technology Conference, OTC 5371, Houston, TX, pp. 143–146.

American Petroleum Institute (API) 1977. API Recommended Practice for Planning, Designing, and Constructing Fixed Offshore Platforms. API RP 2A, American Petroleum Institute Production Department, 300 Corrigan Tower Bldg., Dallas, TX.

American Science and Engineering Company. 1975. Forces acting on unburied offshore pipelines. Project PR-91–68 (revised April 1975).

Angrilli, F., Di Silvio, G., and Zanardo, A. 1972. Hydroelasticity study of a circular cylinder in a water stream. In *Flow Induced Structural Vibrations*. (ed. E. Naudascher), *IUTAM-IAHR Symposium*, pp. 504–512.

Arita, Y., Fujita, H., and Tagaya, K. 1973. A study on the force of current acting on a multitubular column structure. In *Proceedings of the Offshore Technology Conference*, OTC 1815, Houston, TX.

Armitt, J. 1968. The effect of surface roughness and free stream turbulence on the flow around a model cooling tower at critical Reynolds numbers. In *Proceedings of the Symposium on Wind Effects on Buildings and Structures*. Loughborough University of Technology, England.

Atsavapranee, P., Benaroya, H., and Wei, T. 1998. Vortex dynamics in the near wake of a freely-oscillating cylinder. In *Proceedings of FEDSM'98*, pp. 1–6. American Society of Mechanical Engineers, New York. Summer Meeting, Washington, D.C.

Au-Yang, M. K. 1977. Generalized hydrodynamic mass for beam mode vibration of cylinders coupled by fluid gap. *J. Appl. Mech. ASME* **44**, 172–174.

Au-Yang, M. K. 2001. *Flow-Induced Vibrations of Power and Process Plant Components – A Practical Workbook*. American Society of Mechanical Engineers, New York.

Bai, K. J. 1977. The added-mass of two-dimentional cylinders heaving in water of finite depth. *J. Fluid Mech.* **81**, 85–105.

Bai, W. and Eatock Taylor, R. 2006. Higher-order boundary element simulation of fully nonlinear wave radiation by oscillating vertical cylinders. *Appl. Ocean Res.* **28**, 247–265.

Bai, W. and Eatock Taylor, R. 2007. Numerical simulation of fully nonlinear regular and focused wave diffraction around a vertical cylinder using domain decomposition. *Appl. Ocean Res.* **29**, 55–71.

Balasubramanian, S., Skop, R. A., Hann Jr. F. L., and Szewcxyk, A. A. 2000. Vortex-exited vibrations of uniform pivoted cylinders in uniform and shear flow. *J. Fluids Struct.* **14**, 65–86.

Barltrop, N. (ed. and contributing author). 1998. *Floating Structures: A Guide for Design and Analysis*. CMPT/OPL, Ledbury, UK.

Barnouin, B., Mattout, M. R., and Sagner, M. M. 1979. Experimental study of the validity domain of some formulae for hydrodynamic forces for regular and irregular flows. In *Mechanics of Wave-Induced Forces on Cylinders* (ed. T. L. Shaw), pp. 393–405. Pitman, London.

Bartran, D., Kinsey, J. M., Schappelle, R., and Yee, R. 1999. Flow induced vibration of thermowells. *J. Sci. Eng. Meas. Automat.* **38**, 123–132.

Bassett, A. B. 1888. On the motion of a sphere in a viscous liquid. *Philos. Trans. R. Soc. London* **179**, 43–63. (See also *A Treatise on Hydrodynamics* Vol. 2, Chap. 21, 1888. Deighton, Bell, Cambridge; also, Dover, New York, 1961.)

Basu, R. I. 1985. Aerodynamic forces on structures of circular cross-section, part 1. *J. Wind Eng. Indust. Aerodyn.* **21**, 273–294.

Basu, R. I. 1986. Aerodynamic forces on structures of circular cross-section, part 2. The influence of turbulence and three-dimensional effects. *J. Wind Eng. Indust. Aerodyn.* **24**, 33–59.

Batchelor, G. K. 1967. *An Introduction to Fluid Dynamics.* Cambridge University Press, Cambridge.

Batham, J. P. 1973. Pressure distributions on circular cylinders at critical Reynolds numbers. *SFM* **57**, 209–229.

Bearman, P. W. 1984. Vortex shedding from oscillating bluff bodies. *Annu. Rev. Fluid Mech.* **16**, 195–222.

Bearman, P. W. and Macwood, P. R. 1992. Measurements of the hydrodynamic damping of oscillating cylinders. In *Proceedings of the 6th International Conference on the Behavior of Offshore Structures*, BOSS 1992, London, pp. 405–414.

Bearman, P. W. and Obasaju, E. D. 1989. Transverse forces on a circular cylinder oscillating in-line with a steady current. In *Proceedings of the 8th Conference on Offshore Mechanical Arctic Engineering OMAE*, Vol. 2, pp. 253–258.

Bearman, P. W. and Russell, M. P. 1996. Measurements of the hydrodynamic damping of bluff bodies with application to the prediction of viscous damping of TLP hulls. In *Proceedings of the 21st Symposium on Naval Hydrodynamics.* National Academy Press, Washington, D.C.

Bearman, P. W. and Zdravkovich, M. M. 1978. Flow around a circular cylinder near a plane boundary. *J. Fluid Mech.* **89**, pt. 1, 33–47.

Beattie, J. F., Brown, L. P., and Webb, B. 1971. Lift and drag forces on a submerged circular cylinder. In *Proceedings of the Offshore Technology Conference*, OTC 1358, Houston, TX.

Bendat, J. S. and Piersol, A, G. 1971. *Random Data: Analysis and Measurement Procedures.* Wiley-Interscience, New York.

Berger, E. and Wille, R. 1972. Periodic flow phenomena. *Annu. Rev. Fluid Mech.* **4**, 313–340.

Biermann, D. and Herrnstein, Jr. W. H. 1933. The interference between struts in various combinations. NACA Tech. Rep. 468.

Billingham, J. and King, A. C. 2000. *Wave Motion.* Cambridge University Press, New York.

Bingham, H. H., Weimer, D. K., and Griffith, W. 1952. The cylinder and semicylinder in subsonic flow. Princeton University, Department of Physics. Tech. Rep. 11–13.

Birkhoff, G. 1950. *Hydrodynamics – A Study of Logic, Fact and Similitude.* Princeton University Press, Princeton, NJ.

Birkhoff, G. and Zarantonello, E. H. 1957. *Jet, Wakes, and Cavities.* Academic, New York.

Bishop, R. E. D. and Hassan, A. Y. 1963. The lift and drag forces on a circular cylinder in a flowing fluid. *Proc. R. Soc. London Ser. A* **277**, 32–50, 51–75.

Blackburn, H. M., Govardhan, R. N., and Williamson, C. H. K. 2001. A complementary numerical and physical investigation of vortex-induced vibration. *J. Fluids Struct.* **15**, 481–488.

Blackburn, H. M. and Henderson, R. D. 1999. A study of two-dimensional flow past an oscillating cylinder. *J. Fluid Mech.* **385**, 255–286.

Blasius, H. 1908. Grenzchichten in flussigkeiten mit kleiner reibung. *Z. Math. Phys.* **56**, 1–6.

Blevins, R. D. 1977. *Flow Induced Vibration.* Van Nostrand Reinhold, New York.

Blevins, R. D. 1990. *Flow-Induced Vibration.* Van Nostrand Reinhold, New York.

Blevins, R. D. 1999. On vortex-induced fluid forces on oscillating cylinders. In *Proceedings of the ASME Pressure Vessels and Piping Division Publication PVP* **389**, 103–111.

Blevins, R. D. and Coughran, C. S. 2008. Experimental investigation of vortex-induced vibration in two-dimensions. In *Proceedings of the 18th International Offshore Polar Engineering Conference*, Vancouver, BC, Canada, pp. 475–480.

Bloomfield, P. 2000. *Fourier Analysis of Time Series – An Introduction.* John Wiley, Ottawa, Canada.

Bloor, M. S. 1964. The transition to turbulence in the wake of a circular cylinder. *J. Fluid Mech.* **19**, 290–304.

Bloor, M. S. and Gerrard J. H. 1966. Measurements on turbulent vortices in a cylinder wake. *Proc. R. Soc. London Ser. A* **294**, 319–342.

Boltze, E. 1908. Grenzschichten an rotations-körpern. Disscertation University Göttingen, Aerodynamische Versuchsanstatt. AVA.

Borgman, L. E. 1965. The spectral density of ocean wave forces. In *Proceedings of the Coastal Engineering Santa Barbara Specialty Conference ASCE.*

Borgman, L. E. 1967. Spectral analysis of ocean wave forces on piling. *J. Waterways Div. ASCE* **93**, WW2, 129–156.

Borgman, L. E. 1969a. Directional spectra models for design use. In *Proceedings of the Offshore Technology Conference*, OTC 1069, Houston, TX.

Borgman, L. E. 1969b. Ocean wave simulation for engineering design. *J. Waterways Div. ASCE* **95**, WW4, 557–583.

Borgman, L. E. and Yfantis, E. 1979. Three dimensional character of waves and forces. *Civil Eng. Oceans IV, ASCE*, 791–804.

Bostock, B. R. 1972. Slender bodies of revolution at incidence. Ph.D. dissertation submitted to the Department of Engineering, University of Cambridge.

Bouak, F. and Lemay, J. 1998. Passive control of the aerodynamic forces acting on a circular cylinder. *Exp. Thermal Fluid Sci.* **16**, 112–121.

Bouard, R. and Coutanceau, M. 1980. The early stage of development of the wake behind an impulsively started cylinder for $40 < Re < 10^4$. *J. Fluid Mech.* **101**, 583–607.

Boukinda, M. L., Schoefs, F., Quiniou-Ramus, V., Birades, M., and Garretta, R. 2007. Marine growth colonization process in Guinea gulf. *J. Offshore Mech. Arctic Eng.* **129**, 97–101.

Bowie, G. L. 1977. Forces extended by waves on a pipeline at or near the ocean bottom. U.S. Army Corps of Engineers, Coastal Engineering Research Center, Tech. Paper 77-11, Ft. Belvoir, VA.

Bradshaw, P. 2000. A note on "critical roughness height" and "transitional roughness." *Phys. Fluids* **12**, 1611–1614.

Brankovic, M. and Bearman, P. W. 2006. Measurements of transverse forces on circular cylinder undergoing vortex-induced vibrations. *J. Fluids Struct.* **22**, 829–836.

Brannon, H. R. Loftin, T. D., and Whitfield, J. H. 1974. Deep water platform design. In *Proceedings of the Offshore Technology Conference*, OTC 2120, Houston, TX.

Braza, M., Chassaing, P., and Ha Minh, H. 1986. Numerical study and physical analysis of the pressure and velocity fields in the near wake of a circular cylinder. *J. Fluid Mech.* **165**, 79–130.

Breuer, M. 1998. Numerical and modeling influences on large-eddy simulations for the flow past a circular cylinder. *Int. J. Heat Fluid Flow* **19**, 512–521.

Breuer, M. 2000. A challenging test case for large eddy simulation: High Reynolds number circular cylinder flow. *Int. J. Heat Fluid Flow* **21**, 648–654.

Brika, D. and Laneville, A. 1993. Vortex-induced vibrations of a long flexible circular cylinder. *J. Fluid Mech.* **250**, 481–508.

Brooks, J. E. 1976. Added mass of marine propellers in axial translation. David Taylor Naval Ship Research and Development Center, Rep. 76-0079.

Brown, L. J. and Borgman, L. E. 1966. Tables of the statistical distribution of ocean wave forces and methods for the estimation of Cd and Cm. Wave Research Rep. HEL 9-7, University of California, Berkeley.

Bruschi, R. M., Buresti, G., Castoldi, A., and Migliavacca, E. 1982. Vortex shedding oscillations for submarine pipelines: Comparison between full-scale experiments and analytical models. Offshore Technology Conference, OTC 4232, Houston, TX.

Bruun, P. 1976. North Sea offshore structures. *Ocean Eng.* **3**(5), 361–373.

Bryndum, M. B., Jacobsen, V., and Tsahalis, D. T. 1992. Hydrodynamic forces on pipelines: Model tests. *J. Offshore Mech. Arctic Eng.* **114**, 231–241.

Bryson, A. E. 1954. Evaluation of the inertia coefficients of the cross-section of a slender body. *J. Aeronautical Sci.* **21**, 424–427.

Bublitz, P. 1971. Messung der drücke und krafte am ebenen, querangestromten kreiszylinder, Teil I: Untersuchungen am ruhenden kreiszylinder. AVA-Bericht **71** J 11. (See also AVA-Bericht **71** J 20, 1971.)

Burnett, S. J. 1979. OSFLAG Project 4: The effects of marine growth on wave and current loading. *Offshore Res. Focus, CIRIA*, UK, 8–9.

Bushnell, M. J. 1977. Forces on cylinder arrays in oscillating flow. In *Proceedings of the Offshore Technology Conference*, OTC 2903, Houston, TX.

Cao, S., Ozono, S., Hirano, K., and Tamura, Y. 2007. Vortex shedding and aerodynamic forces on a circular cylinder in linear shear flow at subcritical Reynolds number. *J. Fluids Struct.* **23**, 703–714.

Carberry, J. 2002. Wake states of a submerged oscillating cylinder and of a cylinder beneath a free-surface. Ph.D. dissertation, Monash University, Melbourne, Australia.

Cassel, K., Smith, F. T., and Walker, J. D. A. 1996. The onset of instability in unsteady boundary-layer separation. *J. Fluid Mech.* **315**, 223–256.

Catalano, P., Wang, M., Iaccarino, G., and Moin, P. 2003. Numerical simulation of the flow around a circular cylinder at high Reynolds numbers. *Int. J. Heat Fluid Flow* **24**, 463–469.

Chandrasekaran, A. R., Salni, S. S., and Malhotra, M. M. 1972. Virtual mass of submerged structures. *J. Hydraulics ASCE*, Paper 8923.

Chang, P. K. 1970. *Separation of Flow*. Pergamon, New York.

Chaplin, J. R. 1988. Loading on a cylinder in uniform oscillatory flow: Part 1 – Planar oscillatory flow. *Appl. Ocean Res.* **10**(3), 120–128.

Chaplin, J. R. 1993. Orbital flow around a circular cylinder. Part 2. Attached flow at larger amplitudes. *J. Fluid Mech.* **246**, 397–418.

Chaplin, J. R. 2000. Hydrodynamic damping of a cylinder at $\beta \approx 10^6$. *J. Fluids Struct.* **14**, 1101–1117.

Chaplin, J. R. and Kesavan, S. 1993. Velocity measurements in multidirectional waves using a perforated-ball velocity meter. *Appl. Ocean Res.* **16**, 223–234.

Chaplin, J. R., Rainey, R. C. T., and Yemm, R. W. 1997. Ringing of a vertical cylinder in waves. *J. Fluid Mech.* **350**, 119–147.

Chaplin, J. R. and Shaw, T. L. 1971. On the mechanics of flow-induced periodic forces on structures. In *Dynamic Waves in Civil Engineering* (eds. D. A. Howells, I. P. Haigh, and C. Taylor), pp. 73–94. Wiley-Interscience, London.

Chaplin, J. R. and Subbiah, K. 1998. Hydrodynamic damping of a cylinder in still water and in a transverse current. *Appl. Ocean Res.* **20**, 251–254.

Chatelain, P. Curioni, A., Bergdorf, M., Rossinelli, D., Andreoni, W., and Koumoutsakos, P. 2008a. Billion vortex particle direct numerical simulations of aircraft wakes. *Comput. Methods Appl. Mech. Eng.* **197**, 1296–1304.

Chatelain, P. Curioni, A., Bergdorf, M., Rossinelli, D., Andreoni, W., and Koumoutsakos, P. 2008b. *Comput. Methods Appl. Mech. Eng.* **197**, 1296–1304.

Chen, C. F. and Ballengee, D. B. 1971. Vortex shedding from circular cylinders in an oscillating stream. *AIAA J.* **9**, 340–362.

Chen, S. S. 1987. *Flow-Induced Vibration of Circular Cylinder Structures*, Chap. 7. Hemisphere, Springer-Verlag, Washington, D.C.

Chen, S. S. and Jendrzejczyk, J. A. 1979. Dynamic response of a circular cylinder subjected to liquid cross flow. *J. Pressure Vessel Technol.* **101**, 106–112.

Chen, S. S., Wambsganss, M. W., and Jendrzejczyk, J. A. 1976. Added mass and damping of a vibrating rod in confined viscous fluid. *J. Appl. Mech. ASME* **98**, 325–329.

Chen, Y. N. 1972. Fluctuating lift forces of the Kármán vortex streets on single circular cylinders and in tube bundles, Part 3 – lift forces in tube bundles. *J. Eng. Ind. Trans. ASME* **24**, 623.

Chen, Y. N. 1977. The sensitive tube spacing region of tube bank heat exchangers for fluid-elastic coupling in cross flow. ASME Series PVP-PB-026 (eds. M. K. Au-Yang and S. J. Brown).

Cheng, M. and Moretti, P. M. 1991. Lock-in phenomena on a single cylinder with forced transverse oscillation. In *Flow-Induced Vibration and Wear*. Pressure Vessels and Piping, Division of ASME, Vol. 206, pp. 129–133.

Chesnakas, C. J. and Simpson R. L. 1997. Detailed investigation of the three-dimensional separation about a 6:1 prolate spheroid. *AIAA J.* **35**, 990–999.

Chou, P. Y. 1946. On impact of spheres upon water. *Water Entry and Underwater Ballistics of Projectiles.* Chap. 8, Calif. Inst. Tech. OSRD Rep. 2251.

Ciesluk, A. J. and Colonell, J. M. 1974. Experimental determination of hydrodynamic mass effects. AD-781 910.

Clauss, G. F. 2007. The conquest of the inner space – challenges and innovations in offshore technology. *Marine Syst. Ocean Technol.* **3**, 37–50.

Clements, R. R. 1977. Flow representation, including separated regions, using discrete vortices. AGARD Lecture Series No. 86.

Coastal Engineering Manual. 2002. U.S. Army Corps of Engineering.

Coceal, O., Dobre, A., Thomas, T. G., and Belcher, S. E. 2007. Structure of turbulent flow over regular arrays of cubical roughness. *J. Fluid Mech.* **589**, 375–409.

Cohen, K., Siegel, S., McLaughlin, T., and Gilles, E. 2003. Feedback control of a cylinder wake low-dimensional model. *AIAA J.* **4**, 1389–1391.

Cokelet, E. D. 1977. Steep gravity waves in water of arbitrary uniform depth. *Philos. Trans. R. Soc. London Ser. A* **286**, 183–230.

Constantinidis, Y. and Oakley, O. H. 2006. Numerical prediction of bare and straked cylinder VIV. In *Proceedings of OMAE2006*, ASME, Hamburg, OMAE2006-92334. American

Society of Mechanical Engineers, New York.

Constantinidis, Y., Oakley, O. H., Navaro, C., and Holmes, S. 2005. Modeling vortex-induced motions of spars in uniform and stratified motions. In *Proceedings of OMAE2005*. American Society of Mechanical Engineers, New York. Halkidiki, Greece.

Coutanceau, M. and Bouard. R. 1977. Experimental determination of the main features of the viscous flow in the wake of a circular cylinder in uniform translation. Part 1: Steady flow; Part 2: Unsteady flow. *J. Fluid Mech.* **79**, pt. 2, 231–256 and 257–272.

Crandall, S. H., Vigander, S., and March, P. A. 1975. Destructive vibration of trashracks due to fluid-structure interaction. ASME Paper 75-DET-63.

Cummins, W. E. 1957. The force and moment on a body in a time-varying potential flow. *J. Ship Res.* **1**, 7–18.

Dahl, J. M., Hower, F. S., and Triantafyllou, M. S. 2006. Two-degree-of-freedom vortex-induced vibrations using a force assisted apparatus. *J. Fluids Struct.* **22**, 807–818.

Dalheim, J. 1999. Numerical prediction of VIV on deepwater risers subjected to shear currents and waves. Offshore Technology Conference, OTC 10933, Houston, TX.

Dalrymple, R. A. and Dean, R. G. 1975. Waves of maximum height on uniform currents. *J. Waterways Harbors Coastal Eng. Div. ASCE* **101**, No. WW3, 259–268.

Dalton, C. and Helfinstine, R. A. 1971. Potential flow past a group of circular cylinders. *J. Basic Eng. ASME*, 636–642.

Dalton, C. and Szabo, J. M. 1976. Drag on a group of cylinders. ASME 76-Pet-42.

Darbyshire, J. 1952. The generation of waves by wind. *Proc. R. Soc. London Ser. A* **215**, 299–328.

Davenport, W. B. and Root, W. L. 1958. *An Introduction to the Theory of Random Signals and Noise.* McGraw-Hill, New York.

Davis, D. A. and Ciani, J. B. 1976. Wave forces on submerged pipelines – a review with design aids. Civil Engineering Lab. NCBC Tech. Rep. R-844. Port Hueneme, CA.

Davis, J. T., Hover, F. S., Landolt, A., and Triantafyllou, M. S. 2001. Vortex-induced vibrations of cylinders in a tandem arrangement. In *Proceedings of the 4th International Symposium on Cable Dynamics*, Montreal, Q.C., Canada, pp. 109–120.

Dawson, C. and Marcus, M. 1970. DMC – A computer code to simulate viscous flow about arbitrarily shaped bodies. In *Proceedings, 1970 Heat Transfer and Fluid Mechanics Institute* (ed. T. Sarpkaya), pp. 323–338. Stanford University Press, Stanford, CA.

De, S. C. 1955. Contributions to the theory of stokes waves. *Proc. Cambr. Philos. Soc.* **51**, 718–736.

Dean, R. G. 1965. Stream function representation of nuclear ocean waves. *J. Geophys. Res.* **70**, 4561–4572.

Dean, R. G. 1970. Relative validities of water wave theories. *J. Waterways Harbors Coastal Eng. Div. ASCE* **96**, No. WW1, 105–119.

Dean, R. G. 1974. Evolution and development of water wave theories for engineering application, I and II. U.S. Army Coastal Engineering Research Center, Special Rep. 1, Fort Belvoir, VA.

Dean, R. G. and Dalrymple, R. A. 1998. *Water Wave Mechanics for Engineers and Scientists*, Vol. 2. World Scientific, Singapore.

Deardorff, J. W. 1970. A numerical study of three-dimensional turbulent channel flow at large Reynolds numbers. *J. Fluid Mech.* **41**, 453.

Deep Oil Technology, 1992. Joint Industry Program on Vortex Induced Motions of Large Floating Platforms, Houston, TX.

Deniz, S. and Staubli, T. 1997. Oscillating rectangular and octagonal profiles: interaction of leading- and trailing-edge vortex formation. *J. Fluids Struct.* **11**, 3–31.

Deniz, S. and Staubli, T. 1998. Oscillating rectangular and octagonal profiles: modeling of fluid forces. *J. Fluids Struct.* **12**, 859–882.

Dong, S. and Karniadakis, G. E. 2005. DNS of flow past a stationary and oscillating cylinder at Re = 10,000. *J. Fluids Struct.* **20**, 519–531.

Doolittle, R. D. 1974. NSRDC Basin cable strum tests. MAR Inc., Rockville, MD.

Dryden, H. L., Murnaghan, F. D., and Bateman, H. 1956. *Hydrodynamics.* Dover, New York.

Duggal, A. S. and Niedzwecki, J. M. 1995. Dynamic response of a single flexible cylinder in waves. *J. Offshore Mech. Arctic Eng.* **117**, 99–104.

Dütsch, H., Durst, F., Becker, S., and Lienhart, H. 1998. Low-Reynolds number flow around an oscillating circular cylinder at low Keulegan–Carpenter numbers. *J. Fluid Mech.* **360**, 249–271.

Dvorak, R. A. 1969. Calculation of turbulent boundary layers on rough surfaces in pressure gradient. *AIAA J.* **7**, 1752–1759.

Dwyer, H. A. and McCroskey, W. J. 1973. Oscillating flow over a cylinder at large Reynolds number. *J. Fluid Mech.* **61**, pt. 4, 753–767.

Dye, R. C. F. 1978. Photographic evidence of the mechanisms of vortex-exited vibration. *J. Photogr. Sci.* **26**, 203–208.

Dyrbye, C. and Hansen, S. O. 1997. *Wind Loads on Structures.* Wiley, Chichester, UK.

Eatock Taylor, R., Hung, S. M., and Chau, F. P. 1989. On the distribution of second order pressure on a vertical circular cylinder. *Appl. Ocean Res.* **11**, 183–193.

El Baroudi, M. Y. 1960. Measurements of two-point correlations of velocity near a circular cylinder shedding a Kármán vortex street. University of Toronto Institute for Aerospace Studies (UTIAS), Tech. Note 31.

Elston, J. R., Blackburn, H. M., and Sheridan, J. 2006. The primary and secondary instabilities of flow generated by an oscillating circular cylinder. *J. Fluid Mech.* **550**, 359–389.

Ericsson, L. E. and Reding, J. P. 1979. Vortex-induced asymmetric loads on slender vehicles. Lockheed Missiles and Space Company, Inc., Rep. LMSC-D630807. (See also AIAA Paper AIAA-80-0181, 1980.)

Etzold, F. and Fiedler, H. 1976. The near-wake structures of a cantilevered cylinder in a cross flow. *Z. Flugwissenschaften* **24**, 77–82.

Evangelinos, C. and Karniadakis, G. E. 1999. Dynamics and flow structures in the turbulent wake of rigid and flexible cylinders subject to vortex-induced vibrations. *J. Fluid Mech.* **400**, 91–124.

Ewing, J. A. 1973. Mean length of runs of high waves. *J. Geophys. Res.* **78**, 1933–1936.

Fabula, A. G. and Bedore, R. L. 1974. Tow basin tests of cable strum reduction (second series). Naval Undersea Center, Tech. Note TN-1379, San Diego, CA.

Fage, A. 1931. Further experiments on the flow around a circular cylinder. Aeronautical Research Council Reports and Memoranda 1, No. 1369, 186–195.

Fage, A. and Falkner, V. M. 1931. Further experiments on the flow around a circular cylinder. Aeronautical Research Council Reports and Memoranda No. 1369.

Fage, A. and Johansen, R. C. 1928. The structure of vortex sheets. Aeronautical Research Council Reports and Memoranda No. 1143.

Fage, A. and Warsap. J. H. 1930. The effects of turbulence and surface roughness on the drag of a circular cylinder. Aeronautical Research Council Reports and Memoranda No. 1283.

Faltinsen, O. M. 1990. *Sea Loads on Ships and Offshore Structures.* Cambridge University Press, Cambridge.

Faltinsen, O. M., Kjaerland, O., Nøttveit, O., and Vinje, T. 1977. Water impact loads and dynamic response of horizontal circular cylinders in offshore structures. In *Proceedings of the Offshore Technology Conference*, OTC 2741, Houston, TX.

Faltinsen, O. M. and Michaelsen, F. C. 1974. Motions of large structures in waves at zero Froude number. In *Proceedings of the International Symposium on the Dynamics of Marine Vehicles and Structures in Waves*, pp. 91–106. University College, London.

Feng, C.-C. 1968. The measurement of vortex induced effects in flow past stationary and oscillating circular and d-section cylinders. MS thesis, Department of Mechanical Engineering, University of British Columbia, Canada.

Fenton, J. D. 1985. A fifth-order Stokes theory for steady waves. *J. Waterway, Port, Coastal Ocean Eng.* **111**, pt. 2, 216–234.

Fickel, M. G. 1973. Bottom and surface proximity effects on the added mass of rankine ovoids. AD-775 022. Thesis submitted to the Naval Postgraduate School, Monterey, CA.

Field, J. B. 1975. Experimental determination of forces on an oscillating cylinder. MS thesis submitted to the Naval Postgraduate School, Monterey, CA.

Fink, P. T. and Soh, W. K. 1974. Calculation of vortex sheets in unsteady flow and applications in ship hydrodynamics. *Proceedings of the 10th Symposium of Naval Hydrodynamics.* Superintendent of Documents, U.S. Government Printing Service, Washington, D.C.

Fischer, P. F. and Patera, A. T. 1994. Parallel simulation of viscous incompressible flows. *Annu. Rev. Fluid Mech.* **26**, 483–527.

Fitz-Hugh, J. S. 1973. Flow-induced vibration in heat exchangers. Oxford University Rep. RS-57, AERE-P-7238.

Flachsbart, O. 1932. Winddruck auf gasbehilter. Reports of the Aerodynamische Versuchsanstatt in Göttingen, IVth Series, 134–138.

Flagg, C. N. and Newman, J. N. 1971. Sway added-mass coefficients for rectangular profiles in shallow water. *J. Ship Res.* **15**, 257–265.

Frank, W. 1967. Oscillation of cylinder in or below the free surface of deep fluids. David Taylor Naval Ship Research and Development Center, Rep. 2357.

Frédéric, L. and Laneville, A. 2002. Vortex-induced vibrations of a flexible cylinder in a slowly varying flow: Experimental results. In *Proceedings of IMECE 2002.* American Society of Mechanical Engineers, New York. *Intl. Mech. Engng Cong. Expo.* New Orleans, LA.

Fredsøe, J. and Hansen, E. A. 1987. Lift forces on pipelines in steady flow. *J. Waterway, Port Coastal Ocean Eng. OMAE* **1**, 601–609. Also *Trans. ASME J. Offshore Mech. Arctic Eng.* **109**, 52–60.

Fuji, S. and Gomi, M. 1976. A note on the two-dimensional cylinder wake.

J. Fluids Eng. Trans. ASME **98**, Series 1, 318–320.

Fung, Y. C. 1960. Fluctuating lift and drag acting on a cylinder in a flow at supercritical Reynolds numbers. *J. Aerospace Sci.* **27**, 801–814.

Furuya, Y., Miyata, N., and Fujita, H. 1976. Turbulent boundary layer and flow resistance on plates roughened by wires. ASME Paper 76-FE-6.

Gad-el-Hak M. and Bushnell, D. M. 1991. Separation control: Review. *J. Fluids Eng.* **113**, 5–30.

Garrison, C. J. 1978. Hydrodynamic loading of large offshore structures. Three-dimensional source distribution methods. In *Numerical Methods in Offshore Engineering* (eds. O. C. Zienkiewicz, R. W. Lewis, and K. G. Stagg), pp. 97–140. Wiley, Chichester, UK.

Garrison, C. J. 1990. Drag and inertia forces on circular cylinders in harmonic flow. *J. Waterway, Port, Coastal Ocean Div. ASCE* **116**(2), 169–190.

Garrison, C. J., Field, J. B., and May, M. D. 1977. Drag and inertia forces on a cylinder in periodic flow. *J. Waterway, Port, Coastal Ocean Div. ASCE* **103**, WW2 193–204.

Gartshore, I. S. 1984. Some effects of upstream turbulence on the unsteady lift forces imposed on prismatic two dimensional bodies. *J. Fluids Eng.* **106**, 418–424.

Gaston, J. D. and Ohmart, R. D. 1979. Effects of surface roughness on drag coefficients. *Civil Eng. Oceans IV ASCE*, 611–621.

Geernaert, G. L. and Plant, W. J. 1990. *Surface Waves and Fluxes*, Vol. 1, *Current Theory*, p. 337. Kluwer Academic, Dordrecht, The Netherlands.

Gerlach, C. R. and Dodge, F. T. 1970. An engineering approach to tube flow-induced vibrations. In *Proceedings of the Conference on Flow-Induced Vibrations in Reactor System Components*, pp. 205–225. Argonne National Laboratory.

Gerrard, J. H. 1961. An experimental investigation of the oscillating lift and drag of a circular cylinder shedding turbulent vortices. *J. Fluid Mech.* **11**, pt. 2, 215–227.

Gerrard, J. H. 1965. A disturbance-sensitive Reynolds number range of the flow past a circular cylinder. *J. Fluid Mech.* **22**, pt. 1, 187–196.

Gerrard, J. H. 1966. The mechanics of the formation region of vortices behind bluff bodies. *J. Fluid Mech.* **25**, 401–413. Also in Sumer B. M. and Fredsøe, J. 1997. *Hydrodynamics Around Cylindrical Structures*, Vol. 12 of the Advanced Series on Ocean Engineering. World Scientific, Singapore.

Gerrard, J. H. 1978. The wake of cylindrical bluff bodies at low Reynolds number. *Philos. Trans. A* **288**, 351–382.

Gibson, R. J. and Wang, H. 1977. Added mass of pile groups. *J. Waterways ASCE* WW2, 215–223.

Glenny, D. E. 1966. A review of flow around circular cylinders stranded cylinders and struts inclined to the flow direction. Australian Defense Sci. Service Aero. Res. Labs. Mech. Eng. Note 284.

Goda, Y. 1970. A synthesis of breaker indices. *Trans. Jpn. Soc. Civil Eng.* **2**, 227–230.

Goda, Y. 1976. On wave groups. In *Proceedings of the Conference on the Behaviour of Offshore Structures*, *BOSS*, Trondheim, Vol. 1, pp. 115–128.

Goda, Y. 2000. *Random Sea and Design of Maritime Structures*. World Scientific, Singapore.

Goddard, V. P. 1972. Numerical solutions of the drag response of a circular cylinder to stream-wise velocity fluctuations. Ph.D. dissertation,

University of Notre Dame. (See also *Rozpr. Inz.* **33**, 487–508, 1974.)

Goktun, S. 1975. The drag and lift characteristics of a cylinder placed near a plane surface. M.S. thesis submitted to the Naval Postgraduate School, Monterey, CA.

Goldman, R. L. 1958. Karman vortex forces on the vanguard rocket. *Shock Vibration Bull.* **26**, pt. 2, 171–179.

Goldstein, S. 1938. *Modern Developments in Fluid Mechanics.* Clarendon, Oxford.

Goldstein, S. and Rosenhead, L. 1936. Boundary layer growth. *Proc. Cambr. Philos. Soc.* **32**, 392–401.

Gopalkrishnan, R. 1993. Vortex induced forces on oscillating bluff cylinders. Ph.D. dissertation, Department of Ocean Engineering, MIT, Cambridge, MA.

Gopalkrishnan, R., Grosenbaugh, M. A., and Triantafyllou, M. S. 1992. Influence of amplitude modulation on the fluid forces acting on a vibrating cylinder in cross-flow. *Intl. J. Offshore and Polar Eng. Trans. ISOPE* **2**(1), 32–37.

Görtler, H. 1944. Verdrangungswirkung der laminaren grenzschicht und druckwiederstand. *Ing.-Arch.* **14**, 286–305.

Görtler, H. 1948. Grenzschichtentstehung an zylindern bei anfahrt aus der ruhe. *Arch. Math.* **1**, 138–147.

Gosner, K. L. 1978. *A Field Guide to the Atlantic Seashore.* Peterson Field Guide Series, Houghton Mifflin, Boston.

Govardhan, R. and Williamson, C. H. K. 2000. Modes of vortex formation and frequency response of a freely vibrating cylinder. *J. Fluid Mech.* **420**, 85–130.

Grace, R. A. 1973. Available data for the design of unburied submarine pipelines to withstand wave action. In *Proceedings of the 1st Australian Conference on Coastal Engineering,* Sydney, Australia.

Grace, R. A., Castiel, J., Shak, A. T., and Zee, G. T. Y. 1979. Hawaii ocean test pipe project: Force coefficients. *Civil Eng. Oceans IV ASCE* **1**, 99–110.

Grace, R. A. and Nicinski, S. A. 1976. Wave force coefficients from pipeline research in the ocean. In *Proceedings of the Offshore Technology Conference,* OTC 2676, Houston, TX.

Granville, P. S. 1978. Similarity-law characterization methods for arbitrary hydrodynamic roughnesses. David Taylor Naval Ship Research and Development Center, Rep. 78-SPD-815-01.

Graham, D. S. 1979. A bibliography of force coefficients literature germane to unburied pipelines in the ocean. The University of Florida, Department of Civil Engineering. Misc. Rep.

Graham, D. S. and Machemehl, J. L. 1980. Approximate force coefficients for ocean pipelines. ASME 80-Pet-61.

Grass, A. J. and Kemp, P. H. 1979. Flow visualization studies of oscillatory flow past smooth and rough circular cylinders. In *Mechanics of Wave-Induced Forces on Cylinders* (ed. T. L. Shaw), pp. 406–420. Pitman, London.

Grass, A. J., Simons, R. R., and Cavanagh, N. J. 1985. Fluid loading on horizontal cylinders in wave type orbital oscillatory flow. In *Proceedings of the 9th Offshore Mechanical Arctic Engineering Symposium,* Dallas, TX. Vol. 1, pp. 576–583.

Griffin, O. M. 1972. Flow near self exited and forced vibrating circular cylinders. *J. Eng. Industry* **94**, 539–547.

Griffin, O. M. and Koopmann G. H. 1977. The vortex-exited lift and reaction forces on resonantly vibrating cylinders. *J. Sound Vibr.* **54**, 435–448.

Griffin, O. M. and Ramberg, S. E. 1976. Vortex shedding from a cylinder vibrating in line with an incident uniform flow. *J. Fluid Mech.* **75**, pt. 2, 257–271.

Grinstein, F. F. and Karniadakis, G. E. 2002. Alternative LES and hybrid RANS/LES for turbulent flows. *J. Fluids Eng.* **124**, 821–822.

Guilminaeu, E. and Queutey, P. 2001. Numerical simulation in vortex-induced vibrations at low mass-damping. AIAA Paper 2001–2852.

Gurley, K. and Kareem, A. 1993. Gust loading factors for tension leg platforms. *Appl. Ocean Res.* **15**, 137–154.

Guven, O. 1975. An experimental and analytical study of surface-roughness effects on the mean flow past circular cylinders. Ph.D. dissertation submitted to the University of Iowa, Iowa City, IA.

Guven, O., Patel, V. C., and Farrell, C. 1975. Surface roughness effects on the mean flow past circular cylinders. Iowa Institute of Hydraulic Research Rep. 175, Iowa City, IA.

Hafen, B. E., Meggitt, D. J., and Liu, F. C. 1976. Strumming suppression – an annotated bibliography. Civil Engineering Lab. Tech. Rep. N-1456, Port Hueneme, CA.

Halkyard, J. 1996. Status of spar platforms for deepwater production systems. In *Proceedings of the 6th International Offshore and Polar Engineering Conference*, Vol. 1, pp. 262–272.

Hall, P. 1984. On the stability of unsteady boundary layer on a cylinder oscillating transversely in a viscous fluid. *J. Fluid Mech.* **146**, 337–367.

Haller, G. 2004. Exact theory of unsteady separation for two-dimensional flows. *J. Fluid Mech.* **512**, 257–311.

Halse, H. K. 2000. Norwegian deepwater program: Improved predictions of vortex-induced vibrations. *Offshore Technology Conference*, OTC 11996, Houston, TX.

Hansen, E. A. 1990. Added mass and inertia coefficients of groups of cylinders and a cylinder placed near an arbitrarily shaped seabed. In *Proceedings of the 9th Offshore Mechanical Arctic Engineering Symposium*, Vol. 1, pt. A, pp. 107–113.

Hanson, A. R. 1966. Vortex-shedding from yawed cylinders. *AIAA J.* **4**, 738–740.

Hara, T. and Mei, C. C. 1990a. Oscillating flow over periodic ripples. *J. Fluid Mech.* **211**, 183–20.

Hara, T. and Mei, C. C. 1990b. Centrifugal instability of an oscillatory flow periodic ripples. *J. Fluid Mech.* **217**, 1–32.

Haritos, N. 1989. Dynamic response of offshore structures to environmental loading. University of Melbourne, Department of Civil Agriculture Engineering. Research Rep. Struc/01/89.

Harland, L. A. Taylor, P. H., and Vugts, J. H. 1998. The extreme force on an offshore structure and its variability. *Appl. Ocean Res.* **20**(1–2), 3–14.

Hartlen, R. T., Baines, W. D., and Currie, I. G. 1968. Vortex exited oscillations of a circular cylinder. University of Toronto, Rep. UTME-TP 6809. Also see Hartlen, R. T. and Currie, I. G. 1970. Lift oscillation model for vortex-induced vibration. In *Proc. ASCE J. Eng. Mech. Div.* **96**, 577–591.

Hasselmann, K., Barnett, T. P., Bouws, E., Carlson, H., Cartwright, D. E., Enke, K., Ewing, J. A., Gienapp, H., Hasselmann, D. E., Kruseman, P., Meerburg, A., Muller, P., Olbers, D. J., Richter, K., Sell, W., and Walden, H. 1973. Measurements of wind-wave growth and swell decay during the joint North Sea Wave Project (JONSWAP). *Deutsch. Hydrogr. Z. Supply A* **8**, 12, 95.

Havelock, T. H. 1940. The pressure of water waves on a fixed obstacle. *Proc. R. Soc. London Ser. A* **175**, 409–421.

Havelock, T. H. 1963. *Collected Papers.* ONR/ACE-103. U.S. Government Printing Office, Washington. D.C.

Heaf, N. J. 1979. The effect of marine growth on the performance of fixed offshore platforms in the North Sea. In *Proceedings of the Offshore Technology Conference*, OTC 3386, Houston, TX.

Heideman, J. C., Olsen, O. A., and Johansson, P. I. 1979. Local wave force coefficients. *Civil Eng. Oceans IV ASCE*, 684–699.

Heideman, J. C. and Sarpkaya, T. 1985. Hydrodynamic forces on dense arrays of cylinders. In *Proceedings of the Offshore Technology Conference*, OTC 5008, Houston, TX, pp. 421–426.

Helmholtz, Herman, von. 1858. Über integrale der hyrodynamischen gleichungen, Welche den wirbelbewegung entsprechen. *J. Reine Angew. Math. (Crelle's J.)* **55**, 25–55 [also *Ostwald's Klassiker de Ezecken Wiss.* Nr. **79** (1896), 3–37]; trans. by P. G. Tait (with revisions by H. H.) as "On integrals of the hydrodynamical equations, which express vortex motions," *Philos. Mag.* 1867, Ser. 4, **33**, 485–510.

Helmholtz, H. von 1868. "Über Discontinuichiche Flüssigkeitsbewegungen" (On the Discontinuous Motion of Fluids), Monatsbericht der Königlich Preussischen Akedemie der Wissenschaften zu Berlin pp. 215–228; translated and published in *Phil. Mag.* 1868, Ser. 4, **36** (November), 337–345.

Henderson, R. D. 1995. Details of the drag curve near the onset of vortex shedding. *Phys. Fluids* **7**, 2102–2104.

Herbich, J. B. (ed). 1990. *Handbook of Coastal and Ocean Engineering, I, Wave Phenomena and Coastal Structures*, p. 1155. Gulf Publishing Company, Houston, TX.

Higuchi, H., Sawada, H., and Kato, H. 2008. Sting-free measurements on a magnetically supported right circular cylinder aligned with the free stream. *J. Fluid Mech.* **596**, 49–72.

Hoerner, S. F. 1965. *Fluid-Dynamic Drag.* 3rd ed. Book published by the author, New Jersey.

Hogben, N. 1976. Wave loads on structures. In *Proceedings of BOSS '76*, Trondheim, Vol. 1, pp. 187–219.

Hogben, N., Miller, B. L., Searle, J. W., and Ward, G. 1977. Estimation of fluid loading on offshore structures. *Proc. Inst. Civil Eng.* **63**, part 2, 515–562.

Hogben, N. and Standing, R. G. 1974. Wave loads on large bodies. In *Proceedings of the International Symposium on Dynamics of Marine Vehicles and Structures in Waves*, pp. 258–277. University College, London.

Hogben, N. and Standing, R. G. 1975. Experience in computing wave loads on large bodies. In *Proceedings of the Offshore Technology Conference*, OTC 2189, Houston, TX.

Holmes, J. D. 2001. *Wind Loading of Structures.* Spon Press, London, New York.

Holthuijsen, L. H. 2007. *Waves in Oceanic and Coastal Waters.* Cambridge University Press, New York.

Honji, H. 1981. Streaked flow around an oscillating circular cylinder. *J. Fluid Mech.* **107**, 507–520.

Houghton, D. R. 1968. Mechanisms of marine fouling. In *Proceedings of the 1st International Biodeterioration Symposium.* Southampton, U.K.

Houston, J. R. 1981. Combined refraction and diffraction of short waves using the finite element method. *Appl. Ocean Res.* **3**, 163–170.

Hover, F. S., Miller, S. N., and Triantafyllou, M. S. 1997. Vortex-induced vibration of marine cables: Experiments using force feedback. *J. Fluids Struct.* **11**, 306–326.

Hover, F. S., Techet, A. H., and Triantafyllou, M. S. 1998. Forces on oscillating uniform and tapered cylinders in cross flow. *J. Fluid Mech.* **363**, 97–114.

Howe, M. S. 1979. On the added mass of a perforated shell, with application to the generation of aerodynamic sound by a perforated trailing edge. *Proc. R. Soc. London Ser. A* **365**, 209–233.

Huang, W.-X. and Sung, H. J. 2007. Vortex shedding from a circular cylinder near a moving wall. *J. Fluids Struct.* **23**, 1064–1076.

Huarte, F. J. H., Bearman, P. W., and Chaplin, J. R. 2006. On the force distribution along the axis of a flexible circular cylinder undergoing multimode vortex-induced vibrations. *J. Fluids Struct.* **22**, 897–903.

Huerre, P. 2002. Global nonlinear instabilities in wake flows. Presented at the Conference on Bluff Body Wakes and Vortex-Induced Vibrations BBVIV3, Port Douglas, Australia.

Humphreys, J. S. 1960. On a circular cylinder in a steady wind at transition Reynolds numbers. *J. Fluid Mech.* **9**, 603–612.

Humphries, J. A. and Walker, D. H. 1988. Vortex-exited response of large-scale cylinders in sheared flow. *J. Offshore Mech. Arctic Eng.* **110**, 272–277.

Huse, E., Kleiven, G., and Nielsen, F. G. 1998. Large scale model testing of deep sea risers. Offshore Technology Conference, OTC 8701, Houston, TX.

Huse, E. and Muren, P. 1987. Drag in oscillatory flow interpreted from wake considerations. Offshore Technology Conference, OTC 5370, Houston, TX.

Huseby, M. and Grue, J. 2000. An experimental investigation of higher-harmonic wave forces on a vertical cylinder. *J. Fluid Mech.* **414**, 75–103.

Inman, D. L. and Bowen, A. J. 1963. Flume experiments on sand transport by waves and currents. In *Proceedings of the 8th Conference on Coastal Engineering Mexico City, Council on Wave Research*, pp. 137–150.

Iranpour, M. Taheri, F., and Vandiver, J. K. 2008. Structural life assessment of oil and gas risers under vortex-induced vibration. *Marine Struct.* **21**, 353–373.

Isaacson, M. de St. Q. 1978. Vertical cylinders of arbitrary section in waves. *J. Waterway, Port, Coastal Ocean Div. ASCE* **104**, WW4, 309–324.

Isaacson, M. de St. Q. 1979a. Wave forces on large square cylinders. In *Mechanics of Wave-Induced Forces on Cylinders* (ed. T. L. Shaw), pp. 609–622. Pitman, London.

Isaacson, M. de St. Q. 1979b. Wave forces on rectangular caissons. In *Proceedings of Civil Engineering Oceans IV ASCE*, San Francisco, CA, Vol. I, pp. 161–171.

Iwan, W. D. and Jones, N. P. 1987. On the vortex-induced oscillations of long structural elements. *J. Energy Resources Technol.* **109**, 161–167.

Janssen, P. 2004. *The Interaction of Ocean Waves and Wind.* Cambridge University Press, Cambridge.

Jensen, B. L. and Mogensen, B. 1982. Hydrodynamic forces on pipelines placed in a trench under steady current conditions. Progress Rep. 57, pp. 43–50. Institute of Hydrodynamics and Hydraulic Engineering ISVA, Technical University of Denmark.

Jhingran, V. and Vandiver, J. K. 2007. Incorporationg the higher harmonics in VIV fatigue predictions. In *Proceedings of the 26th International Conference on Offshore Mechanical Arctic Engineering OMAE.* San Diego, CA.

Johanning, L. 2003. Hydrodynamic damping of cylinders at high Stokes parameter. Ph.D. dissertation, Faculty of Engineering, University of London.

Johansson, B. and Reinius, E. 1963. Wave forces acting on a pipe at the bottom of the sea. IAHR Congress, Paper No. 1.7, London.

Johnson, R. E. 1970. Regression model of wave forces on ocean outfalls. *J. Waterways Div. ASCE* **96**, 284–305.

Jones, G. W. Jr., Cincotta, J. J., and Walker, R. W. 1969. Aerodynamic forces on a stationary and oscillating circular cylinder at high Reynolds number. NASA Tech. Rep. TR R-300.

Jong, J.-Y. and Vandiver, J. K. 1985. The identification of the quadratic system relating cross-flow and in-line vortex-induced vibration. In *Dynamic System Measurement and Control*, ASME Winter Annual Meeting. American Society of Mechanical Engineers, New York.

Jonsson, I. G. 1976. The dynamics of waves on currents over a weakly varying bed. In *Proceedings of the IUTAM Symposium on Surface Gravity Waves on Water of Varying Depths*, Canberra, Australia, pp. 1–12.

Jonsson, I. G. 1990. Wave current interactions. In *The Sea* (eds. B. Le Méhauté and D. M. Hanes), Chap. 9A, pp. 65–120. Wiley-Interscience, New York.

Jordan, S. K. and Fromm, J. E. 1972. Oscillatory drag, lift and torque on a circular cylinder in a uniform flow. *Phys. Fluids* **15**, 371–376. (See also AIAA 72–111, 1–9).

Kagemoto, H. and Yu, D. K. P. 1986. Interactions among multiple three-dimensional bodies in water waves: an exact algebraic method. *J. Fluid Mech.* **166**, 189–209. (See also 1993, *J. Fluid Mech.* **251**, 687–708.)

Kamphuis, J. W. 2000. *Introduction to Coastal Engineering and Management*, Vol. 16, p. 437. World Scientific, Singapore.

Kaplan, P. 1979. Impact forces on horizontal members. *Civil Eng. Oceans IV ASCE* **11**, 716–731.

Kaplan, P. and Silbert, M. N. 1976. Impact forces on platform horizontal members in the splash zone. In *Proceedings of the Offshore Technology Conference*, OTC 2498, Houston, TX.

Kármán, von, T. L. 1911. Über den mechanismus des widerstandes, den ein bewegter körper in einer flüsigkeit erzeugt. *Nachrichten, gesellschaft der wissenschaften, Göttingen, Math.–Phys. Klasse*, 509–517.

Kármán, von, T. L. 1912. Über den mechanismus des widerstandes, den ein bewegter körper in einer flüsigkeit erzeugt. *Nachrichten, gesellschaft der wissenschaften, Göttingen, Math.–Phys. Klasse*, 547–556.

Karman, Von. T. L. and Wattendorf, F. 1929. The impact on seaplane floats during landing. NACA Tech. Note 321.

Keefe, R. T. 1961. An investigation of the fluctuating forces acting on a stationary circular cylinder in a subsonic stream, and of the associated sound field. University of Toronto Institute for Aerospace Studies (UTIAS), Rep. 76.

Keefe, R. T. 1962. An investigation of the fluctuating forces acting on a stationary cylinder in a subsonic stream, and of the associated sound field. *J. Acous. Soc. Am.* **34**, 1711–1719.

Kennered, Z. H. 1967. Irrotational flow of frictionless fluids. Mostly of invariable density. David Taylor Model Basin, Rep. 2299.

Keulegan, G. H. and Carpenter, L. H. 1956. Forces on cylinders and plates in an oscillating fluid. National Bureau of Standards Rep. 4821.

Keulegan, G. H. and Carpenter, L. H. 1958. Forces on cylinders and plates in an oscillating fluid. *J. Res. Nat. Bureau Standards* **60**, 423–440.

Khalak, A. and Williamson, C. H. K. 1999. Motions, forces and mode transitions in vortex-induced vibrations at low mass-damping. *J. Fluids Struct.* **13**, 813–851.

Kilic, M. S., Haller, G., and Neishtadt, A. 2005. Unsteady fluid flow separation by the method of averaging. *Phys. Fluids* **17**, 067104.

Kim, J. and Moin, P. 1985. Application of a fractional step method to the incompressible Navier–Stokes equations. *J. Comput. Phys.* **59**, 308–323.

Kim, M. H. and Yu, D. K. P. 1989. The complete second-order diffraction solution for an axisymmetric body, Part 1. Monochromatic incident waves. *J. Fluid Mech.* **200**, 235–264.

Kim, W. D. 1966. On a freely floating ship in waves. *J. Ship Res.* **10**, 182–191, 200.

Kim, Y. H., Vandiver, J. K., and Holler R. 1986. Vortex-induced vibration and drag coefficients of long cables subjected to sheared flow. *J. Energy Res. Technol.* **108**, 77–83.

Kim, Y. Y. and Hibbard, H. C, 1975. Analysis of simultaneous wave force and water particle velocity measurements. In *Proceedings of the Offshore Technology Conference*, OTC 2192, Houston, TX.

King, R. 1974. Vortex excited structural oscillations of a circular cylinder in steady currents. Offshore Technology Conference, OTC 1948, Houston, TX. (See also Ph.D. dissertation by R. King, Loughborough University.)

King, R. 1977a. A review of vortex shedding research and its applications. *Ocean Eng.* **4**, 141–171.

King, R. 1977b. Vortex-excited oscillations of yawed circular cylinders. *J. Fluids Eng.* **99**, 495–502.

King, R., Prosser, M. J., and Johns, D. J. 1973. On vortex excitation of model piles in water. *J. Sound Vibr.* **29**, 169–188.

Kinsman, B. 1965. *Wind Waves.* Prentice-Hall, Englewood Cliffs, NJ.

Knight, D. W. and McDonald, J. A. 1979. Hydraulic resistance of artificial strip roughness. *J. Hydraulics Div. ASCE* HY6, 675–689.

Knudsen, J. G. and Katz, D. L. 1958. *Fluid Dynamics and Heat Transfer.* McGraw-Hill, New York.

Komen, G. J., Cavaleri, L., Donelan, M., Hasselmann, K., Hasselmann, S., and Janssen, P. A. E. M. 1994. *Dynamics and Modeling of Ocean Waves*, p. 532. Cambridge University Press, Cambridge.

Koopmann, G. H. 1967a. On the wind-induced vibrations of circular cylinders. MS thesis, Catholic University of America, Washington, D.C.

Koopmann, G. H. 1967b. The vortex wakes of vibrating cylinders at low Reynolds numbers. *J. Fluid Mech.* **28**, 501–512.

Korteweg, D. J., and De Vries, G. 1895. On the change of form of long waves advancing in a rectangular canal, and on a new type of long stationary waves. *Philos. Mag.* Ser. 5, **39**, 422–443.

Koterayama, W. 1979. Wave forces exerted on submerged circular cylinders fixed in deep water. Reports of the Research Institute of Applied Mechanics, Kyushu University, Vol. 27, No. 84, pp. 25–46.

Kozakiewicz, A., Fredsøe J. and Sumer, B. M. 1995. Forces on pipelines in oblique attack. Steady current and waves. In *Proceedings of the 5th International Offshore Polar Engineering Conference.* The Hague, Netherlands, Vol. II, pp. 174–183.

Kravchenko, A. G. and Moin, P. 2000. Numerical studies of flow over a circular cylinder at Re = 3900. *Phys. Fluids* **12**, 403–417.

Kriebel, D. L. 1990. Nonlinear wave interaction with a vertical circular cylinder. Part 1. Diffraction theory. *Ocean Eng.* **17**, 345–377.

Krishnamoorthy, S., Price, S. J., and Paidoussis, M. P. 2001. Cross-flow past an oscillating circular cylinder: Synchronization phenomena in the near wake. *J. Fluids Struct.* **15**, 955–980.

Kumar, S. R., Sharma, A., and Agrawal, A. 2008. Simulation of flow around a row of square cylinders. *J. Fluid Mech.* **606**, 369–397.

Lacasse, S. 1999. Geotechnical contributions to offshore development. OTRC Honors Lecture, Houston, TX, pp. 359–371.

Laguë, F. and Laneville, A. 2002. Vortex-induced vibrations of a flexible cylinder in a slowly varying flow: Experimental results. In *Proceedings of the ASME International Mechanical Engineering Congress and Exposition*, New Orleans, LA, pp. 1–7. American Society of Mechanical Engineers.

Lamb, Sir Horace 1932. *Hydrodynamics.* 6th ed. Dover, New York.

Lambrakos, K. F. and Brannon, H. R. 1974. Wave force calculations for Stokes and non-Stokes waves. In *Proceedings of the Offshore Technology Conference*, OTC 2039, Houston, TX.

Lamont, P. H. 1973. The out-of-plane force on an ogive nosed cylinder at large angles of inclination to a uniform stream. Ph.D. dissertation, University of Bristol, UK.

Lamont, P. H. and Hunt, B. L. 1976. Pressure and force distributions on a sharp-nosed circular cylinder at large angles of inclination to a uniform subsonic steam. *J. Fluid Mech.* **76**, pt. 3, 519–559.

Landweber, L. 1956. On a generalization of Taylor's virtual mass relation for Rankine bodies. *Q. J. Appl. Math.* **14**, 51–62.

Landweber, L. 1967. Vibration of a flexible cylinder in a fluid. *J. Ship Res.* **11**, 143–150.

Landweber, L. and Yih, C.-S. 1956. Forces, moments, and added masses for rankine bodies. *J. Fluid Mech.* **1**, 319–336.

Laneville, A. 2006. On vortex-induced vibration of cylinders describing *X-Y* trajectories. *J. Fluids Struct.* **22**, 773–782.

Larsen, C. M. and Bech, A. 1986. Stress analysis of marine risers under lock-in condition. In *Proceedings of the 5th International Offshore Mechanical Arctic Engineering Symposium,* Tokyo, Japan, pp. 450–457.

Larsen, C. M. and Halse, K. H. 1995. Comparison of models for vortex-induced vibrations of slender marine structures. In *Flow Induced Vibration* (ed. P. W. Bearman), pp. 467–482. Balkema, Rotterdam.

Launder, B. E. and Spalding, D. B. 1972. *Mathematical Models of Turbulence.* Academic, London.

Lavrenov, I. V. 2003. *Wind-Waves in Oceans. Dynamics and Numerical Simulation*, p. 376. Springer-Verlag, Berlin.

Layton, J. A. and Scott, J. L. 1979. Stabilization requirements for submarine pipelines subjected to ocean forces. *Civil Eng. Oceans IV ASCE*, **1**, 60–76.

LeBlond, P. H. and Mysak, L. A. 1978. *Waves in the Ocean.* Elsevier, Amsterdam.

Lecointe, Y. and Piquet, J. 1989. Flow structure in the wake of an oscillating cylinder. *J. Fluids Eng.* **111**, 139–148.

Leehey, P. and Hanson, C. E. 1971. Aeolian tones associated with resonated vibration. *J. Sound Vibr.* **13**, 465–483.

Le Méhauté, B. 1976. *An Introduction to Hydrodynamics and Water Waves.* Springer-Verlag, Dusseldorf.

Lenarts, V., Kerschen, G., and Golival, J. C. 2001. Proper orthogonal decomposition for updating of non-linear mechanical systems. *Mech. Systems Signal Process.* **15**(1), 31–43.

Leonard, A. and Roshko, A. 2001. Aspects of flow-induced vibration. *J. Fluids Struct.* **15**, 415–425.

Leonardi, S., Orlandi, P., and Antonia, R. A. 2007. Properties of d- and k-type roughness in a turbulent channel flow. *Phys. Fluids* **19**, 125101.

Li, B. Y., Lau, S. L., and Ng, C. O. 1991. Second order wave diffraction forces and run up by finite – infinite element method. *Appl. Ocean Res.* **13**, 270–285.

Lie, H. and Kaasen, K. E. 2006. Modal analysis of measurements from a large-scale VIV model test of a riser in linearly sheared flow. *J. Fluids Struct.* **22**, 557–575.

Lienhard, J. H. 1966. Synopsis of lift, drag, and vortex frequency data for rigid circular cylinders. Washington State University College of Engineering, Research Division Bull. 300.

Lighthill, M. J. 1986. Fundamentals concerning wave loading on offshore structures. *J. Fluid Mech.* **173**, 667–681.

Lin, M.-C. and Shieh, L.-D. 1997. Flow visualization and pressure characteristics of a cylinder for water impact. *Appl. Ocean Res.* **19**, 101–112.

Linton, C. M. and Evans, D. V. 1990. The interaction of waves with arrays of vertical circular cylinders. *J. Fluid Mech.* **215**, 549–569.

Linton, C. M. and McIver, P. 1996. The scattering of water waves by an array of circular cylinders in a channel. *J. Eng. Math.* **30**, 661–682.

Littlejohns, P. S. G. 1974. Current induced forces on submarine pipelines. Hydraulic Research Station, Wallingford, England, Rep. INT-138.

Liu, P. L.-F., Lozano, C. J., and Panazaras, N. 1979. An asymmetric theory of combined wave refraction and diffraction. *Appl. Ocean Res.* **1**, 137–146.

Loève, M. 1977. *Probability Theory*. 4th ed. Springer, New York.

Loken, A. E., Torset, O. P., Mathiassen, T., and Arnesen, T. 1979. Aspects of hydrodynamic loading in design of production risers. In *Proceedings of the Offshore Technology Conference*, OTC 3538, Houston, TX.

Longuet-Higgins, M. S. 1952. On the statistical distribution of the periods and amplitudes of sea waves. *J. Geophys. Res.* **80**(8), 2688–2694.

Longuet-Higgins, M. S. 1976. Recent developments in the study of breaking waves. In *Proceedings of the 15th Coastal Engineering Conference*, Honolulu, Vol. I, pp. 441–460.

Longuet-Higgins, M. S. and Stewart, R. W. 1961. The changes in amplitude of short gravity waves on steady non-uniform currents. *J. Fluid Mech.* **10**, 529–549.

Lozano, C. J. and Liu, P. L. F. 1980. Refraction-diffraction model for linear surface water waves. *J. Fluid Mech.* **101**, 705–720.

Lu, X., Dalton, C., and Zhang, J. 1997a. Application of large eddy simulation to flow past a circular cylinder. *J. Offshore Mech. Arctic Eng.* **117**, 219–225.

Lu, X., Dalton, C., and Zhang, J. 1997b. Application of large eddy simulation to an oscillating flow past a circular cylinder. *J. Fluids Eng.* **119**, 519–525.

Lu, X.-Y. and Dalton, C. 1996. Calculation of the timing of vortex formation from an oscillating cylinder. *J. Fluids Struct.* **10**, 527–541.

Lyons, G. J. and Patel, M. H. 1989. Application of a general technique for prediction of riser vortex-induced response in waves and current. *J. Offshore Mech. Arctic Eng.* **111**, 82–91.

MacCamy, R. C. and Fuchs, R. A. 1954. Wave forces on piles: A diffraction theory. U.S. Army Corps of Engineers, Beach Erosion Board, Tech. Memo 69.

Macovsky, M. S. 1958. Vortex-induced vibration studies. David Taylor Model Basin, Rep. 1190.

Mair, W. A. and Maull, D. J. 1971. Bluff bodies and vortex shedding – a report on Euromech 17. *J. Fluid Mech.* **45**, pt. 2, 209–224.

Malenica, S., Clark, P. J., and Molin, B. 1995. Wave and current forces on a vertical cylinder free to surge and sway. *Appl. Ocean Res.* **17**, 79–90.

Malenica, S., Eatock Taylor, R., and Huang, J. B. 1999. Second order water wave diffraction by an array of vertical cylinders. *J. Fluid Mech.* **390**, 349–373.

Maniar, H. D. and Newman. J. N. 1997. Wave diffraction by a long array of cylinders. *J. Fluid Mech.* **339**, 309–330.

Mansy, H., Yang, P.-M., and Williams, D. R. 1994. Quantitative measurements of three-dimensional structures in the wake of a circular cylinder. *J. Fluid Mech.* **270**, 227–296.

Marcollo, H. and Hinwood, J. B. 2002. Vortex-induced-vibration of a long flexible cylinder in uniform flow with both forcing and response. Presented at the Conference on Bluff Body Wakes and Vortex-Induced-Vibrations BBVIV3, Port Douglas, Australia.

Marris, A. W. and Brown, O. G. 1963. Hydrodynamically excited vibrations of cantilever-supported probes. ASME Paper **62**, Hyd-7.

Marshall, P. W. 1976. Failure modes for offshore platforms-fatigue. In *Proceedings of BOSS '76*, Vol. 2, pp. 234–248.

Marxen, O., Lang, M., Rist, U., and Wagner, S. 2003. A combined experimental/numerical study of unsteady phenomena in a laminar seperation bubble. *Flow, Turbulence Combust.* **71**, 133–146.

Massel, S. R. 1996. *Ocean Surface Waves: Their Physics and Prediction.* Vol. 11 of the Advanced Series on Ocean Engineering. World Scientific, Singapore.

Matten, R. B. Hogben, N., and Ashley, R. M. 1979. A circular cylinder oscillating in still water, in waves and in currents. In *Mechanics of Wave-Induced Forces on Cylinders* (ed. T. L. Shaw), pp. 475–489. Pitman, London.

Maull, D. J. and Milliner, M. G. 1978. Sinusoidal flow past a circular cylinder. *Coastal Eng.* **2**, 149–168.

May, M. D. 1976. Oscillatory flow past a circular cylinder. MS thesis submitted to the Naval Postgraduate School., Monterey, CA.

McConnell, K. G. and Young, D. F. 1965. Added mass of a sphere in a bounded fluid. *J. Eng. Mech. ASCE* **91**, EM4, 147–164.

McGregor, D. M. 1957. An experimental investigation of the oscillating pressures on a circular cylinder in a fluid stream. University of Toronto Institute for Aerospace Studies (UTIAS), Tech. Note 14.

McIver, M. 1992. Second-order oscillatory forces on a body in waves. *Appl. Ocean Res.* **14**, 325–331.

McIver, P. 2002. Wave interaction with arrays of structures. *Appl. Ocean Res.* **24**(3), 121–126.

McLachlan, N. W. 1932. The accession to inertia of flexible discs vibrating in a fluid. *Proc. Phys. Soc. London* **44**, 546–555.

Mei, C. C. 1978. Numerical methods in water wave diffraction and radiation. *Annu. Rev. Fluid Mech.* **10**, 393–416.

Mei, C. C. 1989. *The Applied Dynamic of Ocean Surface Waves.* World Scientific, Singapore.

Mei, C. C., Stiassnie, M., Yue, D. K.-P. 2005. *Theory and Applications of Ocean Surface Waves, Part I: Linear Aspects, Part 2: Nonlinear Aspects.* Vol. 23 of the Advanced Series on Ocean Engineering, p. 1071. World Scientific, Singapore.

Meskell, C., Fitzpatrick, J. A., and Rice, H. J. 2001. Application of force-state mapping to a non-linear fluid elastic system. *Mech. Systems Signal Process.* **15**(1), 75–85.

Meyer, R. E. 1979. Theory of water-wave refraction. *Adv. Appl. Mech.* **19**, 53–141.

Meyerhoff, W. K. 1970. Added masses of thin rectangular plates calculated from potential theory. *J. Ship Res.* **14**, 100–111.

Miau, J. J., Tu, J. K., Chou, J. H., and Lee, G. B. 2006. Sensing flow separation on a circular cylinder by micro-electrical-mechanical-system thermal-film sensors. *AIAA J.* **44**, 2224–2230.

Michel, W. H. 1967. Sea spectra simplified. Presented at the Society of Naval Architects Marine Engineering, Meeting of the Gulf Section.

Michelassi, V., Wissink, J. G., Fröhlich, J., and Rodi, W. 2003. Large-eddy simulation of flow around low-pressure turbine blade with incoming wakes. *AIAA J.* **41**, 2143–2156.

Miller, B. L. 1977. The hydrodynamic drag of roughened circular cylinders. The Naval Architect. *J. R. Inst. Nav. Arch.* **119**, 55–70.

Miller, B. L. 1980. Wave slamming on offshore structures. National Maritime Institute Rep. NMI-R81.

Milne-Thomson, L. M. 1960. *Theoretical Hydrodynamics*. 4th ed. Macmillan, New York.

Mittal, R. and Moin, P. 1997. Suitability of upwind-biased finite-difference schemes for large-eddy simulation of turbulent flows. *AIAA J.* **35**, 1415–1417.

Mobarek, I. 1965. Directional spectra of laboratory wind waves. *J. Waterways Div. ASCE* **91**, No. WW3.

Moe, G. and Gudmestad, O. T. 1998. Predictions of Morrison-type force in irregular waves at high Reynolds number. *Int. J. Offshore Polar Eng.* **8**, 273–279.

Moe, G., Holden, K., and Yttervoll, P. O. 1994. Motion of spring supported cylinders in subcritical and critical water flows. In *Proceedings of the 4th International Offshore Polar Engineering Conference*, Vol. 3, pp. 468–475, Osaka, Japan.

Moe, G. and Wu, Z.-J. 1990. The lift force on a cylinder vibrating in a current. *J. Offshore Mech. Arctic Eng.* **112**, 297–303.

Moeller, M. J. 1982. Measurement of unsteady forces on a circular cylinder in cross flow at subcritical Reynolds numbers. Ph.D. dissertation, Massachusetts Institute of Technology, Cambridge, MA.

Mogridge, G. R. and Jamieson, W. 1975. Wave forces on a circular caisson: theory and experiment. *Can. J. Civil Eng.* **2**, 540–548.

Mogridge, G. R. and Jamieson, W. 1976. Wave forces on a square caissons. In *Proceedings of the 15th Coastal Engineering Conference*, Honolulu, Vol. III, pp. 2271–2289.

Mohr, K. H. 1981. Messungen Instationärer Drucke by Queranströmung von Kreiszylindern unter Beräcksichtigung Fluidlastischer Effecte. Ph.D. dissertation, KFA Jülich, Germany Rep. July–1732.

Molin, B. 1979. Second-order diffraction loads upon three-dimensional bodies. *Appl. Ocean Res.* **1**, 197–202.

Molin, B. 1992. Motion damping of slotted structures. In *Hydrodynamics: Computations, Model Tests and Reality* (ed. H. J. J. van den Boom), pp. 297–303. Elsevier Science, Amsterdam.

Molin, B. 2001. On the added mass and damping of periodic arrays of fully or partially porous disks. *J. Fluids Struct.* **15**, 275–290.

Molin, B. 2002. *Hydrodynamique des Structures Offshore*. Technip, Paris.

Moore, F. K. 1957. On the separation of the unsteady laminar boundary layer. *Boundary Layers IUTAM Symp.* 296–311. Freiberg, Berlin.

Moran, J. P. 1965. On the hydrodynamic theory of water-exit and entry. Thermal Advanced Research Inc., Rep. TAR-TR-6501.

Moretti, P. M. and Lowery, R. L. 1975. Hydrodynamic inertia coefficients for a tube surrounded by a rigid tube. *J. Pressure Vessel Technol.* **97**, 345–351.

Morison, J. R., O'Brien, M. P., Johnson, J. W., and Schaaf, S. A. 1950. The force exerted by surface waves on piles. Petroleum Transactions. *Am. Inst. Mining Eng.* **189**, 149–157.

Morkovin, M. V. 1964. Flow around circular cylinder – A kaleidoscope of challenging phenomena. In *Proceedings of the Symposium on Fully Separated Flows*, pp. 102–118. ASME, New York.

Mujumdar, A. S. and Douglas, W. J. M. 1970. The unsteady wake behind a group of three parallel cylinders. ASME 70-Pet-8.

Munk, M. 1934. Fluid Mechanics, Part 2. In *Aerodynamic Theory*, Vol. I. (ed. W. F. Durand), Berlin, Springer, p. 224.

Munk, W. H. 1949. The solitary wave theory and its application to surf problems. *Ann NY Acad. Sci.* **51**, 376–423.

Muto, K., Kuroda, K. K. and Kasai, Y. 1979. Forced vibration test of 1/5 scale Model of CANDU core. In *Proceedings of the 5th International Conference on Structural Mechanics in Reactor Technology*, Berlin, Germany, Vol. 123, pp. 13–17.

Myers, J., Holm, C. H., and McAllister, R. F. 1969. *Handbook of Ocean Engineering*. McGraw-Hill, New York.

Nagai, K. 1973. Runs of the maxima of the irregular sea. *Coastal Eng. Jpn.* **16**, 13–18.

National Physical Laboratory. 1969. Strouhal number of model stacks free to oscillate. NPL Aero Rep. 1257.

Navier, M. 1827. Mémoire sur les lois du movement des fluids. *Mem. Acad. Sci.* **6**, 389–416.

Nedergaard, H., Ottersen, Hansen, N.-E., and Fines, S. 1994. Response of free hanging tethers. In *Proceedings of the 7th International Conference on the Behaviour of Offshore Structures*, Boston, MA, pp. 315–326.

Nehari, D., Armenio, V., and Ballio, F. 2004. Three-dimentional analysis of the unidirectional oscillatory flow around a circular cylinder at low Keulegan–Carpenter and β numbers. *J. Fluid Mech.* **520**, 157–186.

Nelson, R. C. 1996. Hydraulic roughness of coral reef platforms. *Appl. Ocean Res.* **18**(5), 265–274.

Neumann, G. 1953. On ocean wave spectra and a new method of forecasting wind-generated sea. U.S. Army Corps of Engineers, Beach Erosion Board, Tech. Memo 43.

Newman, D. J. and Karniadakis G. E. 1997. Simulations of flow past a freely vibrating cable. *J. Fluid Mech.* **344**, 95–136.

Newman, J. N. 1962. The exciting forces on fixed bodies in waves. *J. Ship Res.* **6**, 10–17.

Newman, J. N. 1977. *Marine Hydrodynamics.* MIT Press, Cambridge, MA.

Newman, J. N. 1996. The second-order wave force on a vertical cylinder. *J. Fluid Mech.* **320**, 417–443.

Nielsen, J. N. 1960. *Missile Aerodynamics.* McGraw-Hill, New York.

Nolte, K. G. and Hsu, F. H. 1973. Statistics of ocean wave groups. In *Proceedings of the Offshore Technology Conference*, OTC 1688, Houston, TX, Vol. II, pp. 637–656.

Norberg, C. 1994. An experimental investigation of the flow around a circular cylinder: Influence of aspect ratio. *J. Fluid Mech.* **258**, 287–316.

Norberg, C. 2000. Flow around a circular cylinder: Aspects of fluctuating lift. *J. Fluids Struct.* Vol. **15**, 459–469.

Norberg, C. 2001. Flow around a circular cylinder: Aspects of fluctuating lift. *J. Fluids Struct.* **15**, 459–469.

Norberg, C. 2003. Fluctuating lift on a circular cylinder: review and new measurements. *J. Fluids Struct.* **17**, 57–96.

Novak, M. and Tanaka, H. 1975. *Pressure correlations on a vibrating cylinder.* In *Proceedings of the 4th International Conference on Wind Effects on Buildings Structures* (ed. K. J. Eaton), pp. 227–232, 273. Cambridge University Press, Cambridge.

Ochi, M. K. and Tsai, C.-H. 1984. Prediction of impact pressure induced by breaking waves on vertical cylinders in random seas. *Appl. Ocean Res.* **6**, 157–165.

Ogilvie, T. F. 1964. Recent progress toward the understanding and prediction of ship motions. In *Proceedings of the Fifth Symposium on Naval Hydrodynamics*, pp. 3–128. U.S. Government Printing Office, Washington, D.C.

Ohl, C. O. G., Eatock Taylor, R., Taylor, P. H., and Borthwick, A. G. L. 2001. Water wave diffraction by a cylinder array. Part 1. Regular waves. *J. Fluid Mech.* **442**, 1–32.

Ohmart, R. D. and Gratz, R. L. 1979. Drag coefficients from hurricane wave data. *Civil Eng. Oceans IV ASCE*, 260–272.

Okajima, A., Nagamori, T., Matsunaga, F., and Kiwata, T. 1999. Some experiments on flow-induced vibration of a circular cylinder with surface roughness. *J. Fluids Struct.* **13**, 853–864.

Okajima, A., Nakamura, A., Kosugi, T., and Uchida, H. 2002. Flow-induced in-line oscillation of a circular cylinder. Presented at the Conference on Bluff Body Wakes and Vortex-Induced Vibrations, BBVIV3, Port Douglas, Australia.

Olinger, D. J. and Sreenivasan, K. R. 1988. Nonlinear dynamics of the wake of an oscillating cylinder. *Phys. Rev. Lett.* **60**, 797–800.

Olunloyo, V. O. S., Osheku, C. A., and Oyediran, A. A. 2007. Dynamic response interaction of vibrating offshore pipeline on moving seabed. *J. Offshore Mech. Arctic Eng.* **129**, 107–119.

Omer, G. C. and Hall, H. H. 1949. The scattering of a tsunami by a cylindrical island. *J. Seismol. Soc. Am.* **39**, 257–260.

Otter, A. 1990. Damping forces on cylinder oscillating in a viscous fluid. *Appl. Ocean Res.* **12**, 153–155.

Pantazopoulos, M. S. 1994. Vortex-induced vibration parameters: critical review. *13th Inter. Conf. Offshore Mech. Arctic Eng.* **1**, 199–255.

Parkinson, G. V. 1974. Mathematical models of flow-induced vibrations of bluff bodies. In *Flow-Induced Structural Vibrations* (ed. E. Naudascher), pp. 81–127. Springer-Verlag, Berlin.

Parkinson, G. V. 1989. Phenomena and modeling of flow-induced vibrations of bluff bodies. *Progr. Aerosp. Sci.* **26**, 169–224.

Patrikalakis, N. M. and Chryssostomidis, C. 1986. Vortex induced response of a flexible cylinder in a sheared current. *J. Energy Res. Technol. Trans. ASME* **108**, 59–64.

Patton, K. T. 1965. Tables of hydrodynamic mass factors for translational motion. American Society of Mechanical Engineers, Paper 65-WA-UNT.

Peregrine, D. H. 1976. Interaction of water waves and currents. In *Advances in Applied Mechanics*, Vol. 16, pp. 9–117. Academic, New York.

Peters, A. S. and Stoker, J. J. 1957. The motion of a ship as a floating rigid body in a seaway. *Commun. Pure Appl. Math.* **10**, 399–490.

Phillips, O. M. 1956. The intensity of aeolian tones. *J. Fluid Mech.* **1**, 607–624.

Phillips, O. M. 1977. *The Dynamics of the Upper Ocean.* Cambridge, University Press, Cambridge.

Phillips, O. M. 1985. Spectral and statistical properties of the equilibrium range in wind-generated gravity waves. *J. Fluid Mech.* **156**, 505–531.

Pierson, W. J. and Holmes, P. 1965. Irregular wave forces on a pile. *J. Waterways Div. ASCE* **91**, WW4, 1–10.

Pierson, W. J. and Moskowitz, L. 1964. A proposed spectral form for fully developed wind seas based on the similarity theory of C. A. Kitaigorodskii. *J. Geophys. Res.* **69**, 5181–5190.

Poisson, S. D. 1831. *Nouvell theorie de l'action capillaire.* Bachelier, Paris.

Prandtl, L. 1904. Über flüssigkeitsbewegung bei schr kleiner Reibung. In *Proceedings of the Third International Mathematical Congress.* Heidelberg.

Prastianto, R. W. Otsuka, K., and Ikeda, Y. 2008. Experimental study on two flexible hanging-off circular cylinders undergoing wake interference. In *Proceedings of OCEAN'08 MTS/ EEE KOBE-TECHNO OCEAN'08 Conf.* Kobe, Japan.

Price, S. J. and Paidoussis, M. P. 1984. The aerodynamic forces acting on groups of two and three circular cylinders when subject to cross-flow. *J. Wind Eng. Ind. Aerodyn.* **17**, 329–347.

Price, S. J., Sumner, D., Smith, J. G., Leung, K., and Paidoussis, M. P. 2002. Flow visualization around a circular cylinder near to a plane wall. *J. Fluids Struct.* **16**, 175–191.

Priest, M. S. 1971. Wave forces on exposed pipelines on the ocean bed. In *Proceedings of the Offshore Technology Conference*, OTC 1383, Houston, TX.

Protos, A., Goldschmidt, V. W., and Toebes, G. H. 1968. Hydroelastic forces on bluff cylinders. *J. Basic Eng.* **90**, 378–386.

Quadflieg, H. 1977. Vortex induced load on the cylinders pair at high Reynolds numbers [in German]. *Forsch. Ingenieur.* **43**(1), 9–18.

Rahman, M. 1984. Wave diffraction by large offshore structures: An exact second theory. *Appl. Ocean Res.* **6**, 90–100.

Rahman, M. 1995. *Water Waves: Relating Modern Theory to Advanced Engineering Applications*, p. 343. Clarendon, Oxford.

Raman, H. G. V., Prabhakara, R., and Venkatanarasaiah, P. 1975. Diffraction of nonlinear waves by circular cylinder. *Acta Mech.* **23**, 145–158.

Ramberg, S. E. and Griffin, O. M. 1976. The effects of vortex coherence, spacing and circulation on the flow-induced forces on vibrating cables and bluff structures. Naval Research Laboratory, Rep. 7945, Washington, D.C.

Rance, P. J. 1969. Wave forces on cylindrical members of structures. Hydraulic Research Station Annual Report.

Raudkivi, A. J. and Small, A. F. 1974. Hydroelastic excitation of cylinders. *J. Hydraulic Res.* **12**(1), 99–131.

Reid, R. O. 1958. Correlation of water level variations with wave forces on a vertical pile for non-periodic waves. In *Proceedings of the 6th Coastal Engineering Conference*, pp. 749–786. Council for Wave Research, Berkeley, CA.

Reid, R. O. and Bretschneider, C. L. 1953. Surface waves and offshore structures. Texas A&M Research Foundation Report.

Rice, S. O. 1944–1945. Mathematical analysis of random noise. *Bell Sys. Tech. J.* **23**, 282–332; *Bell Sys. Tech. J.* **24**, 46–156; also in *Noise and Stochastic Processes* (ed. N. Wax), pp. 133–294. Dover, New York.

Robertson, J. M. 1965. *Hydrodynamics in Theory and Application*. Prentice-Hall, Englewood Cliffs, NJ.

Robinson, S. K. 1991. Coherent motions in the turbulent boundary layer. *Annu. Rev. Fluid Mech.* **23**, 601–639. (See also NASA Tech. Memo. 103859, 1991, and *J. Fluid Mech.* **412**, 355–378).

Robinson, S. K. and Kline, S. J. 1990. Turbulent boundary layer structure: Progress, status, and challenges. In *Structure and Drag Reduction* (ed. A. Gyr), pp. 3–32. Springer-Verlag, Berlin.

Rodriguez, O. and Pruvost, J. 2000. Wakes of an oscillating cylinder. In *Proceedings of the IUTAM Symposium*, Marseille, France.

Roos, F. W. and Willmarth, W. W. 1971. Some experimental results on sphere and disk drag. *AIAA J.* **9**, 2, 285–291.

Rosenhead, L. 1929. The Kármán street of vortices in a channel of finite breadth. *Phil. Trans. R. Soc. London Ser. A* **228**, 275–329.

Rosenhead, L. 1963. *Laminar Boundary Layers*. Clarendon, Oxford.

Roshko, A. 1953. On the development of turbulent wakes from vortex streets. NACA Tech. Note 2913.

Roshko, A. 1961. Experiments on the flow past a circular cylinder at very high Reynolds number. *J. Fluid Mech.* **10**, 345–356.

Roshko, A. and Fiszdon, W. 1969. On the persistence of transition in the near wake. In *Problems of Hydrodynamics and Continuum Mechanics*,

pp. 606–616. Society of Industrial and Applied Mathematics, Philadelphia.

Rott, N. 1956. Diffraction of a weak shock with vortex generation. *J. Fluid Mech.* **1**, 111–128.

Rott, N. 1964. Theory of time-dependent laminar flows. In *Theory of Laminar Flows* (ed. F. K. Moore), pp. 395–438. Princeton University Press, Princeton, NJ.

Ruhl, J. A. 1976. Offshore platforms: Observed behavior and comparisons with theory. In *Proceedings of the Offshore Technology Conference*, OTC 2553, Houston, TX.

Rye, H. 1974. Wave group formation among storm waves. In *Proceedings of the 14th Coastal Engineering Conference*, Copenhagen, pp. 164–183.

Sagatun, S. I., Herfjord, K., and Holmâs 2002. Dynamic simulation of marine risers moving relative to each other due to vortex and wake effects. *J. Fluids Struct.* **16**, 375–390.

Sainsbury, R. N. and King, D. 1971. The flow-induced oscillations of marine structures. *Proc. Inst. Civil Eng.* **49**, 269–302.

Saint-Venant, B. de 1843. Note a joindre un memoire sur la dynamique des fluids. *C. R.* **17**, 1240–1244.

Sakai, T. and Battjes, J. A. 1980. Wave shoaling calculated from Cokelet's theory. *Coastal Eng.* **4**, 65–84.

Sakamoto, H. and Hsniu H. 1994. Optimum suppression of fluid forces acting on a circular cylinder. *ASME J. Fluid Eng.* **116**, 221–227.

Saltara, F., Meneghini, J. R., Siqueira, C. R., and Bearman, P. W. 1998. The simulation of vortex shedding from an oscillating circular cylinder with turbulence modeling. American Society of Mechanical Engineers, Paper FEDSM98–5189, in CD-ROM.

Sarkar, A. and Païdoussis, M. P. 2003. A compact limit-cycle oscillation model of a cantilever conveying fluid. *J. Fluids Struct.* **17**, 525–539.

Sarpkaya, T. 1955. Oscillatory gravity waves in flowing water. *J. Eng. Mech. Div. ASCE* **87**(815), 1–33.

Sarpkaya, T. 1957. Oscillatory gravity waves in flowing water. *Trans. Am. Soc. Civil Eng.* **122**, 564–586.

Sarpkaya, T. 1959. Oblique impact of a bounded stream on a plane lamina. *J. Franklin Inst.* **267**, 229–242.

Sarpkaya, T. 1960. Added mass of lenses and parallel plates. *J. Eng. Mech. ASCE* **86**, EM3, 141–152.

Sarpkaya, T. 1961. Torque and cavitation characteristics of butterfly valves. *J. Appl. Mech.* **28**, 511–518.

Sarpkaya, T. 1962. Unsteady flow of fluids in closed systems. *J. Eng. Mech. ASCE* **88**, 1–5.

Sarpkaya, T. 1963. Lift, drag, and added-mass coefficients for a circular cylinder immersed in a time-dependent flow. *J. Appl. Mech. Trans. ASME* **85**, 13–15.

Sarpkaya, T. 1966. Separated flow about lifting bodies and impulsive flow about cylinders. *AIAA J.* **17**, 1193–1200.

Sarpkaya, T. 1975. An inviscid model of two-dimensional vortex shedding for transient and asymptotically steady flow over an inclined plate. *J. Fluid Mech.* **68**, 109–130.

Sarpkaya, T. 1976a. Forces on cylinders near a plane boundary in a sinusoidally oscillating fluid. *J. Fluids Eng.* **98**, 499–505.

Sarpkaya, T. 1976b. In-line and transverse forces on smooth and sand-roughened cylinders in oscillatory flow at high Reynolds numbers. Naval Postgraduate School, Tech. Rep. NPS-69SL76062, Monterey, CA.

Sarpkaya, T. 1976c. In-line and transverse forces on cylinders in oscillatory flow at high Reynolds numbers. In *Proceedings of the Offshore Technology Conference*, OTC 2533, Vol, 11, pp. 95–108.

Sarpkaya, T. 1976d. Vortex shedding and resistance in harmonic flow about smooth and rough circular cylinders. In *Proceedings of the International Conference on Behavior of Offshore Structures BOSS'76*, Vol. 1, pp. 220–235. The Norwegian Institute of Technology.

Sarpkaya, T. 1976e. Vortex shedding and resistance in harmonic flow about smooth and rough circular cylinders at high Reynolds numbers. Naval Postgraduate School, Rep. NPS-59SL76021, Monterey, CA.

Sarpkaya, T. 1977a. In-line and transverse forces on cylinders in oscillatory flow at high Reynolds numbers. *J. Ship Res.* **21**, 200–216.

Sarpkaya, T. 1977b. Unidirectional periodic flow about bluff bodies. Final Tech. Rep. to NSF, Naval Postgraduate School, Rep. NPS-69SL77051, Monterey, CA.

Sarpkaya, T. 1977c. Transverse oscillations of a circular cylinder in uniform flow. Part 1. Naval Postgraduate School, Tech. Rep. NPS-69SL77071, Monterey, CA.

Sarpkaya, T. 1977d. In-line and transverse forces on cylinders near a wall in oscillatory flow at high Reynolds numbers. In *Proceedings of the Offshore Technology Conference*, OTC 2898, Houston, TX.

Sarpkaya, T. 1978a. Fluid forces on oscillating cylinders. *J. Waterway, Port, Coastal Ocean Div. ASCE* **104**, WW4, 275–290.

Sarpkaya, T. 1978b. Hydrodynamic resistance of roughened cylinders in harmonic flow. *J. R. Inst. Naval Arch.* **120**, 41–55.

Sarpkaya, T. 1978c. Wave impact loads on cylinders. In *Proceedings of the Offshore Technology Conference*, OTC 3065, Houston, TX.

Sarpkaya, T. 1979a. Hydrodynamic forces on various multiple-tube riser configurations. In *Proceedings of the Offshore Technology Conference*, OTC 3539, Houston, TX.

Sarpkaya, T. 1979b. Vortex-induced oscillations-a selective review. *J. Appl. Mech. Trans. ASME* **46**, 241–258.

Sarpkaya, T. 1980. Hydroelastic response of cylinders in harmonic flow. *J. R. Inst. Naval Arch.* **3**, 103–110.

Sarpkaya, T. 1985. Past progress and outstanding problems in time-dependent flows about ocean structures. In *Proceedings of the International Symposium on Separated Flow Around Marine Structures*, Norwegian Institute of Technology, Trondheim, Norway, pp. 1–36.

Sarpkaya, T. 1986a. Force on a circular cylinder in viscous oscillatory flow at low Keulegan–Carpenter numbers. *J. Fluid Mech.* **165**, 61–71.

Sarpkaya, T. 1986b. In-line transverse forces on smooth and rough cylinder in oscillatory flow at high Reynolds numbers. Naval Postgraduate School, Tech. Rep. NPS-69-86-003, Monterey, CA.

Sarpkaya, T. 1987. Oscillating flow about cylinders: Experiments and analysis, *Forum on Unsteady Flow Separation, ASME-FED* **52**, 139–146.

Sarpkaya, T. 1989. Computational methods with vortices – The 1988 Freeman Scholar Lecture. *J. Fluids Eng.* **111**, 5–51.

Sarpkaya, T. 1990a. On the effect of roughness on cylinders. *J. Offshore Mech. Arctic Eng.* **112**, 334–340.

Sarpkaya, T. 1990b. Wave forces on cylindrical piles. In *The Sea, Ocean Engineering Science* (eds. B. Le Méhauté and D. M. Haynes), Vol. 9, pt. A, pp. 169–195. Wiley, New York.

Sarpkaya, T. 1991a. Comments on "the accurate calculation of vortex shedding." *Phys. Fluids* **A3**, 2013.

Sarpkaya, T. 1991b. Hydrodynamic lift and drag on rough circular cylinders. Offshore Technology Conference, OTC 6518, Houston, TX.

Sarpkaya, T. 1991c. Nonimpulsively started steady flow about a circular cylinder. *AIAA J.* **29**, 1283–1289.

Sarpkaya, T. 1991d. Recent progress in basic numerical and physical experiments on oscillating flow about cylinders. Presented at the Second Osaka International Colloquium on Viscous Fluid Dynamics in Ship and Ocean Technology, Osaka, Japan.

Sarpkaya, T. 1992. Brief reviews of some time dependent flows. *J. Fluids Eng.* **114**, 283–298.

Sarpkaya, T. 1993a. Coherent structures in oscillatory boundary layers. *J. Fluid Mech.* **253**, 105–140.

Sarpkaya, T. 1993b. On the Instability of the Stokes boundary layer. In *Near Wall Turbulence Flows* (eds. R. M. C. So, C. G. Speziale, and B. E. Launder). Elsevier Science, New York.

Sarpkaya, T. 1994. Vortex element methods for flow simulation. In *Advances in Applied Mechanics* (eds. Th. Wu and A. Hutchinson), Vol. 31, pp. 113–247. Academic, London.

Sarpkaya, T. 1995. Hydrodynamic damping, flow-induced oscillations, and biharmonic response. *J. Offshore Mech. Arctic Eng.* **117**, 232–238.

Sarpkaya, T. 1996a. Interaction of vorticity, free-surface, and surfactants. *Annu. Rev. Fluid Mech.* **28**, 83–128.

Sarpkaya, T. 1996b. Unsteady flows. In *Handbook of Fluid Dynamics and Fluid Machinery* (eds. J. A. Schetz and A. E. Fuhs), Vol. 1, pp. 697–732. Wiley, New York.

Sarpkaya, T. 1997. Discussion of dynamics of a hydroelastic cylinder with very low mass and damping (by

A. Khalak and C. H. Williamson). *J. Fluids Struct.* **11**, 549–552.

Sarpkaya, T. 2000. Resistance in unsteady flow: Search for an in-line force model. *Int. J. Offshore Polar Eng.* **10**(4), pp. 249–255.

Sarpkaya, T. 2001a. Hydrodynamic damping and quasi-coherent structures at large Stokes numbers. *J. Fluids Struct.* **15**, 909–928.

Sarpkaya, T. 2001b. On the force decompositions of Lighthill and Morison. *J. Fluids Struct.* **15**, 227–233.

Sarpkaya, T. 2002a. Experiments on the stability of sinusoidally oscillating flow over a circular cylinder. *J. Fluid Mech.* **457**, 157–180.

Sarpkaya, T. 2002b. Taylor–Gortler instability and separation on a cylinder in sinusoidally oscillating flow. In *Proceedings of IUTAM Symposium on Unsteady Flows*. Toulouse, France.

Sarpkaya, T. 2004. A critical review of the intrinsic nature of vortex-induced vibrations. *J. Fluids Struct.* **19**, 389–447.

Sarpkaya, T. 2005. On the parameter $\beta = Re/KC = D^2/\nu T$. *J. Fluids Struct.* **21**, 435–440.

Sarpkaya, T. 2006a. Hydrodynamic damping: theoretical, numerical, and experimental facts. Naval Postgraduate School, Tech. Rep. NPS/TS-0406, Monterey, CA. Also OMAE 2006 Keynote Address, Hamburg [see the movie; it is on the web (www.omae2006.com)].

Sarpkaya, T. 2006b. Structures of separation on a circular cylinder in periodic flow. *J. Fluid Mech.* **567**, 281–297.

Sarpkaya, T. and Butterworth, W. 1992. Separation points on a cylinder in oscillating flow. *J. Offshore Mech. Arctic Eng.* **114**, 28–35.

Sarpkaya, T. and Cakal, I. 1983. A comprehensive sensitivity analysis of the OTS data. In *Proceedings of the Offshore Technology Conference*, OTC 4616, Houston, TX.

Sarpkaya, T., Cinar, M., and Ozkaynak, S. 1980. Hydrodynamic interference of two cylinders in harmonic flow. In *Proceedings of the Offshore Technology Conference*, OTC 3775, Houston, TX.

Sarpkaya, T. and Dalton, C. 1992. Analysis of wave plus current-induced forces on cylinders. In *Proceedings of the Offshore Technology Conference*, OTC 6815, Houston, TX.

Sarpkaya, T., de Angelis, M., and Hanson, C. 1997. Oscillating turbulent flow with or without a current about a circular cylinder. *J. Offshore Mech. Arctic Eng.* **119**, 73–78.

Sarpkaya, T. and Ihrig, C. J. 1986. Impulsively-started flow about rectangular prisms – experiments and discrete vortex analysis. *J. Fluids Eng.* **108**, 47–54.

Sarpkaya, T. and Kline, H. K. 1982. Impulsively started flow about four types of bluff body. *J. Fluids Eng.* **104**, 207–213.

Sarpkaya, T. and O'Keefe, J. L. 1996. Oscillating flow about two- and three-dimensional bilge keels. *J. Offshore Mech. Arctic Eng.*, 1–6.

Sarpkaya, T., Putzig, C., Gordon, D., Wang, X., and Dalton, C. 1992. Vortex trajectories around a circular cylinder in oscillatory plus mean flow. *J. Offshore Mech. Arctic Eng.* **114**, 291–298.

Sarpkaya, T., Raines, S., and Trytten, D. O. 1982. Wave forces on inclined smooth and rough circular cylinders. *Proceedings of the Offshore Technology Conference*, OTC 4227, Houston, TX, Vol. 1, pp. 731–736.

Sarpkaya, T. and Rajabi, F. 1980. Hydrodynamic drag on bottom-mounted smooth and rough cylinders in periodic flow. In *Proceedings of* the Offshore Technology Conference, OTC 3761, Houston, TX.

Sarpkaya, T. and Shoaff, R. L. 1979a. A discrete-vortex analysis of flow about stationary and transversely oscillating circular cylinders. Naval Postgraduate School, Tech. Rep. NPS-69SL79011, Monterey, CA.

Sarpkaya, T. and Shoaff, R. L. 1979b. Inviscid model of two-dimensional vortex shedding by a circular cylinder. *AIAA J.* **17**(11), 1193–1200.

Sarpkaya, T. and Shoaff, R. L. 1979c. Numerical modeling of vortex-induced oscillations. In *Proceedings of the Specialty Conference on Civil Engineering of Oceans IV*, Vol. 1, pp. 504–515.

Sarpkaya, T. and Storm, M. A. 1985. In-line force on a cylinder translating in oscillatory flow. *Appl. Ocean Res.* **7**(4), 188–196.

Savkar, S. D. and Litzinger, T. A. 1982. Buffeting forces induced by cross flow through staggered arrays of cylinders. General Electric, Corporate Research and Development, Rep. 82RD238.

Sawaragi, T. 1995. *Coastal Engineering – Waves, Beaches Wave-Structure Interactions*. Elsevier, Amsterdam.

Sawaragi, T. and Nakamura, T. 1979. An analytical study of wave force on a cylinder in oscillatory flow. *Proceedings of the Specialty Conference on Coastal Structures ASCE* **79**, 154–173.

Scandura, P., Vittori, G., and Blondeaux, P. 2000. Three dimensional oscillatory flow over steep ripples. *J. Fluid Mech.* **412**, 355–378.

Schewe, G. 1983a. On the force fluctuation acting on a circular cylinder in crosflow from subcritical up to transcritical Reynolds numbers. *J. Fluid Mech.* **133**, 265–285.

Schewe, G. 1983b. On the structure and resolution of wall pressure fluctuations associated with turbulent

boundary-layer flow. *J. Fluid Mech.* **134**, 311–328.

Schiller, L. and Linke, W. 1933. Druck und reibungswiderstand des zylinders bei Reynol's-schen zahlen 5000 bis 40000. *Zeitschrift für Mathematische* **24**, 193–198.

Schlichting, H. 1932. Berechnung ebener periodischer Grenzschicht-strömungen. *Phys. Z.* **33**, 327–335.

Schlichting, H. 1968. *Boundary-Layer Theory*. 6th ed. McGraw-Hill, New York.

Schlichting, H. 1979. *Boundary-Layer Theory*. 7th ed. McGraw-Hill, New York.

Schmidt, D. W. von and Tilmann, P. M. 1972. Über die zirkulationsentwicklung in nachlaufen von rundstaben. *Acoustics* **27**, 14–22.

Schmidt, L. V. 1965. Measurements of fluctuating air loads on a circular cylinder. *J. Aircraft* **2**(1), 49–55.

Schnitzer, E. and Hathaway, M. E. 1953. Estimation of hydrodynamic impact loads and pressure distributions on bodies approximating elliptical cylinders with special reference to water landings of helicopters. NACA Tech. Note TN-2889.

Schuh, H. 1953. Calculation of unsteady boundary layers in two-dimensional laminar flow. *Zeitschrift für Wirtschaftlichen* **1**, 122–131.

Schwabe, M. 1935. Über druckermittlung in der instationären ebenen Strömung. *Ing.-Arch.* **6**, 34–50; NACA Tech. Memo TM-1039 (1943).

Sears, W. R. 1956. Some recent developments in airfoil theory. *J. Aerospace Sci.* **23**, 490–499.

Sears, W. R. and Telionis, D. P. 1975. Boundary-layer separation in unsteady flow. *J. Appl. Math.* **28**, 215–235.

Sedov, L. I. 1965. *Two-Dimensional Problems in Hydrodynamics and Aerodynamics*, translated from Russian (ed. C. K. Chu *et al.*), Interscience Publishers, New York.

Sekita, K. 1975. Laboratory experiments on wave and current forces acting on a fixed platform. In *Proceedings of the Offshore Technology Conference*, OTC 2191, Houston, TX.

Shih, W. C. L. and Hove, D. T. 1977. High Reynolds number flow considerations for the OTEC cold water pipe. Science Applications, Inc., Rep. SAI-78-607-LA, El Segundo, CA.

Shih, W. C. L., Wang, W. C. L., Coles, D., and Roshko, A. 1992. Experiments on flow past rough circular cylinders at large Reynolds numbers. Presented at the 2nd International Colloquium on Bluff Body Aerodynamics and Applications, Melbourne, Australia.

Siefert, W. 1976. Consecutive high waves in coastal waters. In *Proceedings of the 15th Coastal Engineering Conference*, Vol. 1, pp. 171–182.

Shore Protection Manual. 1973, 1984. U.S. Army Coastal Engineering Research Center, Vol. I.

Simmons, J. E. L. 1977. Similarities between two dimensional and axisymmetric vortex wakes. *Aeronaut. Q.* **28**, pt. 1, 15–20.

Simpson, R. L. 1973. A generalized correlation of roughness density effects on the turbulent boundary layer. *AIAA J.* **11**, 242–244.

Simpson, R. L. 1989. Turbulent boundary-layer separation. *Annu. Rev. Fluid Mech.* **21**, 205–234.

Sin, V. K. and So, R. M. C. 1987. Local force measurements on finite span cylinders in cross flow. *J. Fluids Eng.* **109**, 136–143.

Sinha, J. K. and Moorthy, R. I. K. 1999. Added mass of submerged perforated tubes. *Nucl. Eng. Des.* **193**, 23–31.

Skarecky, R. 1975. Yaw effects on galloping instability. *Proc. Eng. Mech. Div. ASCE*, 11759.

Skjelbreia, L. and Hendrickson, J. A. 1960. Fifth order gravity wave theory. In *Proceedings of the 7th Coastal Engineering Conference*, The Hague, pp. 184–196.

Skomedal, N. G., Teigen, P., and Vada, T. 1989. Computation of vortex induced vibration and the effect of correlation on circular cylinders in oscillatory flow. In *Proceedings of the 8th International Conference on Offshore Mechanical Arctic Engineering*, Vol. **11**, The Hague, pp. 327–334.

Skomedal, N. G., Vada, T., and Sortland, B. 1989. Viscous forces on one and two circular cylinders in planar oscillatory flow. *Appl. Ocean Res.* **11**, 114–134.

Skop, R. A. and Balasubramanian, S. 1997. A new twist on an old model for vortex-exited vibrations. *J. Fluids Struct.* **11**, 395–412.

Skop, R. A., Ramberg, S. E., and Ferer, K. M. 1976. Added mass and damping forces on circular cylinders. ASME 76-Pet-3.

Skourup, J. 1994. Diffraction of 2-D and 3-D irregular seas around a vertical circular cylinder. In *Proceedings of the Offshore Mechanical Arctic Engineering Conference ASME*, Vol. 1, pp. 293–300. American Society of Mechanical Engineers, New York.

Smith, C. R., Walker, J. D. A., Haidari, A. H., and Sorbun, U. 1991. On the dynamics of near-wall turbulence. *Philos. Trans. R. Soc. London Ser. A* **336**, 131–175.

Sommerfeld, A. 1896. Mathematiche theory der diffraction, *Math. Ann.* **47**, 317–374.

Sommerfeld, A. 1949. *Partial Differential Equations in Physics.* Academic, New York.

Sonneville, P. 1976. Ètude de la Structure tridimensionelle des écoulements autour d'un cylindre circulaire. Bulletin de la Direction des Ètudes et

Recherches, Série A No. 3-1976, Electricité de France.

Sorensen, R. M. 1993. *Basic Wave Mechanics: For Coastal and Ocean Engineers*, p. 284. Wiley, New York.

Soylemez, M. and Goren, O. 2004. Diffraction of oblique waves by thick rectangular barriers. *Appl. Ocean Res.* **25**, 345–353.

Spalart, P. R. and Almaras, S. R. 1992. A one-equation turbulence model for aerodynamic flows. AIAA Paper 92-0439.

Spring, B. H. and Monkmeyer, P. L. 1974. Interaction of plane waves with vertical cylinders. In *Proceedings of the 14th Coastal Engineering Conference*, Vol. 111, pp. 1828–1847.

St. Dennis, M. 1973. Some cautions on the employment of the spectral technique to describe the waves of the sea and the response there to of ocean systems. Offshore Technology Conference, OTC 1819, Houston, TX.

Stansby, P. K. 1974. The effect of end plates on the base pressure coefficient of a circular cylinder. *Aeronaut. J.* **78**(757), 36–37.

Stansby, P. K. 1976. The locking-on of vortex shedding due to the cross-stream vibration of circular cylinders in uniform and shear flows. *J. Fluid Mech.* **74**, 641–665.

Stansby, P. K. 1977. An inviscid model of vortex shedding from a circular cylinder in steady and oscillatory far flows. *Proc. Inst. Civil Eng.* **63**, pt. 2, 865–880.

Stansby, P. K. 1979. Mathematical modeling of vortex shedding from circular cylinders in planar oscillatory flows, including effects of harmonics. In *Mechanics of Wave-Induced Forces on Cylinders* (ed. T. L. Shaw), pp. 450–460. Pitman, London.

Stansby, P. K. and Star, P. 1992. On a horizontal cylinder resting on a sand bed under waves and current. *Int. J. Offshore Polar Eng.* **2**, 262–266.

Stappenbelt, B. and Lalji, F. 2008. Investigation on vortex induced oscillation and helical strakes effectiveness at very high incidence angles. *Int. J. Offshore Polar Eng.* **18**(2), 99–105.

Staubli, T., 1983. Calculation of the vibration of an elastically mounted cylinder using experimental data from forced oscillation. *J. Fluids Eng.* **105**, 225–229.

Stelson, T. E. and Mavis, F. T. 1955. Virtual mass and acceleration in fluids. *Proc. ASCE* **81**, Separate No. 670, 1–9.

Stoker, J. J. 1957. *Water Waves*. Interscience, New York.

Stokes, Sir George G. 1845. On the theories of internal friction of fluids in motion. *Trans. Cambr. Philos. Soc.* **8**, 287–305.

Stokes, G. G. 1847. On the theory of oscillatory waves. *Trans. Cambr. Philos. Soc.* **8**, 441–455 [reprinted in *Mathematical and Physical Papers* (1880), London, Vol. 1, pp. 314–326].

Stokes, Sir George G. 1851. On the effect of the internal friction of fluids on the motion of pendulums. *Cambr. Phil. Trans.* **IX** (also *Mathematical and Physical Papers*, Vol. 3. Cambridge University Press, Cambridge).

Stokes, G. G. 1880. On the effect of the internal friction of fluids on the motion of pendulums. *Trans. Cambr. Philos. Soc.* **IX**, 8.

Stokes, G. G. Sir 1898. "Mathematical proof of the identity of the streamlines obtained by means of a viscous film with those of a perfect fluid moving in two dimensions. British Association for the advancement of Science, Report of 78th Annual Meeting, Bristol, pp. 143–144. (Also Collected Papers of G. G. Stokes, Vol. 5, p. 278).

Strouhal, V. 1878. Uber eine besondere art der tonerregung. *Ann. Phys. Chem.* New Series **5**, 216–251.

Sugimoto, T., Saito, S., Matsuda, K., Okajima, A., Kiwata, T., and Kosugi,

T. 2002. Water tunnel experiments on in-line oscillation of a circular cylinder with a finite span length. Presented at the Conference on Bluff Body Wakes VIV, BBVIV3, Port Douglas, Australia.

Sumer, B. M. and Fredsøe, J. 1988. Transverse vibrations of an elastically mounted cylinder exposed to an oscillating flow. *J. Offshore Mech. Arctic Eng.* **110**, 387–394.

Sumer, B. M. and Fredsøe, J. 1997. Hydrodynamics around cylindrical structures. Vol. 12 of the Advanced Series on Coastal Engineering **12**, World Scientific, Singapore.

Sumer, B. M., Jensen, B. L., and Fredsøe, J. 1992. Pressure measurements around a pipeline exposed to combine waves and current. In *Proceedings of the 11th Offshore Mechanical Arctic Engineering Conference*, Calgary, Canada, V-A pp. 113–121.

Sun, H. and Faltinsen, O. M. 2006. Water impact of horizontal circular cylinders and cylindrical shells. *Appl. Ocean Res.* **28**, 5, 299–311.

Surry, D. 1969. The effect of high intensity turbulence on the aerodynamics of a rigid circular cylinder at subcritical Reynolds numbers. University of Toronto Institute for Aerospace Studies (UTIAS), Rep. 142.

Suthon, P. 2009. Instability and quasi-coherent structures in sinusoidal flow over a circular cylinder. Ph.D. dissertation, University of Houston, Houston, TX.

Svendsen, I. A. 2006. *Introduction to Nearshore Hydrodynamics*. Advanced Series on Ocean Engineering. World Scientific, Singapore.

Sverdrup, H. U. and Munk, W. H. 1947. Wind, sea and swell; theory of relations for forecasting. U.S. Navy Hydrographic Office, Pub. 601.

Szebehely, V. G. 1959. Hydrodynamic impact. *Appl. Mech. Rev.* **12**, 297–300.

Szechenyi, E. 1975. Supercritical Reynolds number simulation for two-dimensional flow over circular cylinders. *J. Fluid Mech.* **70**, pt. 3, 529–542. (See also *La Recherche Aerospatiale*, May–June 1974.)

Szepessy, S. and Bearman, P. W. 1992. Aspect ratio and end plate effects on vortex shedding from a circular cylinder. *J. Fluid Mech.* **234**, 191–217.

Takaki, M. 1977. On the hydrodynamic forces and moments acting on the two-dimensional bodies oscillating in shallow water. *Rep. Res. Inst. Appl. Mech. Kyushu Univ.* **25**, 1–64.

Taneda, S. 1972. The development of the lift of an impulsively started elliptic cylinder at incidence. *J. Phys. Soc. Jpn.* **33**, 1706–1711.

Taneda, S. and Honji, H. 1971. Unsteady flow past a flat plate normal to the direction of motion. *J. Phys. Soc. Jpn.* **30**, 262–272.

Tani, I. 1964. Low-speed flows involving bubble separations. In *Progress in Aeronautical Sciences*, Vol. 5, pp. 70–103. Pergamon, New York.

Tanida, Y., Okajima, A., and Watanabe, Y. 1973. Stability of a circular cylinder oscillating in uniform flow or in a wake. *J. Fluid Mech.* **61**, pt. 4, 769–784.

Taylor, Sir Geoffrey I. 1928a. The energy of a body moving in an infinite fluid, with an application to airships. *Proc. R. Soc. London Ser. A* **120**, 13–20.

Taylor, Sir Geoffrey I. 1928b. The forces on a body placed in a curved or converging stream of fluid. *Proc. R. Soc. London Ser. A* **120**, 260–283.

Taylor, Sir Geoffrey I. 1974. The interaction between experiment and theory in fluid mechanics. *Annu. Rev. Fluid Mech.* **6**, 1–16.

Taylor, J. Lockwood 1930. Some hydrodynamical inertia coefficients. *Philos. Mag.* **9**, 161–183.

Terzaghi, K. 1927. The science of foundations – its power and future. *Trans. Am. Soc. Civil Eng.* **93**, 270–405.

Thom, A. 1928. The boundary layer of the front portion of a cylinder. Aeronautical Research Council Reports and Memoranda No. 1176.

Thomas, G. P. 1979. *Water Wave-Current Interactions: A Review. Mechanics of Wave-Induced Forces on Cylinders* (ed. T. L. Shaw), pp. 179–203. Pitman, London.

Thompson, W. T. 1988. *Theory of Vibration with Applications*, 3rd ed. Prentice-Hall, Englewood Cliffs, NJ.

Thomson, K. D. 1972. The estimation of viscous normal force, pitching moment side force and yawing moment on bodies of revolution of incidences up to 90″. Australian Defense Scientific Service, WRERep-782 (WR & D), Melbourne, Australia.

Thomson, K. D. and Morrison, D. F. 1969. The spacing, position and strength of vortices in the wake of slender cylindrical bodies at large incidence. Australian Dept. of Supply, W.R.E. HSA 25.

Thomson, K. D. and Morrison, D. F. 1971. The spacing, position and strength of vortices in the wake of slender cylindrical bodies at large incidence. *J. Fluid Mech.* **50**, pt. 4, 751–783.

Thomson, William (Lord Kelvin) 1849. On the vis-viva of a liquid motion. *Cambr. Dublin Math. J. Papers* **I**, 107.

Thrasher, L. W. and Aagaard, P. M. 1969. Measured wave force data on offshore platforms. In *Proceedings of the Offshore Technology Conference*, OTC 1007, Houston, TX.

Toebes, G. H. 1969. The unsteady flow and wake near an oscillating cylinder. *J. Basic Eng. Trans. ASME* **91**, 493–502.

Tournier, C. and Py, B. 1978. The behavior of naturally oscillating three-dimensional flow around a cylinder. *J. Fluid Mech.* **85**, 161–186.

Triantafyllou, M. S. Gopalkrishnan, R., and Grosenbaugh, M. A. 1994. Vortex induced-vibrations in a sheared flow: a new predictive method. In *Proceedings of the International Conference on Hydroelasticity Marine Technology*, pp. 31–34.

Triantafyllou, M. S. and Grosenbaugh, M. A. 1995. Prediction of vortex induces vibrations on sheared flows. In *Proceedings of the 6th International Conference on Flow-induced Vibrations*, pp. 73–82. Balkema, Rotterdam.

Triantafyllou, M. S., Techet, A. H., Hover, F. S., and Yue, D. K. P. 2003. VIV of slender structures in shear flow. In *Proceedings of the IUTAM Symposium on Flow-Structure Interactions*, Rutgers State University.

Troesch, A. W. and Kim, S. K. 1991. Hydrodynamic forces acting on cylinders oscillating at small amplitudes. *J. Fluids Struct.* **5**, 113–126.

Tsuchiya, Y. and Yamaguchi, M. 1972. Some considerations on water particle velocities of finite amplitude wave theories. *Coastal Eng. Jpn.* **15**, 43–57.

Tutar, M. and Holdø, A. E. 2000. Large eddy simulation of a smooth circular cylinder oscillating normal to a uniform flow. *Trans. ASME* **122**, 694–701.

van Atta, C. W. 1968. Experiments in vortex shedding from yawed circular cylinders. *AIAA J.* **6**, 931–933.

van Koten, H. 1976. Fatigue analysis of marine structures. In *Proceedings of BOSS '76*, Vol. 1, pp. 653–678.

Van Oortmerssen, G. 1972. Some aspects of very large offshore structures. In *Proceedings of the 9th Symposium on Naval Hydrodynamics*, pp. 957–1001.

Vandiver, J. K. 1993. Dimensionless parameters important to the prediction of vortex-induced vibration of long, flexible cylinders in ocean currents. *J. Fluids Struct.* **7**, 423–455.

Vandiver, J. K. and Jong, J.-Y. 1987. The relationship between in-line and cross-flow vortex-induced vibration of cylinders. *J. Fluids Struct.* **1**, 381–399.

Vandiver, J. K. and Li, L. 1994. SHEAR 7 Program Theoretical Manual. Department of Ocean Engineering. MIT, Cambridge, MA.

Vengatesan, V., Varyani, K. S., and Barltrop. N. D. P. 2000. Wave-current forces on rectangular cylinder at low KC numbers. *Int. Offshore Polar Eng.* **10**, 276–284.

Verley, R. L. P. 1982. A simple model of vortex-induced forces in waves and oscillating currents. *Appl. Ocean Res.* **4**, 117–120.

Verley, R. L. P. and Moe, G. 1979. The forces on a cylinder oscillating in a current. River and Harbour Laboratory, The Norwegian Institute of Technology, Rep. STF60 A79061.

Versteeg, H. K. and Malalasekera, W. 2007. *An Introduction to Computational Fluid Dynamics: The Finite-Volume Method.* 2nd ed. Pearson, Prentice-Hall, Essex, UK.

Vickery, B. J. and Watkins, R. D. 1962. Flow-induced vibrations of cylindrical structures. In *Proceedings of the First Australiasian Conference*, pp. 213–241.

Vikestad, K., Vandiver, J. K., and Larsen, C. M. 2000. Added mass and oscillation frequency for a circular cylinder subjected to vortex-induced vibrations and external disturbance. *J. Fluids Struct.* **14**, 1071–1088.

von Muller, W. 1929. Systeme von doppelquellen in der ebener strömung, insbesondere die strömung um zwei dreiszylinder. *ZAMM* **9**(3), 200–213.

Voorhees, A. and Wei, T. 2002. Three-dimensionality in the wake of a surface piercing cylinder mounted as an inverted pendulum. Presented at the Conference on Bluff Body Wakes Vortex-Induced Vibration BBVIV3, Port Douglas, Australia.

Vugts, J. H. 1968. The hydrodynamic coefficients for swaying, heaving and rolling cylinders in a free surface. *Int. Shipbuild. Progr.* **15**, 251–276.

Wade, B. G. and Dwyer, M. 1976. On the application of Morison's equation to fixed offshore platforms. In *Proceedings of the Offshore Technology Conference*, OTC 2723, Houston, TX.

Walker, J. D. A. 2003. Unsteady separation processes at high Reynolds number and their control. In *Proceedings of International Union of Theoretical & Applied Mechanics – 2002 on Unsteady Separated Flows*, Toulouse, France, 1–12.

Wang, C.-Y. 1968. On the high frequency oscillating viscous flows. *J. Fluid Mech.* **32**, 55–68.

Wang, C. Z. and Wu, G. X. 2007. Time domain analysis of second order wave diffraction by an array of vertical cylinders. *J. Fluids Struct.* **23**, 605–631.

Wardlaw, A. D. Jr. 1974. Prediction of yawing force at high angle of attack. *AIAA J.* **12**, 1142–1144. (See also NOLTR 73–209, Oct. 1973.)

Wardlaw, R. L. and Cooper, K. R. 1973. A wind tunnel investigation of the steady aerodynamic forces on smooth and stranded twin bundled power conductors for the aluminum company of America. National Aeronautics Establishment, Canada, LTR-LA-117.

Warren, W. F. 1962. An experimental investigation of fluid forces on an oscillating cylinder. Ph.D. dissertation, University of Maryland, College Park, MD.

Watson, E. J. 1955. Boundary layer growth. *Proc. R. Soc. London Ser. A* **231**, 104–116.

Watt, B. J. 1978. Basic structural systems – A review of their design and analysis requirements. In *Numerical Methods in Offshore Engineering* (eds. O. C. Zienkiewicz, R. W. Lewis, and K. G. Stagg), pp. 1–42. Wiley, Chichester, UK.

Waugh, J. G. and Ellis, A. T. 1969. Fluid-free-surface proximity effect on a sphere vertically accelerated from rest. *J. Hydronaut.* **3**, 175–179.

Weaver, W. Jr. 1961. Wind-induced vibrations in antenna-members. *Proc. ASCE* **87**, EMI, 141–149.

Wehausen, J. V. 1971. The motion of floating bodies. *Annu. Rev. Fluid Mech.* **3**, 237–268.

Wehausen, J. V. and Laitone, E. V. 1960. Surface waves. In *Handbook der Physic*, Vol. 9, pp. 446–778. Springer-Verlag, Berlin.

Wei, T. and Smith, C. R. 1986. Secondary vortices in the wake of circular cylinders. *J. Fluid Mech.* **169**, 513–533.

Weihs, D. 1977. Resistance of a closely spaced row of cylinders. *J. Appl. Mech. Trans. ASME* **44**, Series E, 177–178.

Weinblum, G. 1952. On hydrodynamic masses. David Taylor Model Basin, Rep. 809.

Wendel, K. 1950. Hydrodynamische masses und hydrodynamische masses-tragheits-momente. *Jahrbuch der Schiffbautechnischen Gesellschaft*, **44**, Translation 260, David Taylor Naval Ship Research and Development Center, 1956.

West, G. S. and Apelt, C. J. 1993. Measurements of fluctuating pressures and forces on a circular cylinder in the Reynolds number range 10^4 to 2.5×10^5. *J. Fluids Struct.* **7**, 227–244.

Whitham, G. H. 1974. *Linear and Nonlinear Waves*. Wiley, New York.

Wiegel, R. L. 1964. *Oceanographical Engineering*. Prentice-Hall, Englewood Cliffs, NJ.

Wiegel, R. L. 1969. Waves and their effects on pile supported structures.

Rep. HEL-9-15, Hydraulic Engineering Lab., University of California, Berkeley.

Wiegel, R. L., Beebe, K. E., and Moon, J. 1957. Ocean wave forces on circular cylindrical piles. *J. Hydraulics Div. ASCE* **83**, HY2, 1199.1–1199.36.

Wilkinson, R. H., Chaplin, J. R., and Shaw, T. L. 1974. On the correlation of dynamic pressure on the surface of a prismatic bluff body. In *Flow-Induced Structural Vibrations* (ed. E. Naudascher), pp. 471–487. Springer-Verlag, Berlin.

Wille, R. 1974. Generation of oscillatory flows. In *Flow-Induced Structural Vibrations* (ed. E. Naudascher), pp. 1–16. Springer-Verlag, Berlin.

Williams, A. N., Li, W., and Wang, K.-H. 2000. Water wave interaction with a floating porous cylinder. *Ocean Eng.* **27**, 1–28.

Williams, J. C. 1977. Incompressible boundary layer separation. *Annu. Rev. Fluid Mech.* **9**, 113–144.

Williamson, C. H. K. and Roshko, A. 1988. Vortex formation in the wake of an oscillating cylinder. *J. Fluids Struct.* **2**, 355–381.

Wilson, B. W. 1965. Analysis of wave forces on a 30-inch diameter pile under confused sea conditions. Coastal Engineering Research Center, U.S. Army Tech. Memo 15.

Wilson, J. F. and Caldwell, H. M. 1971. Force and stability measurements of submerged pipelines. *J. Eng. Ind. Trans. ASME*, 1290–1298.

Wissink J. G. and Rodi, W. 2003. DNS of a laminar separation bubble in the presence of oscillating external flow. *Flow, Turbulence Combust.* **71**, 311–331.

Wolfram, J., Javidan, P., and Theophanatos, A. 1989. The loading of heavily roughened cylinders in waves and linear oscillatory flow. In *Proceedings of the 8th Offshore*

Mechanical Arctic Engineering Conference, pp. 183–190.

Wolfram, J., Jusoh, I., and Al Sell, D. 1993. Uncertainty in the estimation of the fluid loading due to the effects of marine growth. In *Proceedings of the 12th Offshore Mechanical Arctic Engineering Conference*, Glasgow, Vol. II, pp. 219–228.

Wolfram, J. and Naghipour, M. 1999. On the estimation of Morison force coefficients and their predictive accuracy for very rough circular cylinder. *J. Appl. Res.* **21**, 311–328.

Wootton, L. R., Warner, M. H., and Cooper, D. H. 1974. Some aspects of the oscillations of full-scale piles. In *Proceedings of the IUTAM-IAHR Symposium on Flow-Induced Structural Vibrations* (ed. E. Naudascher), pp. 587–661. Springer-Verlag, Berlin. (See also 1972 CIRIA Rep. 40.)

Wright, J. C. and Yamamoto, T. 1979. Wave forces on cylinders near plane boundaries. *J. Waterways Div ASCE* **105**, WWl, 1–13.

Wu, G. X. and Hu, Z. Z. 2004. Simulation of nonlinear interactions between waves and floating bodies through a finite element based numerical tank. *Proc. R. Soc. London Ser. A* **460**, 2797–2817.

Wu, Z.-J. 1989. Current induced vibrations of a flexible cylinder. Ph.D. dissertation, Department of Civil Engineering, The Norwegian Institute of Technology.

Wundt, H. 1955. Wachstum der laminaren grenzschicht an schräg angeströmten zylindern bei anfahrt aus der ruhe. *Ing-Arch.* **23**, 212–230.

Yamaguchi, T. 1971. On the vibration of the cylinder by Kármán vortex. *Mitsubishi Heavy Industries Tech. J.* **8**(1), 1–9.

Yamamoto, T. 1976. Hydrodynamic forces on multiple circular cylinders. *J. Hydraul. Div. ASCE* **102**, HY9, 1193–1210.

Yamamoto, T. and Nath, J. H. 1976. Forces on many cylinders near a plane boundary. ASCE National Water Resources and Ocean Engineering Convention, Preprint 2633.

Yamamoto, T., Nath, J. H., and Slotta, L. S. 1974. Wave forces on cylinders near plane boundary. *J. Waterways Div. ASCE* **100**, WW4, 345–359.

Young, I. R. 1999. *Wind Generated Ocean Waves*, Vol. 2, p. 288. Elsevier, Amsterdam.

Yu, Y. T. 1945. Virtual masses of rectangular plates and parallel pipes in water. *J. Appl. Mech.* **16**, 724–729.

Zdravkovich, M. M. 1969. Smoke observations of the formation of a Kármán vortex street. *J. Fluid Mech.* **37**, 491–496.

Zdravkovich, M. M. 1977. Review of flow interference between two circular cylinders in various arrangements. *J. Fluids Eng. Trans. ASME* **99**, Series 1, 618–633.

Zdravkovich, M. M. 1981. Review and classification of various aerodynamic and hydrodynamic means for suppressing vortex shedding. *J. Wind Eng. Industr. Aerodyn.* **7**, 145–189.

Zdravkovich, M. M. 1982. Modification of vortex shedding in the synchronization range. *J. Fluids Eng.* **104**, 513–517.

Zdravkovich, M. M. 1990. On origins of hysteretic responses of a circular cylinder induced by vortex shedding. *Z. Flugwissenchaft Weltraumforsch.* **14**, 47–58.

Zdravkovich, M. M. 1996a. Different modes of vortex shedding: An overview. *J. Fluids Struct.* **10**, 427–437.

Zdravkovich, M. M. 1996b. Inadequacy of a conventional Keulegan-number for wave and current combination. *J. Offshore Mech. Arctic Eng.* **118**, 309–311.

Zdravkovich, M. M. 1997. *Flow Around Circular Cylinders, Vol. 1: Fundamentals.* Oxford Scientific, Oxford.

Zdravkovich, M. M. 2003. *Flow Around Circular Cylinders, Vol. 2: Applications.* Oxford Scientific, Oxford.

Zhang, J. and Dalton, C. 1996. Interaction of vortex-induced vibration of a circular cylinder and a steady approach flow at a Reynolds number of 13,000. *Comput. Fluids* **25**, 283–294.

Zhang, J. and Dalton, C. 1999. The onset of three-dimensionality in an oscillating flow past a fixed circular cylinder. *Int. J. Numer. Meth. Fluids* **30**, 19–42.

Zhou, C. Y., So, R. M., and Lam, K. 1999. Vortex-induced vibrations of elastic circular cylinders. *J. Fluids Struct.* **13**, 165–189.

Zhou, D., Chan, E. S., and Melville, W. K. 1991. Wave impact pressures on vertical cylinders. *Appl. Ocean Res.* **13**, 220–234.

Zhou, Y. and Kareem, A. 2001. Gust loading factor: New model. *J. Struct. Eng.* **127**, 168–175.

Zhou, Y., Kareem, A., and Gu, M. 2000. Equivalent static buffeting wind loads on structures. *J. Struct. Eng. ASCE* **126**, 989–992.

Zhou, Y., Kareem, A., Gu, M., and Moran, R. M. 2002. Mode shape corrections for wind load effects. *J. Eng. Mech.* **128**, 15–23.

Zhu, X., Faltinsen, O. M., and Hu, C. 2007. Water entry and exit of a horizontal circular cylinder. *J. Offshore Mech. Arctic Eng.* **129**, 253–264.

Index

Printed in the United States
By Bookmasters